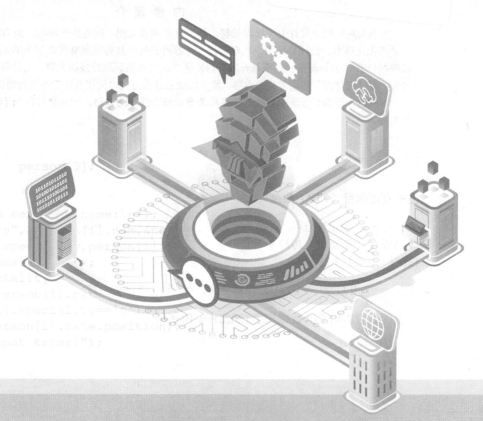

云计算平台
运维与应用（双语）

安 宁◎主 编

彭天炜 张纯容 宋 牧◎副主编

中国铁道出版社有限公司
CHINA RAILWAY PUBLISHING HOUSE CO., LTD.

内 容 简 介

本书为高等职业教育中英双语教材。基于真实项目案例，贴近生产环境，具有较强的实践指导意义，包括虚拟化技术、云计算技术、私有云技术、容器云技术及容器集群管理五个单元的内容，并融入职业技能等级证书"云计算平台运维与开发"1+X 证书、全国高职院校技能大赛"云计算"赛项、金砖国家职业技能大赛"云计算"赛项等相关内容，紧紧围绕云计算方向"岗课赛证"各方面知识、技能建设开发。

本书适合作为高等职业教育云计算相关专业双语课程的教材，同时也可作为云计算运维技术人员的指导教程。

图书在版编目（CIP）数据

云计算平台运维与应用:汉文、英文/安宁主编 .—北京：中国
铁道出版社有限公司，2023.9
ISBN 978-7-113-30309-9

Ⅰ.①云… Ⅱ.①安… Ⅲ.①云计算 - 汉、英 Ⅳ.① TP393.027

中国国家版本馆 CIP 数据核字 (2023) 第 103646 号

书　　名：云计算平台运维与应用（双语）
作　　者：安　宁

策　　划：潘晨曦　祁　云　　　　　　　编辑部电话：(010) 63549458
责任编辑：祁　云　绳　超
封面设计：刘　颖
责任校对：刘　畅
责任印制：樊启鹏

出版发行：中国铁道出版社有限公司（100054，北京市西城区右安门西街 8 号）
网　　址：http://www.tdpress.com/51eds/
印　　刷：番茄云印刷（沧州）有限公司
版　　次：2023 年 9 月第 1 版　2023 年 9 月第 1 次印刷
开　　本：850 mm×1 168 mm 1/16　印张：23.25　字数：504 千
书　　号：ISBN 978-7-113-30309-9
定　　价：79.80 元

前 言

近年来，互联网、大数据、云计算、人工智能、区块链等技术加速创新，日益融入经济社会发展各领域全过程。数字经济发展速度之快、辐射范围之广、影响程度之深前所未有，正在成为重组全球要素资源、重塑全球经济结构、改变全球竞争格局的关键力量。国家"十二五"规划确定信息技术为国家七大战略性新兴产业之一，是国家未来重点扶持的产业，而云计算正是新一代信息技术当中重要内容之一。党的二十大报告明确提出，构建新一代信息技术、人工智能等一批新的增长引擎。云计算作为 IT 领域的新技术、新规范，目前市面上与之相配套的教材尤其面向职业教育的教材较少，现有教材内容也大多停留在对云计算基础知识讲解的层面，技术内容相对滞后，缺少实战内容或实战内容不够深入，与生产实践、市场需求相距甚远。

本书为高等职业教育中英双语教材，立足于项目实战，贴近生产环境，没有通篇讲解深奥晦涩的理论知识，也没有追求对云计算各方面技术讲解面面俱到，而是注重如何让一名对云计算感兴趣的学习者能够快速使用云计算搭建自己的业务系统。本书包括虚拟化技术、云计算技术、私有云技术、容器云技术及容器集群管理五个单元，每个单元中结合相应内容以真实生产场景中的项目需求为背景设计实战项目。本书以双语、活页式呈现，不仅适用于云计算相关专业课程、双语课程的教学使用，同时也可作为云计算运维技术人员的指导教程。

　　本书由成都职业技术学院计算机网络技术专业教师安宁任主编，彭天炜、张纯容、宋牧任副主编。编写团队成员具有丰富的企业从业经历、项目实战经验，其中，安宁编写单元 3 和单元 4，彭天炜编写单元 1，张纯容编写单元 2，宋牧编写单元 5。

　　在本书编写过程中，编者查阅并借鉴了诸多资料，在此向相关作者表示由衷的感谢。

　　尽管在编写过程中，编者投入了大量的时间与精力，但鉴于水平与时间有限，书中不足之处在所难免，欢迎广大读者批评指正。

编　者

2023 年 5 月

目　录

单元 1

虚拟化技术

在传统的 IT 系统中，所有的业务都要依赖于物理服务器或数据中心，而物理服务器或数据中心最大的问题就是能源使用率的问题。有关统计数据表明，服务器和数据中心能源使用率仅有 5%，远没有被充分利用。与此同时，物理主机硬件成本高，不够灵活，并且单台主机同一时刻只能运行单一的操作系统。

虚拟化技术出现以后，整个 IT 系统的能源使用率及灵活度得到了大大的提升。有了虚拟化技术，单台主机可以同时运行多种操作系统，各操作系统从内核级别隔离，安全性高。结合云计算技术，还可以实现虚拟主机的灵活调度、动态分配、弹性伸缩。在许多类似于银行外包、呼叫中心、机房网吧等应用场景中，使用桌面虚拟化技术，仅电费这一项的成本就得以大大缩减。

1.1 虚拟化类型

物理主机依赖一种称为 Hypervisor 的软件实现物理资源的虚拟化，提供给虚拟机使用。按照 Hypervisor 在主机中所处位置的不同，将虚拟化分为两种类型：Ⅰ型虚拟化和Ⅱ型虚拟化。

1. Ⅰ型虚拟化

Hypervisor 直接安装在物理硬件（physical hardware）上，在此之上运行虚拟机（virtual machine，VM）。此时 Hypervisor 充当了类似于操作系统的角色，主机通常成为一台专门用来做虚拟化的主机，这种虚拟化方式称为Ⅰ型虚拟化。典型的Ⅰ型虚拟化技术有 VMware 的 ESX 和 ESXi，以及 Xen。Ⅰ型虚拟化系统架构图如图 1-1 所示。

2. Ⅱ型虚拟化

在物理硬件上首先安装操作系统，例如 Redhat、Ubuntu、Windows 等，在此基础上再安装

Hypervisor，此时 Hypervisor 作为主机当中的一个应用程序来运行，并提供虚拟化及虚拟机的管理功能。这种虚拟化称为Ⅱ型虚拟化。在 Windows 操作系统中比较常用的 VMware Workstation 以及 VirtualBox 都属于Ⅱ型虚拟化。此外，还有 Linux 操作系统中常用的 KVM 虚拟化技术。Ⅱ型虚拟化系统架构图如图 1-2 所示。

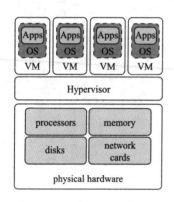

图 1-1　Ⅰ型虚拟化系统架构图

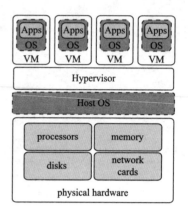

图 1-2　Ⅱ型虚拟化系统架构图

1.2　常见的虚拟化软件

目前的虚拟化软件产品繁多，除 VMware 提供基于桌面的 VMware Workstation 外，还有针对服务器的 VMware vSphere（VMware ESXI）和安装在 Mac 计算机上的 VMware Fusion（Mac），另外，还有 Oracle VM VirtualBox、XenServer、Microsoft Hyper-V、KVM、华为 Fusion Sphere 等。

随着新技术的不断发展，虚拟化应用与虚拟化技术正与当今时代下的云计算紧密结合，提供更加灵活、自助服务式的 IT 基础架构。图 1-3 ～图 1-7 为常见的几款虚拟化软件。

图 1-3　VMware Fusion

图 1-4　VM VirtualBox

图 1-5　XenServer

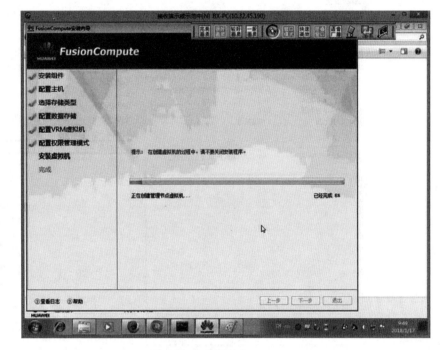

图 1-6　Microsoft Hyper-V

图 1-7　华为 Fusion Sphere

1.3　项目实战——基于 KVM 虚拟化技术的主机管理

项目背景

　　某公司因业务拓展需要在 Linux 系统中进行开发及测试工作。为了降低成本、充分利用现有硬件资源，决定使用 KVM 虚拟化技术搭建虚拟机并安装 Linux 操作系统满足开发测试环境。

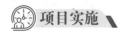 项目实施

任务 1　搭建一台 KVM 虚拟机

1. 环境准备

（1）虚拟化环境准备

在 VMware Workstation 中创建虚拟机，打开 CPU 设置，在"虚拟化引擎"中选中"虚拟化 Intel VT-x/EPT 或 AMD-V/RVI(V)"复选框，以支持 CPU 虚拟化，这是使用 KVM 的前提条件，如图 1-8 所示。之后为虚拟机安装 CentOS7 操作系统。

图 1-8　虚拟机设置

在系统中验证 CPU 是否支持虚拟化。使用 grep 命令过滤 cpuinfo 文件内容，查看是否包含 svm 或 vmx 关键字段，如包含，代表当前系统支持 CPU 虚拟化。

```
[root@localhost ~]# grep -o -E 'svm|vmx' /proc/cpuinfo
```

验证是否加载 KVM 模块。使用 lsmod 命令查看系统中加载的系统模块，如包含 KVM 模块代表当前系统已加载了 KVM 模块。

```
[root@localhost ~]# lsmod|grep kvm
```

```
kvm_intel            174841  0
kvm                  578518  1 kvm_intel
irqbypass             13503  1 kvm
```

如果返回为空，则需载入 KVM 模块。

```
modprobe kvm_intel
```

（2）搭建 yum 源，关闭防火墙及 selinux

yum 源可以使用镜像自带的仓库，也可以选择阿里云等仓库。这里以阿里云镜像仓库为例。

```
[root@localhost ~]# curl -o /etc/yum.repos.d/CentOS-Base.repo https://
mirrors.aliyun.com/repo/Centos-7.repo
yum repolist
Loaded plugins: fastestmirror
Loading mirror speeds from cached hostfile
 * base: mirrors.aliyun.com
 * extras: mirrors.aliyun.com
 * updates: mirrors.aliyun.com
repo id              repo name                                  status
base/7/x86_64        CentOS-7 - Base - mirrors.aliyun.com       10,072
extras/7/x86_64      CentOS-7 - Extras - mirrors.aliyun.com     509
local                local                                      3,971
updates/7/x86_64     CentOS-7 - Updates - mirrors.aliyun.com 3,728
repolist: 18,280
[root@localhost ~]# systemctl stop firewalld
[root@localhost ~]# cat /etc/selinux/
config         final/          semanage.conf targeted/      tmp/
[root@localhost ~]# cat /etc/selinux/config

# This file controls the state of SELinux on the system.
# SELINUX= can take one of these three values:
#     enforcing - SELinux security policy is enforced.
#     permissive - SELinux prints warnings instead of enforcing.
#     disabled - No SELinux policy is loaded.
SELINUX=disabled
# SELINUXTYPE= can take one of three two values:
#     targeted - Targeted processes are protected,
#     minimum - Modification of targeted policy. Only selected processes are
protected.
#     mls - Multi Level Security protection.
```

```
SELINUXTYPE=targeted
```

（3）安装相关应用程序

KVM 是一种基于 Linux 内核实现的虚拟化技术，通过加载内核模块 kvm.ko 实现对虚拟 CPU 及内存的管理，而对于 I/O、存储及网络等的虚拟化需要借助 qemu 应用程序来实现。因此，使用 KVM 虚拟化技术，需要安装 qemu-kvm、libvirt 及 virt-install 应用程序。其中，qemu-kvm 是 KVM 的核心包，libvirt 提供虚拟化服务，可以用来管理 KVM 等多种 Hypervisor。virt-install 是安装虚拟机的实用工具。相关应用程序的示意图如图 1-9 所示。

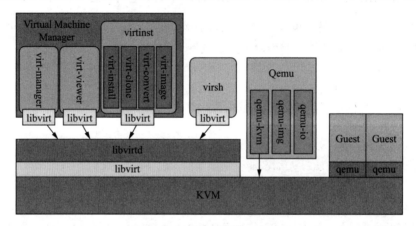

图 1-9　相关应用程序的示意图

使用 yum install 命令安装 qemu-kvm、libvirt 及 virt-install。

```
[root@localhost ~]# yum -y install qemu-kvm libvirt virt-install
```

2. 启动服务

启动 libvirtd 并设置开机自启动。使用 ip address 查看系统网络信息，可以看到新增虚拟网桥 virbr0 及虚拟网卡 virbr0-nic。

```
[root@localhost ~]# systemctl start libvirtd
[root@localhost ~]# systemctl enable libvirtd
[root@localhost ~]# ip address
1: lo: <LOOPBACK,UP,LOWER_UP> mtu 65536 qdisc noqueue state UNKNOWN
    link/loopback 00:00:00:00:00:00 brd 00:00:00:00:00:00
    inet 127.0.0.1/8 scope host lo
      valid_lft forever preferred_lft forever
    inet6 ::1/128 scope host
      valid_lft forever preferred_lft forever
 2: eno16777736: <BROADCAST,MULTICAST,UP,LOWER_UP> mtu 1500 qdisc pfifo_fast
state UP qlen 1000
    link/ether 00:0c:29:2f:24:43 brd ff:ff:ff:ff:ff:ff
```

```
    inet 192.168.134.128/24 brd 192.168.134.255 scope global dynamic
eno16777736
       valid_lft 1188sec preferred_lft 1188sec
    inet6 fe80::20c:29ff:fe2f:2443/64 scope link
       valid_lft forever preferred_lft forever
  3: virbr0: <NO-CARRIER,BROADCAST,MULTICAST,UP> mtu 1500 qdisc noqueue state
DOWN
     link/ether 52:54:00:6b:8e:4a brd ff:ff:ff:ff:ff:ff
     inet 192.168.122.1/24 brd 192.168.122.255 scope global virbr0
       valid_lft forever preferred_lft forever
  4: virbr0-nic: <BROADCAST,MULTICAST> mtu 1500 qdisc pfifo_fast master virbr0
state DOWN qlen 500
     link/ether 52:54:00:6b:8e:4a brd ff:ff:ff:ff:ff:ff
```

服务启动成功后，查看网络信息，这里出现了一个新的网络设备 virbr0，这是虚拟桥设备，其地址为 192.168.122.1，后续创建 KVM 虚拟机默认使用的网络就是以 virbr0 为网关的 NAT 网络。

3. 使用 cirros 磁盘镜像

cirros 可在其官方网站上下载。将下载的镜像放在主机 /opt 目录下。

4. 管理虚拟机

（1）创建虚拟机

系统镜像使用 cirros，虚拟机名称为 cirros，内存为 256 MB，网络模式使用默认的 NAT 网络。

```
[root@localhost ~]#  virt-install --virt-type kvm --name cirros --ram
256 --boot hd --disk path=/opt/cirros-0.3.4-x86_64-disk.img --network
network=default --graphics vnc,listen=0.0.0.0 --noautoconsole
[root@localhost ~]# virsh list --all
 Id    Name                           State
----------------------------------------------------------
 3     cirros                         running
```

（2）登录虚拟机控制台

使用 virsh console 命令进入虚拟机控制台，进入后查看虚拟机的 IP 地址，可以看到虚拟机获取到的 IP 地址与 virbr0 同网段，cirros 虚拟机就是以 virbr0 为网关的。

```
[root@localhost opt]# virsh console cirros
=== network info ===
if-info: lo,up,127.0.0.1,8,::1
if-info: eth0,up,192.168.122.40,24,fe80::5054:ff:fec4:bc
```

```
ip-route:default via 192.168.122.1 dev eth0
ip-route:192.168.122.0/24 dev eth0  src 192.168.122.40
=== datasource: None None ===
=== cirros: current=0.3.4 uptime=277.42 ===

  ____               ____  ____
 / __/ __ ____ ____ / __ \/ __/
/ /__ / // __// __// /_/ /\ \
\___//_//_/  /_/   \____/___/
   http://cirros-cloud.net


login as 'cirros' user. default password: 'cubswin:)'. use 'sudo' for root.
cirros login: cirros
Password:
$ ip a
1: lo: <LOOPBACK,UP,LOWER_UP> mtu 16436 qdisc noqueue
    link/loopback 00:00:00:00:00:00 brd 00:00:00:00:00:00
    inet 127.0.0.1/8 scope host lo
    inet6 ::1/128 scope host
     valid_lft forever preferred_lft forever
2: eth0: <BROADCAST,MULTICAST,UP,LOWER_UP> mtu 1500 qdisc pfifo_fast qlen
1000
    link/ether 52:54:00:c4:00:bc brd ff:ff:ff:ff:ff:ff
    inet 192.168.122.40/24 brd 192.168.122.255 scope global eth0
    inet6 fe80::5054:ff:fec4:bc/64 scope link
     valid_lft forever preferred_lft forever
```

在 NAT 模式下，如果虚拟机需要访问外部网络，需开启宿主机的路由转发功能。

```
[root@localhost ~]# vi /etc/sysctl.conf
net.ipv4.ip_forward=1
[root@localhost ~]# sysctl -p
net.ipv4.ip_forward = 1
[root@localhost ~]# cat /proc/sys/net/ipv4/ip_forward
1
```

在虚拟机中验证访问外网的情况。

```
$ ping www.baidu.com
PING www.a.shifen.com (14.215.177.39) 56(84) bytes of data.
64 bytes from 14.215.177.39 (14.215.177.39): icmp_seq=1 ttl=128 time=32.8 ms
64 bytes from 14.215.177.39 (14.215.177.39): icmp_seq=2 ttl=128 time=32.9 ms
```

```
64 bytes from 14.215.177.39 (14.215.177.39): icmp_seq=3 ttl=128 time=33.1 ms
64 bytes from 14.215.177.39 (14.215.177.39): icmp_seq=4 ttl=128 time=33.3 ms
^C
```

任务 2 为 KVM 虚拟机创建桥接网络

在本单元任务 1 中，使用的网络模式是默认的 NAT 网络，在有些场景中，可能需要虚拟机与宿主机在同一子网，这时需要为虚拟机创建虚拟桥接网络。

创建一台 centos7 虚拟机，虚拟机名称为 guest01，内存 1 GB，网络模式使用桥接模式，创建完成后使用 VNC 远程桌面协议登录虚拟机访问其桌面。

1. 创建虚拟桥接网络

添加虚拟网桥 br0。

```
[root@localhost ~]# brctl addbr br0
```

修改网卡配置文件。

```
[root@localhost ~]# cat /etc/sysconfig/network-scripts/ifcfg-ens33
TYPE=Ethernet
BOOTPROTO=static
NAME=ens33
DEVICE=ens33
ONBOOT=yes
BRIDGE=br0
[root@localhost ~]# cat /etc/sysconfig/network-scripts/ifcfg-br0
TYPE=Bridge
BOOTPROTO=none
NAME=br0
DEVICE=br0
ONBOOT=yes
IPADDR=192.168.134.128
NETMASK=255.255.255.0
GATEWAY=192.168.134.2
DNS1=192.168.134.2
```

重启网络服务。

```
[root@localhost ~]# systemctl restart network
[root@localhost ~]# ip a
1: lo: <LOOPBACK,UP,LOWER_UP> mtu 65536 qdisc noqueue state UNKNOWN group
default qlen 1000
    link/loopback 00:00:00:00:00:00 brd 00:00:00:00:00:00
```

```
        inet 127.0.0.1/8 scope host lo
            valid_lft forever preferred_lft forever
        inet6 ::1/128 scope host
            valid_lft forever preferred_lft forever
    2: ens33: <BROADCAST,MULTICAST,UP,LOWER_UP> mtu 1500 qdisc pfifo_fast master
br0 state UP group default qlen 1000
        link/ether 00:0c:29:e7:4d:35 brd ff:ff:ff:ff:ff:ff
    3: virbr0: <BROADCAST,MULTICAST,UP,LOWER_UP> mtu 1500 qdisc noqueue state UP
group default qlen 1000
        link/ether 52:54:00:91:4f:b7 brd ff:ff:ff:ff:ff:ff
        inet 192.168.122.1/24 brd 192.168.122.255 scope global virbr0
            valid_lft forever preferred_lft forever
    4: virbr0-nic: <BROADCAST,MULTICAST> mtu 1500 qdisc pfifo_fast master virbr0
state DOWN group default qlen 1000
        link/ether 52:54:00:91:4f:b7 brd ff:ff:ff:ff:ff:ff
    12: vnet0: <BROADCAST,MULTICAST,UP,LOWER_UP> mtu 1500 qdisc pfifo_fast
master virbr0 state UNKNOWN group default qlen 1000
        link/ether fe:54:00:ed:80:55 brd ff:ff:ff:ff:ff:ff
        inet6 fe80::fc54:ff:feed:8055/64 scope link
            valid_lft forever preferred_lft forever
    13: br0: <BROADCAST,MULTICAST,UP,LOWER_UP> mtu 1500 qdisc noqueue state UP
group default qlen 1000
        link/ether 00:0c:29:e7:4d:35 brd ff:ff:ff:ff:ff:ff
        inet 192.168.134.128/24 brd 192.168.134.255 scope global noprefixroute br0
            valid_lft forever preferred_lft forever
        inet6 fe80::20c:29ff:fee7:4d35/64 scope link
            valid_lft forever preferred_lft forever
[root@localhost ~]# brctl show
bridge name        bridge id            STP enabled      interfaces
br0                8000.000c29e74d35    no               ens33
virbr0             8000.525400914fb7    yes              virbr0-nic
                                                         vnet0
```

2. 创建虚拟机

添加镜像文件 centos7-01.raw，模拟 VM 磁盘。

```
[root@localhost ~]# qemu-img create -f raw /opt/ guest01.raw 5G
```

启动虚拟机。

```
[root@localhost ~]# virt-install --virt-type kvm --name guest01 --ram
1024 --cdrom=/opt/CentOS-7-x86_64-DVD-1908.iso --disk path=/opt/ guest01.raw
--network bridge=br0 --graphics vnc,listen=0.0.0.0  --noautoconsole
```

3. 搭建 VNC 远程桌面服务

在 CentOS 中安装 VNC 服务端程序并开启服务。

```
[root@localhost opt]# yum -y install tigervnc-server
[root@localhost opt]# vncserver
Password:
Verify:
Would you like to enter a view-only password (y/n)? n
A view-only password is not used
```

在 VNC 客户端软件中输入宿主机 IP 及 VNC 连接的端口号，端口号默认为 5900。如果有多个 VNC 连接，端口会从 5900 开始依次开启可用端口，这里可以使用 ps aux 和 netstat-ntpl 命令确认每个 VNC 连接对应的端口号。

```
[root@localhost ~]# ps aux|grep vnc
root        4125  2.5  0.4 207884 37304 pts/0     Sl   10:58    0:00 /bin/
Xvnc :1 -auth /root/.xauthmzSuvW -desktop localhost.localdomain:1 (root) -fp
catalogue:/etc/X11/fontpath.d -geometry 1024x768 -pn -rfbauth /root/.vnc/passwd
-rfbport 5901 -rfbwait 30000
root        4132  0.0  0.0 113280  1204 pts/0     S    10:58    0:00 /bin/sh /
root/.vnc/xstartup
root        4313  2.3  0.2 193464 23520 pts/0     Sl   10:58    0:00 /bin/
Xvnc :2 -auth /root/.xauthmzSuvW -desktop localhost.localdomain:2 (root) -fp
catalogue:/etc/X11/fontpath.d -geometry 1024x768 -pn -rfbauth /root/.vnc/passwd
-rfbport 5902 -rfbwait 30000
root        4532  0.0  0.0 113280  1204 pts/0     S    10:58    0:00 /bin/sh /
root/.vnc/xstartup
root        4943  0.0  0.0 112828   984 pts/0     R+   10:58    0:00 grep
--color=auto vnc
[root@localhost ~]# netstat -ntpl
Active Internet connections (only servers)
Proto Recv-Q Send-Q Local Address            Foreign Address         State
PID/Program name
tcp        0      0 127.0.0.1:631            0.0.0.0:*               LISTEN
1176/cupsd
tcp        0      0 127.0.0.1:25             0.0.0.0:*               LISTEN
1394/master
tcp        0      0 0.0.0.0:5901             0.0.0.0:*               LISTEN
4125/Xvnc
tcp        0      0 0.0.0.0:5902             0.0.0.0:*               LISTEN
4313/Xvnc
```

访问 centos7 虚拟机桌面，并查看虚拟机的 IP 地址，可以看到虚拟机的 IP 地址与宿主机 ens33 网卡的 IP 地址在同一网段，如图 1-10 所示。

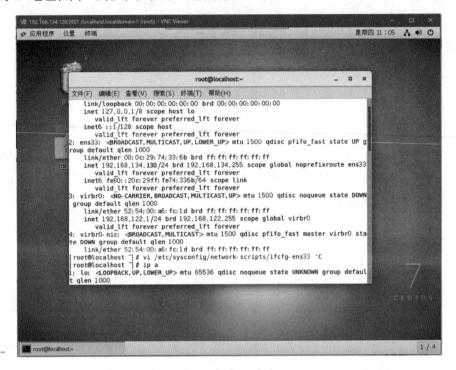

图 1-10　查看 IP 地址

如果想要将原有的虚拟机的网络模式从 NAT 转换为桥接模式，可以通过修改虚拟机 xml 配置文件的方式实现。在 /etc/libvirt/qemu/ 目录下，以虚拟机名称命名的配置文件为虚拟机的 xml 配置文件，定义了关于虚拟机各方面的属性参数，将网络配置部分改为桥接。

```
# virsh edit /etc/libvirt/qemu/ guest01
        <interface type='bridge'>
            <source bridge='br0'>
# virsh shutdown guest01
# virsh start guest01
```

1.4 项目实战——管理 Linux 服务器

项目背景

某公司为提高服务器工作效率，决定将原有部分 Windows 服务器中的业务迁移至 Linux 服务器。公司决定在 Windows 10 操作系统上安装虚拟软件 VMware Workstation，并安装 CentOS 7 操作系统的 Linux 服务器。

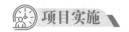 项目实施

任务 1 获取 Linux 操作系统镜像

对于技术人员来说，希望可以方便地找到各个发行版的操作系统镜像进行学习和研究，最基本的方式当然是通过官网获得。除此之外，在许多大型社区中，也提供了包含各种类型、各种版本的操作系统的镜像站，对学习者或者技术人员十分友好。例如，在阿里云开发者社区的镜像站中，可以方便地获取常用的操作系统镜像。下面以下载 CentOS7-1804 版本操作系统镜像完整版为例进行说明。选择镜像站，如图 1-11 所示。

图 1-11 选择镜像站

选择相应的操作系统，例如 centos，如图 1-12 所示。

图 1-12 选择 centos

进入下载目录，如图 1-13 所示。

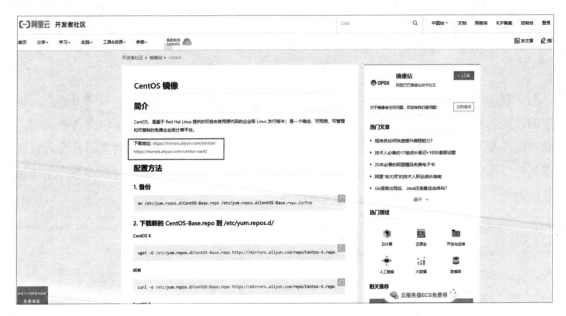

图 1-13　进入下载目录

选择相应版本的镜像进行下载，如图 1-14 ～图 1-17 所示。

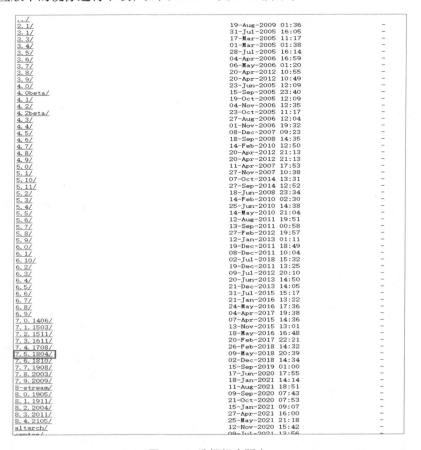

图 1-14　选择相应版本

```
Index of /centos-vault/7.5.1804/

../
atomic/                          05-Jun-2015 11:33        -
centosplus/                      10-May-2018 12:48        -
cloud/                           03-Nov-2015 11:59        -
configmanagement/                06-Oct-2017 15:49        -
cr/                              09-May-2018 20:43        -
dotnet/                          29-Sep-2017 12:33        -
extras/                          10-May-2018 04:12        -
fasttrack/                       01-Sep-2017 11:08        -
isos/                            04-May-2018 12:20        -
nfv/                             26-Feb-2018 14:32        -
opstools/                        13-Sep-2017 12:54        -
os/                              09-May-2018 14:54        -
paas/                            18-May-2016 15:36        -
rt/                              10-Feb-2017 21:18        -
sclo/                            04-Nov-2015 10:27        -
storage/                         13-Nov-2015 17:33        -
updates/                         13-Aug-2018 17:04        -
virt/                            12-Nov-2015 12:07        -
```

图 1-15　选择 isos 目录

```
Index of /centos-vault/7.5.1804/isos/

../
x86_64/                          11-May-2018 16:27
```

图 1-16　选择 x86_64/ 目录

```
Index of /centos-vault/7.5.1804/isos/x86_64/

../
0_README.txt                            09-May-2018 20:16         2495
CentOS-7-x86_64-DVD-1804.iso            03-May-2018 21:07   4470079488
CentOS-7-x86_64-DVD-1804.torrent        11-May-2018 15:43        85846
CentOS-7-x86_64-Everything-1804.iso     07-May-2018 12:55   9397338112
CentOS-7-x86_64-Everything-1804.torrent 11-May-2018 15:43       179848
CentOS-7-x86_64-LiveGNOME-1804.iso      02-May-2018 18:21   1388314624
CentOS-7-x86_64-LiveGNOME-1804.torrent  11-May-2018 15:43        53563
CentOS-7-x86_64-LiveKDE-1804.iso        02-May-2018 18:28   1890582528
CentOS-7-x86_64-LiveKDE-1804.torrent    11-May-2018 15:43        72717
CentOS-7-x86_64-Minimal-1804.iso        03-May-2018 21:07    950009856
CentOS-7-x86_64-Minimal-1804.torrent    11-May-2018 15:43        36836
CentOS-7-x86_64-NetInstall-1804.iso     03-May-2018 20:34    519045120
CentOS-7-x86_64-NetInstall-1804.torrent 11-May-2018 15:43        20405
sha1sum.txt                             09-May-2018 20:02          454
sha1sum.txt.asc                         11-May-2018 15:12         1314
```

图 1-17　选择 CentOS-7 镜像

任务 2　创建新的虚拟机

选择使用 VMware Workstation 软件提供虚拟化环境。VMware Workstation 软件是一款功能强大的桌面虚拟计算机软件，适合单台计算机使用，用户能够在单一的桌面上同时运行不同的操作系统，以及能够进行开发、测试 、部署新的应用程序。VMware Workstation 软件以较好的灵活性与先进的技术成为市面上主流的桌面虚拟计算机软件之一。

选择"新建虚拟机"命令，如图 1-18 所示。单击"创建新的虚拟机"按钮。

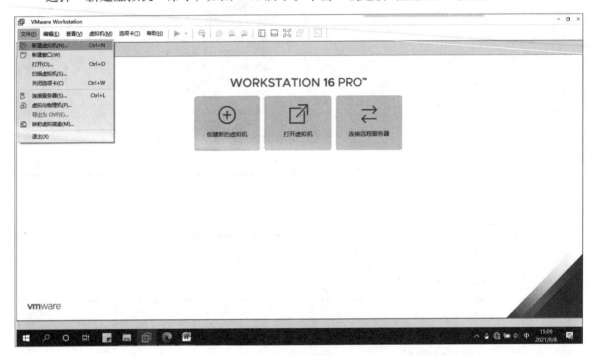

图 1-18　新建虚拟机

选中"自定义（高级）"单选按钮，如图 1-19 所示 。单击"下一步"按钮。

图 1-19　选中"自定义（高级）"单选按钮

选择虚拟机硬件兼容性，使用默认选项即可，如图 1-20 所示。单击"下一步"按钮。

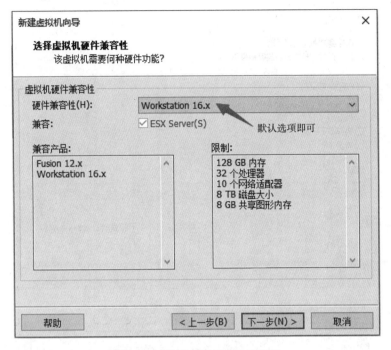

图 1-20　选择虚拟机硬件兼容性

选择"稍后安装操作系统"命令。在虚拟机创建完成后再加载镜像文件进行系统安装，如图 1-21 所示。单击"下一步"按钮。

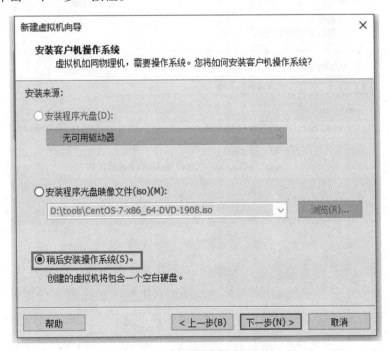

图 1-21　选择稍后安装操作系统

选择操作系统类型为 Linux，版本为 CentOS 7，如图 1-22 所示。单击"下一步"按钮。

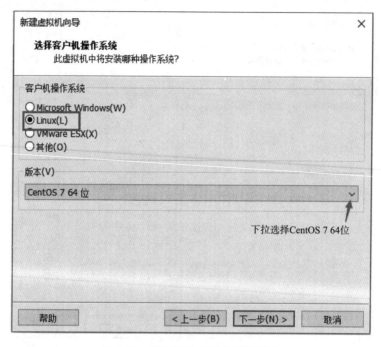

图 1-22 选择操作系统类型及版本

定义虚拟机名称及虚拟机安装位置，如图 1-23 所示。

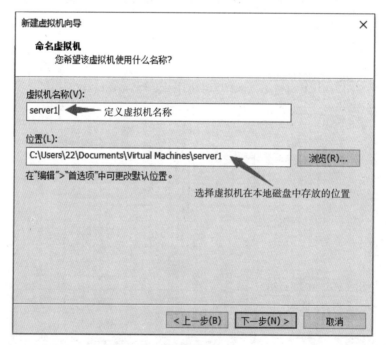

图 1-23 定义虚拟机名称及虚拟机安装位置

对处理器做相关配置，原则是虚拟机资源配置不超过物理宿主机资源配置的上限。对于其他硬件的相关配置都遵循这一原则。例如，物理宿主机为双核处理器，那么虚拟机处理器内核数量不能超过 2，如图 1-24 所示。单击"下一步"按钮。

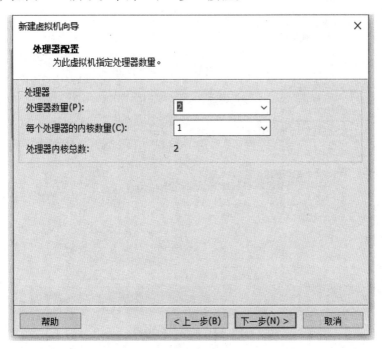

图 1-24　配置处理器数量及内核数量

设置虚拟机内存为 2 048 MB，如图 1-25 所示。单击"下一步"按钮。

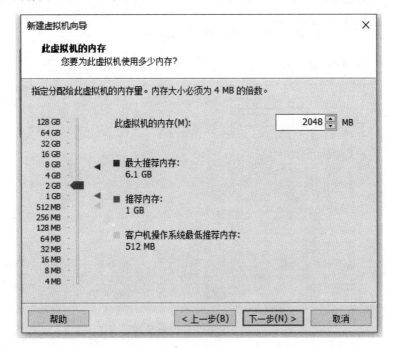

图 1-25　设置虚拟机内存

选中"使用网络地址转换（NAT）"单选按钮，如图 1-26 所示。单击"下一步"按钮。

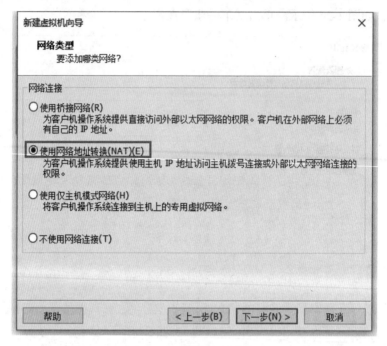

图 1-26 网络类型选择

选择 I/O 控制器类型为 LSI Logic（L），如图 1-27 所示。单击"下一步"按钮。

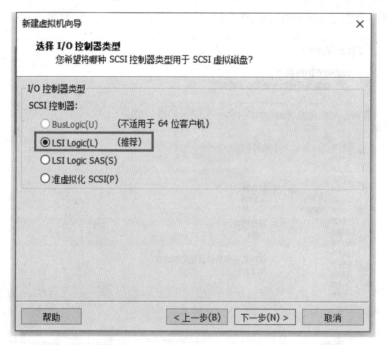

图 1-27 选择 I/O 控制器类型

选择磁盘类型为 SCSI（S），如图 1-28 所示。单击"下一步"按钮。

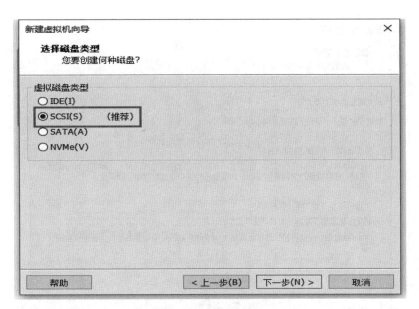

图 1-28　选择磁盘类型

选择磁盘为"创建新虚拟磁盘",如图 1-29 所示。单击"下一步"按钮。

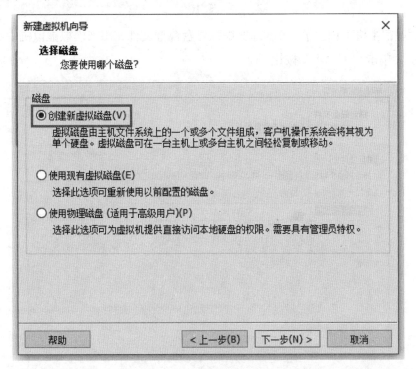

图 1-29　选择磁盘

在配置磁盘时请勿勾选"立即分配所有磁盘空间"复选框,否则物理宿主机的磁盘将被直接占用相应空间,而相反,物理宿主机的存储空间将随着虚拟机的使用动态增加,更加节约资源。设置磁盘大小为 100 GB,如图 1-30 所示。单击"下一步"按钮。

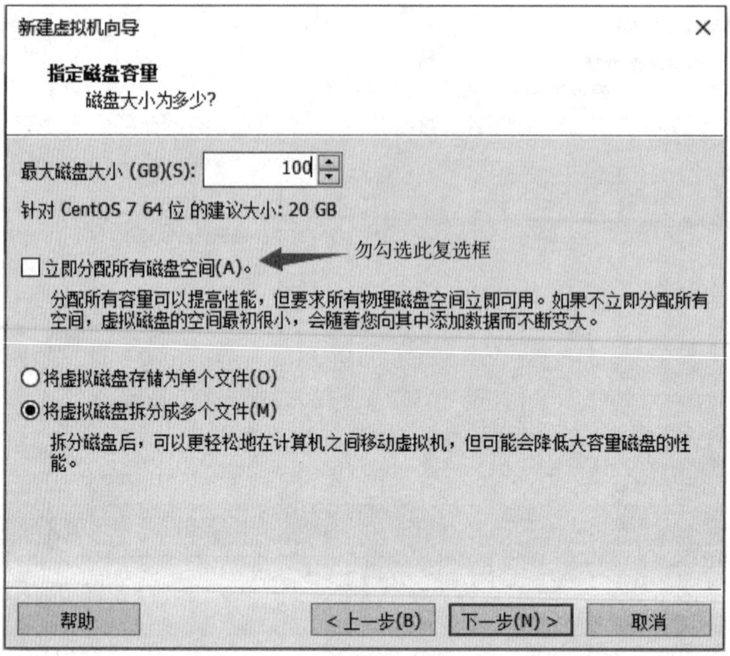

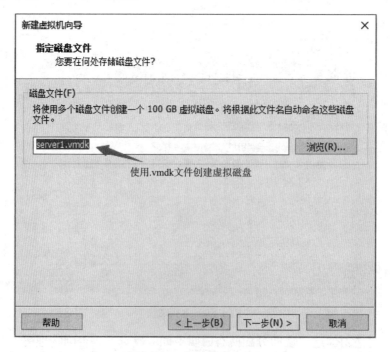

图 1-30　设置磁盘大小

这里在物理宿主机上创建了一个 vmdk 虚拟磁盘镜像文件，用来模拟虚拟机中的磁盘设备，如图 1-31 所示。单击"下一步"按钮。

图 1-31　指定磁盘文件

在虚拟机创建过程中，可以根据需要对硬件设备进行调整，选择自定义硬件，如图 1-32 所示。

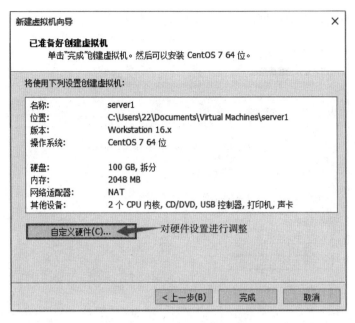

图 1-32　选择自定义硬件

指定虚拟光驱中使用的镜像文件为 CentOS 7 镜像文件，如图 1-33 所示。

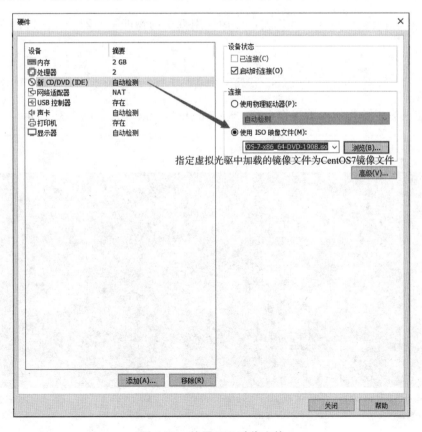

图 1-33　使用 ISO 映像文件

单击"完成"按钮，虚拟机创建成功，如图 1-34 所示。

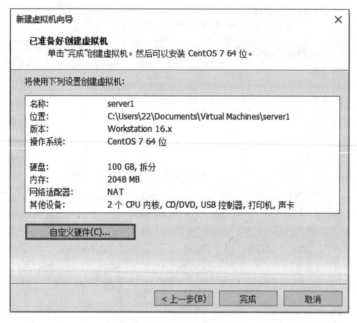

图 1-34 虚拟机创建成功

至此，虚拟机已创建完成，我们就可以像使用真实的物理服务器一样，进行安装操作系统等其他操作了。

任务 3 安装 CentOS 7 操作系统

下面给创建好的虚拟机安装操作系统，选择 Install CentOS 7，如图 1-35 所示。

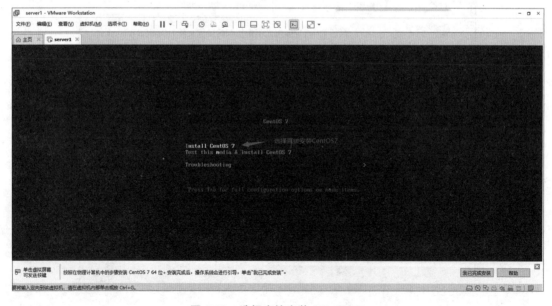

图 1-35 选择直接安装 CentOS 7

选择语言及时区。在安装系统过程中，使用手动方式进行分区，将整个磁盘分为根分区（/），交换分区（swap），以及启动分区（/boot），并设置分区大小。过程如图 1-36～图 1-41 所示。

图 1-36　对磁盘进行手动分区

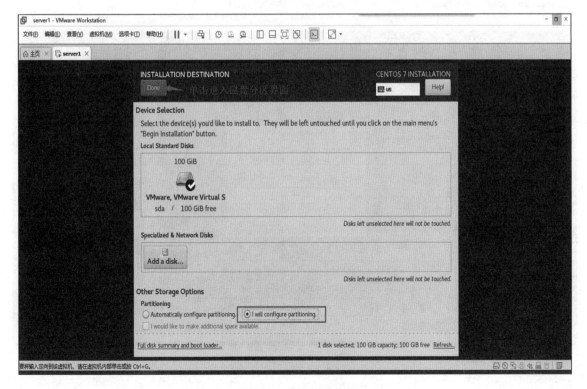

图 1-37　选择手动分区

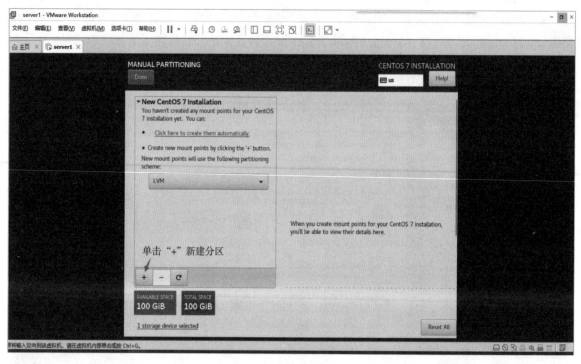

图 1-38　新建分区

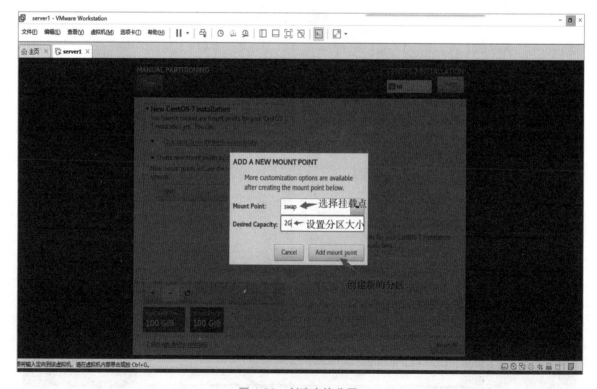

图 1-39　创建交换分区

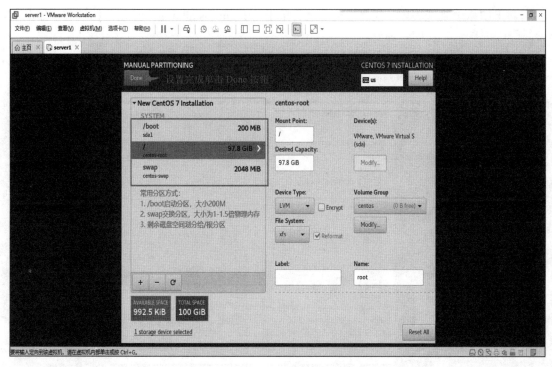

图 1-40　创建根分区

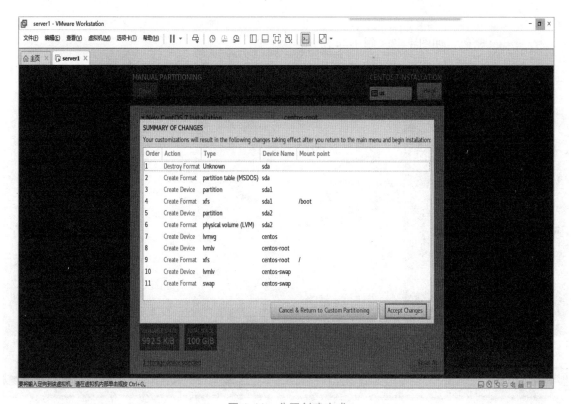

图 1-41　分区创建完成

可将 KDUMP 功能关闭，提高性能。之后单击"Begin Installation"按钮开始安装，如图 1-42 所示。

图 1-42　开始安装

设置 root 密码，如图 1-43、图 1-44 所示。

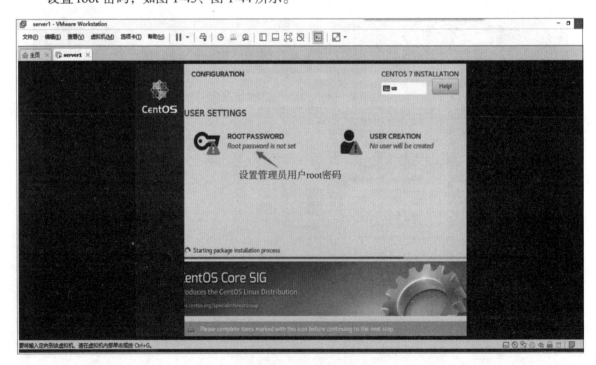

图 1-43　设置 root 密码

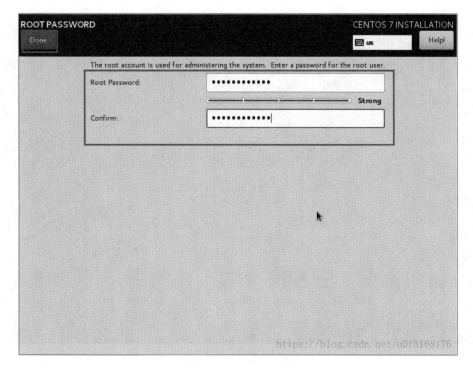

图 1-44　密码设置完成

安装完成后，单击 Reboot 按钮重启系统，并使用 root 用户及其密码登录，如图 1-45～图 1-47 所示。

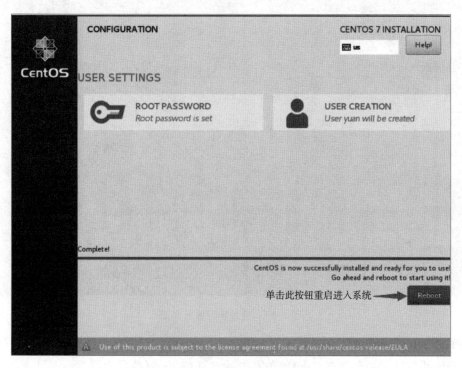

图 1-45　重启系统

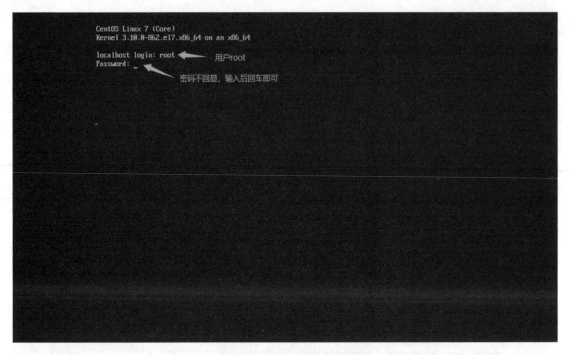

图 1-46　输入用户名和密码

图 1-47　进入字符终端界面

　　当看到图 1-47 所示的命令提示符"[root@localhost ~]#"时，代表使用 root 登录系统成功。至此，已经成功安装了一台具有 CentOS 7 操作系统的 Linux 服务器主机。

单元 2

云计算技术

在虚拟化技术的基础上，云计算技术飞速发展，越来越多的公司开始将业务迁移上云。与传统 IT 系统相比，云计算就是将计算资源放在云端，通过网络访问提供给用户，用户按使用量付费，并且无须关心运维问题。用户可以随时随地访问云端的资源，也无须担心容量不够、硬件损坏等问题。云计算是一种将可伸缩、弹性、共享的物理和虚拟资源池，以按需自服务的方式供应和管理，并提供网络访问的模式。

2.1 云计算的五个特征

1. 自助服务

用户在使用云端资源时，根据自身需求配置获取所需的计算能力。这一特点降低了用户的时间成本和操作成本，用户无须进行额外的人工交互，就能具备在需要的时候按照自身需求实现相应业务的能力。

2. 广泛的网络接入

用户可以从任何网络覆盖的地方，使用各种客户端设备，包括移动电话、平板计算机、笔记本计算机和工作站访问资源。

3. 资源池化

计算资源可以是物理服务器或虚拟计算资源，将它们作为一个集合，这个集合称为资源池。资源池化后，云计算提供的计算资源可以不受物理硬件的限制，根据用户的需求以资源池为整体进行资源的调度。并且，当虚拟资源所依托的底层物理硬件出现故障，对用户来说也是透明的，云服务会自动将虚拟资源迁移到资源池中可用的底层硬件之上，用户无须进行运维，并且不会中断业务。

4. 快速弹性

对于用户的需求，云计算可以提供快速、弹性、可伸缩的服务。在一些大型电商网站的应用场景中，云计算的这一特性优势尤为突出。例如淘宝等，当购物节到来时，站点的访问量会急剧增加，这时，如果基于传统 IT 系统，就需要通过增加物理服务器的节点数来应对，而购物节过后，这些资源又会被闲置，而部署在云中的业务，可以随着访问量的增加和减少动态地增加资源或释放资源，对计算资源获得最高效的利用。物理或虚拟资源能够快速、弹性，有时是自动化地供应，以达到快速增减资源的目的。并且在云计算中，对用户来说可以获取的计算资源是没有限制的，可以在任何时间购买任意数量的资源。云计算的快速弹性对用户来说非常友好，用户无须提前规划资源量和容量，需要扩大规模或缩减都可以即时调整。

5. 按需计费

在云计算中，用户无须提前购置硬件，在不使用计算资源时也不需要支付额外的费用。只需要根据资源使用量、使用时间来付费。

2.2 云计算的三种服务模式

IT 系统从底层依次分为基础层、网络层、存储层、服务器层、操作系统层、数据库层、中间件层、应用软件层，在此基础上是数据信息层。其中，基础层、网络层、存储层、服务器层属于基础设施层；操作系统层、数据库层、中间件层构成平台软件层，为上层应用软件提供操作系统、库函数、数据库等运行环境。根据云服务器提供商提供给用户对应 IT 系统层次的不同，云计算可以分为三种服务模式，分别是 IaaS（infrastructure as a service，基础设施即服务），PaaS（platform as a service，平台即服务），SaaS（software as a service，软件即服务）。IT 系统体系结构如图 2-1 所示。

1. IaaS

将 IT 系统的基础设施架构层作为服务提供给用户使用。云服务提供商进行机房基础设施建设，并且部署网络、存储及物理服务器，将物理计算资源进行池化，将硬件服务器或虚拟服务器、存储、网络（包括公网 IP、DNS 服务、负载均衡服务、防火墙等）提供给用户。用户获取 IaaS 服务后，需要自己安装操作系统，部署数据库、中间件，安装应用软件。使用 IaaS 的用户通常为系统技术人员。

2. PaaS

云服务提供商在建设好基础设施后，在服务器上安装操作系统、数据库和中间件，此时所提供的资源已具备了上层应用的运行环境，因此使用 PaaS 的用户通常是开发技术人员，在 PaaS 基础上进行开发工作。

3. SaaS

在平台软件层基础上安装好各种应用系统软件，用户通过网络访问并使用这些软件。对于用户来说，不需要做任何维护或管理，只需要使用软件即可。目前，许多常用的软件，如微信小程序、支付宝等都属于 SaaS 产品。

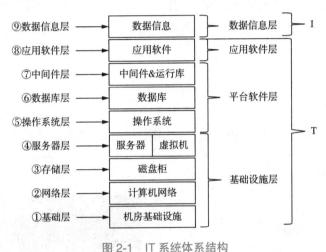

图 2-1　IT 系统体系结构

2.3　云计算的四种部署模型

1. 私有云

私有云是指使用云计算服务的用户是来自一个单位组织的人员，例如一个学校或者一个企事业单位。私有云通常为本地部署，但也可以托管在公有云上，单位组织使用 VPN 进行访问。目前使用最广泛的私有云操作系统为 OpenStack，是一种开源的云操作系统，对数据中心的计算、存储和网络资源进行统一的管理。很多公司的私有云平台产品都是基于 OpenStack 的。

2. 社区云

社区云是指使用云计算服务的用户是来自多个单位组织的人员。

3. 公有云

公有云是指云计算服务的用户是来自公共大众。通常公有云由云计算公司通过互联网的方式提供给用户使用。

4. 混合云

云计算资源来自多种部署方式的云，例如同时包含私有云及公有云，就称为混合云。

2.4　项目实战——使用云计算服务搭建博客系统

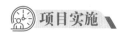 项目背景

某技术小组现要搭建自己的博客系统来进行技术分享，但由于没有现成合适的物理服务器，因此决定使用公有云资源来完成站点的搭建。各公有云中的产品类型不同、特点不同、功能不同，但基本使用方式相同、该项目的运维人员选择的公有云资源是腾讯云。

项目实施

任务 1　获取云服务器

获取云服务器作为网站的 Web 服务器。使用浏览器访问并登录腾讯云，如图 2-2 所示。

图 2-2　访问腾讯云

根据业务的需求配置云服务器并确认选购信息购买。

单击产品，在产品列表中选择计算服务中的"云服务器"，单击"立即选购"按钮。如图 2-3、图 2-4 所示。

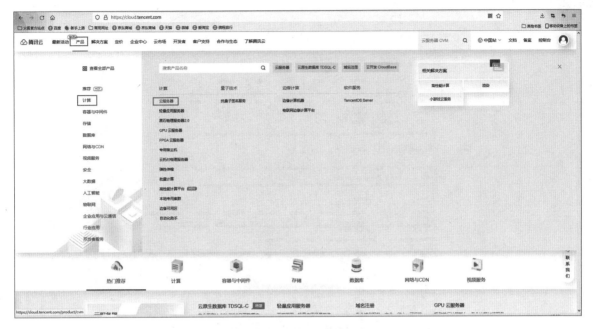

图 2-3　选择计算服务中的"云服务器"

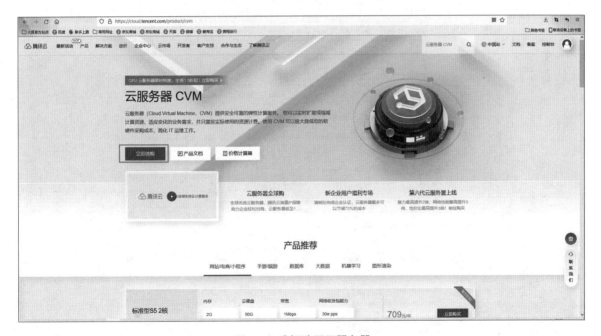

图 2-4　选择购买云服务器

在云服务器配置界面，根据业务选购相应的服务。本站点需要在 Linux 操作系统中搭建
Web 服务器，并需要公有 IP 地址以便用户访问。在本任务中，演示步骤基于实验环境，以最低
成本完成实验为准则，因此计算资源及网络都选择按使用量计费，实例配置选择较低的配置。
在实际使用时，可根据具体需求选择包年、包月及更高的配置，如图 2-5 所示。

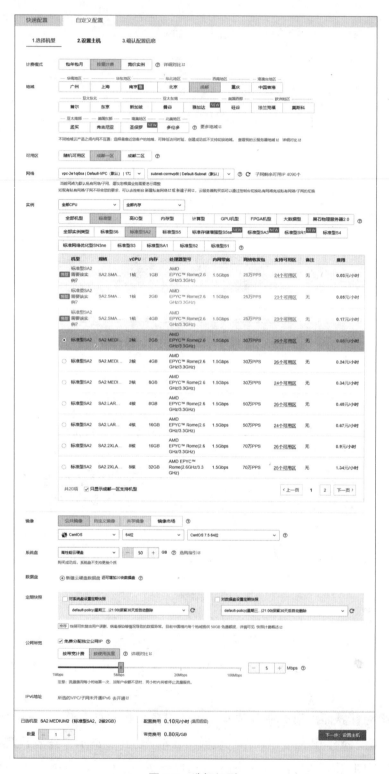

图 2-5　选择机型

设置主机安全组，新建安全组，放行常用端口，为主机设置名称、登录密码，如图 2-6 所示。

图 2-6　设置主机

确认配置信息，完成选购，如图 2-7 所示。

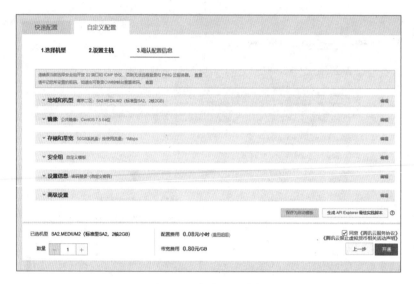

图 2-7　确认配置信息

单击"开通"按钮，这时就获取了一台拥有公有 IP 地址的云服务器，如图 2-8 所示。

图 2-8　云服务器构建完成

任务 2　获取云数据库产品

获取云数据库为论坛网站提供数据库服务，如图 2-9、图 2-10 所示。

图 2-9　选择数据库服务

图 2-10 选购云数据库

后续在搭建站点的过程中，云服务器需要连接数据库，因此在购置数据库时选择与云服务器同区、同一私有网络的同一子网，如图 2-11 ～图 2-13 所示。

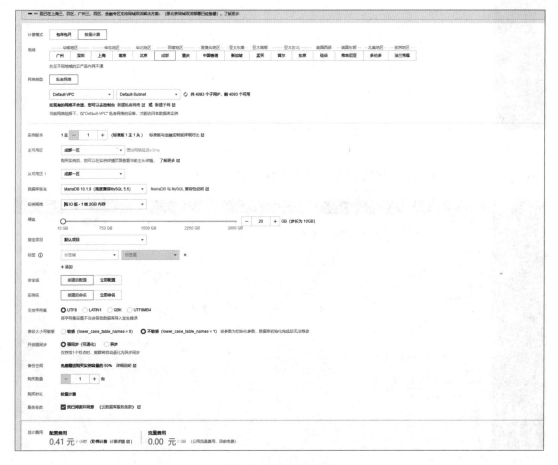

图 2-11 配置云数据库

图 2-12　确认购买实例

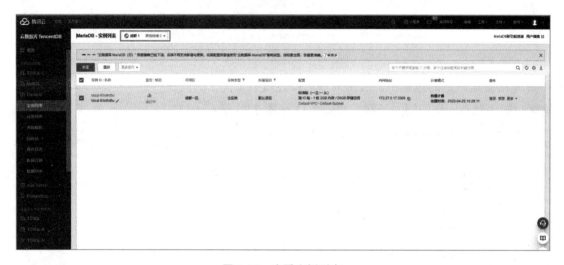

图 2-13　查看实例列表

创建一个账号用来远程连接数据库，如图 2-14、图 2-15 所示。

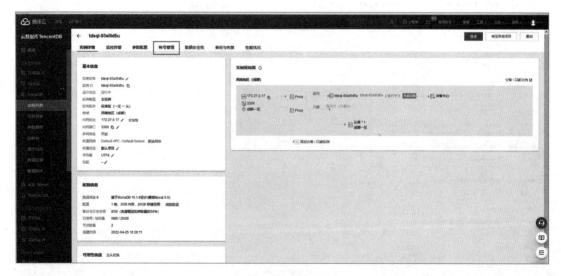

图 2-14　管理账号

图 2-15　新建账号

主机配置为 "%"，表示所有 IP 地址均可使用用户 user 访问数据库，访问密码为 User123456#，如图 2-16 所示。

图 2-16　配置账号信息

开放数据库的所有权限给用户 user，以便对库和表进行操作，如图 2-17 所示。

账号创建完成后，使用用户 user 及相应密码登录数据库，如图 2-18、图 2-19 所示。

选择 "新建" → "新建库" 命令，新建名为 wordpress 的数据库，作为博客系统站点的后台数据库，如图 2-20 ～图 2-22 所示。

图 2-17　修改账号权限

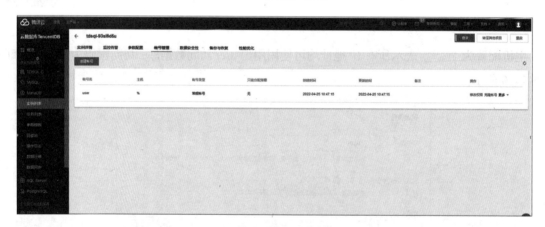

图 2-18　使用账号登录

图 2-19　输入账号信息

图 2-20　选择新建库

图 2-21　新建数据库

图 2-22　配置数据库

任务3   搭建 wordpress 论坛系统

登录云服务器进行 Web 服务器的环境搭建，如图 2-23 所示。

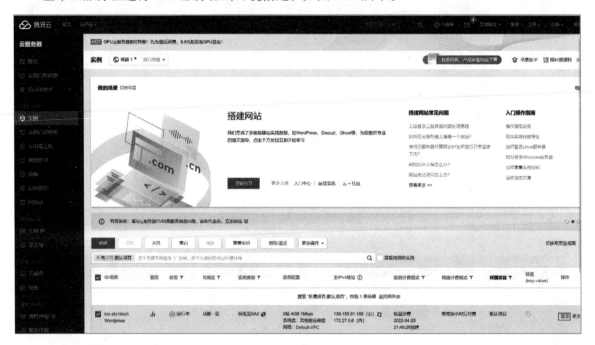

图 2-23   登录云服务器

使用设置的用户名及密码，使用 ssh 协议远程登录云主机，如图 2-24、图 2-25 所示。

图 2-24   使用 ssh 协议远程登录云主机

图 2-25　服务器终端

在本任务中，使用 LNMP 架构，因此需要安装 php、nginx 服务。数据库使用在任务 2 中购置的服务器。

使用 yum 命令安装 php、php-fpm、php-mysql，nginx 仅支持 php 工作在 fast-cgi 模式下。安装完成后，使用 vi 编辑器修改 php-fpm 配置文件，设置用户及用户组为 nginx，使 php-fpm 接受 nginx 的调用。

```
[root@VM-0-6-centos ~]# yum -y install php php-fpm
[root@VM-0-6-centos ~]# vi /etc/php-fpm.d/www.conf
...
user = nginx
group = nginx
...
```

启动 php-fpm 服务，php-fpm 服务工作在 9000 端口，使用 netstat 命令查看当前系统正在工作的端口，可以看到 9000 端口已经处于侦听状态，说明 php-fpm 服务已正常工作。

```
[root@VM-0-6-centos ~]# systemctl start php-fpm              // 启动 php 服务
[root@VM-0-6-centos ~]# systemctl enable php-fpm

[root@VM-0-6-centos ~]# netstat -nptl
Active Internet connections (only servers)
Proto Recv-Q Send-Q Local Address      Foreign Address      State   PID/Program name
tcp      0      0 127.0.0.1:25        0.0.0.0:*            LISTEN 1032/master
tcp      0      0 127.0.0.1:9000      0.0.0.0:*            LISTEN 15455/php-fpm: mast
tcp      0      0 0.0.0.0:3306        0.0.0.0:*            LISTEN 15300/mysqld
tcp      0      0 0.0.0.0:80          0.0.0.0:*            LISTEN 2248/nginx: master
tcp      0      0 0.0.0.0:22          0.0.0.0:*            LISTEN 867/sshd
tcp6     0      0 ::1:25              :::*                 LISTEN 1032/master
tcp6     0      0 :::22               :::*                 LISTEN 867/sshd
```

安装 nginx。配置 nginx 的 yum 仓库，使用 yum 安装 nginx 服务。登录 nginx 官网，选择 RHEL/CentOS 系统，编写 yum 配置文件 /etc/yum.repos.d/nginx.repo，如图 2-26 ～图 2-28 所示。

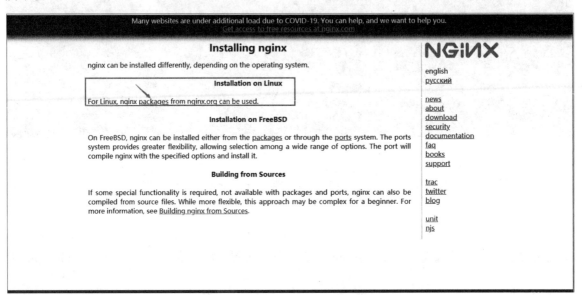

图 2-26　选择软件包安装 nginx

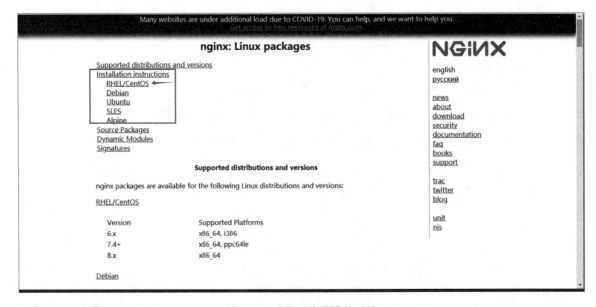

图 2-27　选择对应的操作系统

```
                              RHEL/CentOS

Install the prerequisites:

  sudo yum install yum-utils
            非 root 用户使用 sudo 提权，root 用户执行时不需加 sudo
To set up the yum repository, create the file named /etc/yum.repos.d/nginx.repo with the following
contents:

  [nginx-stable]
  name=nginx stable repo
  baseurl=http://nginx.org/packages/centos/$releasever/$basearch/
  gpgcheck=1
  enabled=1
  gpgkey=https://nginx.org/keys/nginx_signing.key
  module_hotfixes=true

  [nginx-mainline]
  name=nginx mainline repo
  baseurl=http://nginx.org/packages/mainline/centos/$releasever/$basearch/
  gpgcheck=1
  enabled=0
  gpgkey=https://nginx.org/keys/nginx_signing.key
  module_hotfixes=true

By default, the repository for stable nginx packages is used. If you would like to use mainline nginx
packages, run the following command:

  sudo yum-config-manager --enable nginx-mainline

To install nginx, run the following command:

  sudo yum install nginx
```

图 2-28　配置 yum 源并安装 nginx

安装完成后，开启 nginx 服务。

```
[root@VM-0-6-centos ~]# systemctl start nginx
[root@ VM-0-6-centos ~]# systemctl enable nginx
```

配置 nginx，修改 nginx 虚拟主机配置文件，添加索引文件 index.php，打开 location ~\.php$ 配置段的注释，做如下配置。修改完成后重新加载 nginx。

```
[root@VM-0-6-centos ~]# vi /etc/nginx/conf.d/default.conf
server {
    listen       80;
    server_name  localhost;

    #charset koi8-r;
    #access_log  /var/log/nginx/host.access.log  main;

    location / {
        root   /usr/share/nginx/html;
        index  index.php index.html index.htm;
    }
```

```
#error_page   404              /404.html;

# redirect server error pages to the static page /50x.html
#
error_page   500 502 503 504   /50x.html;
location = /50x.html {
    root   /usr/share/nginx/html;
}

# proxy the PHP scripts to Apache listening on 127.0.0.1:80
#
#location ~ \.php$ {
#    proxy_pass   http://127.0.0.1;
#}

# pass the PHP scripts to FastCGI server listening on 127.0.0.1:9000
#
location ~ \.php$ {
    root           /usr/share/nginx/html;
    fastcgi_pass   127.0.0.1:9000;
    fastcgi_index  index.php;
    fastcgi_param  SCRIPT_FILENAME  $document_root$fastcgi_script_name;
    include        fastcgi_params;
}

# deny access to .htaccess files, if Apache's document root
# concurs with nginx's one
#
#location ~ /\.ht {
#    deny  all;
#}
}

[root@VM-0-6-centos ~]# nginx -s reload
```

部署 wordpress。下载 wordpress 站点代码，解压文件。

```
[root@VM-0-6-centos ~]# wget https://cn.wordpress.org/wordpress-5.0.3-zh_
CN.tar.gz--2022-04-25 23:20:22-- https://cn.wordpress.org/wordpress-5.0.3-zh_
CN.tar.gz
```

```
Resolving cn.wordpress.org (cn.wordpress.org)... 198.143.164.252
Connecting to cn.wordpress.org (cn.wordpress.org)|[198.143.164.252|:443...
connected.HTTP request sent,awaiting response... 200 OK
Length: 11098483 (11M) [application/octet-stream]
Saving to: "wordpress-5.0.3-zh_CM.tar.gz"
[root@VM-0-6-centos ~]# tar -zxvf latest-zh_CN.tar.gz
[root@VM-0-6-centos ~]# ll
total 19012
-rw-r--r-- 1 root root 19463922 Apr6 21:00 latest-zh_CN.tar.gz
drwxr-xr-x 5 1006 10064096 Apr6 21:00 wordpress
```

删除 nginx 工作目录下的默认文件，将站点代码复制到工作目录下，并进行权限设置。

```
[root@VM-0-6-centos ~]# rm -rf /usr/share/nginx/html/*
[root@VM-0-6-centos ~]# cp -r wordpress/* /usr/share/nginx/html/
[root@VM-0-6-centos ~]# chmod -R 777 /usr/share/nginx/html
```

使用浏览器访问云主机公网 IP 地址，如果有可用域名，也可将其与云服务器绑定，并通过域名访问，进行站点初始化，如图 2-29 所示。

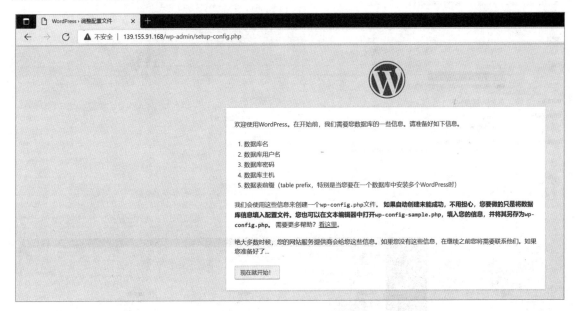

图 2-29　wordpress 安装界面

进行数据库连接信息的配置，数据库名为本单元任务 2 中创建的数据库 wordpress；用户名、密码为本单元任务 2 为数据库创建的远程访问账户；数据库主机为本单元任务 2 中购置的与云服务器同子网的数据库服务器内网 IP 地址，如图 2-30、图 2-31 所示。

设置站点管理员信息，如图 2-32、图 2-33 所示。

安装完成，此时使用管理员登录就可以访问博客系统站点了，如图 2-34、图 2-35 所示。

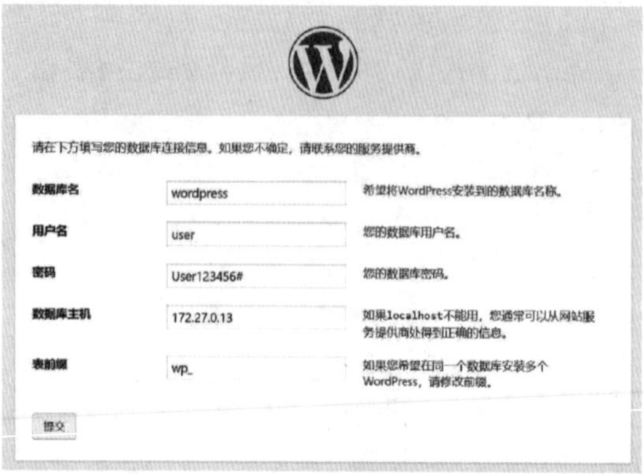

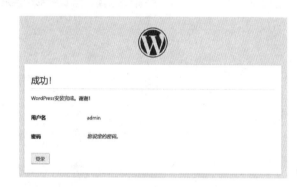

图 2-30　配置环境参数

图 2-31　开始安装

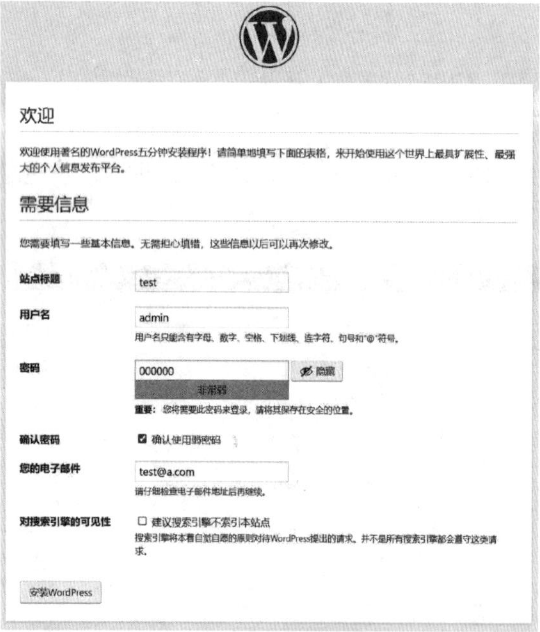

图 2-32　配置站点管理员信息

图 2-33　安装完成

图 2-34　登录 wordpress

图 2-35　wordpress 仪表盘

单元 3

私有云技术——OpenStack

3.1　OpenStack 云操作系统

　　OpenStack 是一个开源的云操作系统，最早是由 Rackspace 和 NASA（美国国家航空航天局）一同成立的开源单元。OpenStack 可统一管理大型资源池 [包括裸机（Bare Metal）、虚拟机（Virtual Machines）及容器（Containers）] 中的计算资源、存储资源和网络资源。所有这些资源都通过统一的身份验证机制的 API（OpenStack SDK）进行管理和配置。此外，OpenStack 还提供了仪表板 dashboard，使管理员能够通过 Web 界面（Horizon Web UI）进行云平台的管理，同时可以赋予用户权限查看相应的资源。除了标准的基础架构即服务功能外，还可以部署第三方服务组件，例如提供编排（Terraform）、故障管理和服务管理（例如管理容器的 Kubernetes 及 CloudFoundry）以及其他服务，以确保用户应用程序的高可用性。图 3-1 为 OpenStack 示意图。

　　OpenStack 开源单元的官方网站界面如图 3-2 所示。

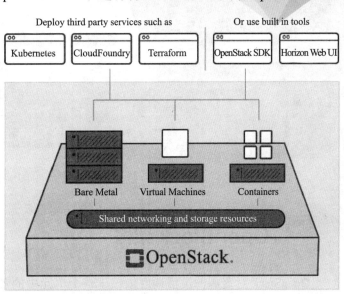

图 3-1　OpenStack 示意图

图 3-2　OpenStack 官方网站界面

OpenStack 平均每半年更新一次版本，版本号从 A 开始，依次使用 B、C、D……开头的单词作为版本名称，当前使用 Y 版本，Z 版本正在开发中，如图 3-3 所示。

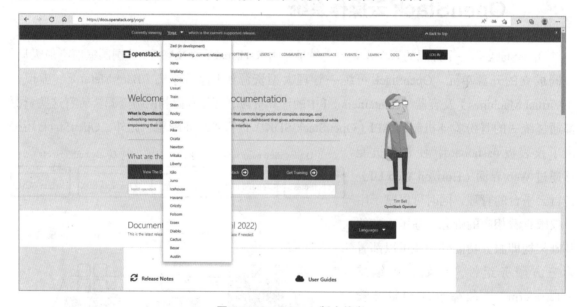

图 3-3　OpenStack 版本信息

3.2　OpenStack 架构

OpenStack 对资源的灵活调度依赖于整个单元中的各个组件，或者说各个服务，这些组件可以根据应用场景和需要实现即插即用。图 3-4 中展示的就是 OpenStack 的各个组件及功能。

OpenStack 的这种架构保证了其高度的灵活性，及其分布式的部署架构，以查看这些服务的适用位置以及它们如何协同工作。

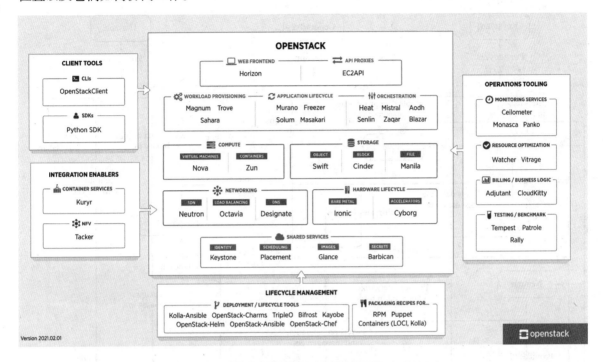

图 3-4 OpenStack 组件示意图

在众多组件中，必备的核心组件有以下六个：

① Nova，计算服务。提供对计算资源（包括裸机、虚拟机和容器）的大规模可扩展、按需自助访问。

② Glance，镜像服务。镜像服务负责发现、注册和搜索虚拟机的镜像。

③ Cinder，块存储服务。提供块存储设备的虚拟化管理，提供自助服务 API，用户通过 API 进行存储资源的请求和使用。

④ Swift，对象存储服务。提供内置冗余及高容错机制的对象存储服务，可以作为 Cinder 的备份存储，为 Glance 提供存储服务。

⑤ Keystone，身份认证服务。负责客户端身份认证、服务发现和令牌管理等。

⑥ Neutron，网络服务。提供网络虚拟化服务，为用户提供 API 接口，可以自定义网络、子网、路由等，配置 DHCP、DNS、负载均衡等服务，支持 VLAN、FLAT、GRE、VXLAN 等网络模式。

除了必备核心组件外，还有常用组件 Horizon 仪表盘，为平台提供 dashboard 管理界面，可以在浏览器可视化环境中对 OpenStack 进行运维管理。图 3-4 为 OpenStack 组件示意图。

OpenStack 最重要的工作就是管理虚拟机的生命周期，以虚拟机为中心，其他各个组件分别为虚拟机 VM 提供相应的服务。Glance 为虚拟机 VM 提供镜像；Nova 管理实例的生命周期；Neutron 为实例提供网络连接服务；Cinder 为实例提供存储卷，来存储实例数据。整个 OpenStack 各组件体系结构如图 3-5 所示。

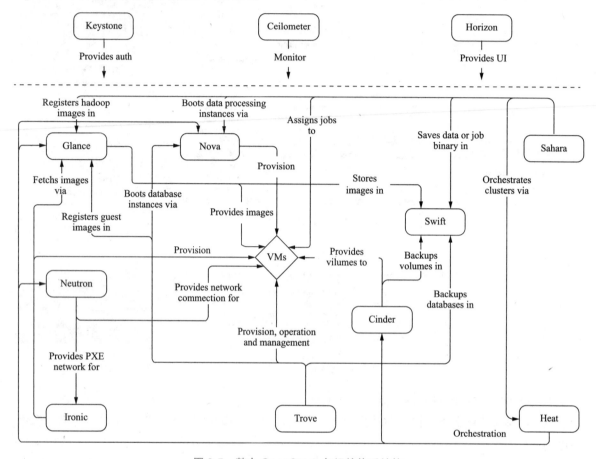

图 3-5 整个 OpenStack 各组件体系结构

每个组件都是由多个服务协同工作完成功能提供的。图 3-6 为各服务之间的逻辑体系结构图。

这里以 Nova 组件为例，阐述组件内部各服务是如何协同工作的。在 Nova 管理实例生命周期的过程中，主要涉及以下服务：

nova-api，负责接收和响应客户 API 调用。除了自身的 API，还兼容对接 Amazon EC2 API，也就是说，可以在 OpenStack 中调用亚马逊的 EC2 实例 API。

nova-scheduler，负责实施调度服务，决定将虚拟机调度在哪个计算节点上运行。

nova-compute，管理虚拟机的核心服务，通过调用 Hypervisor API 实现虚拟机生命周期管理。

nova-conductor，负责为 nova-compute 服务提供数据库交互。

Hypervisor，虚拟化管理程序。常用的 Hypervisor 包括 KVM、Xen、VMware 等。

nova-console，提供虚拟机控制台访问服务。默认使用的访问方式为 nova-novncproxy，是一种基于浏览器的 VNC 远程桌面访问服务。

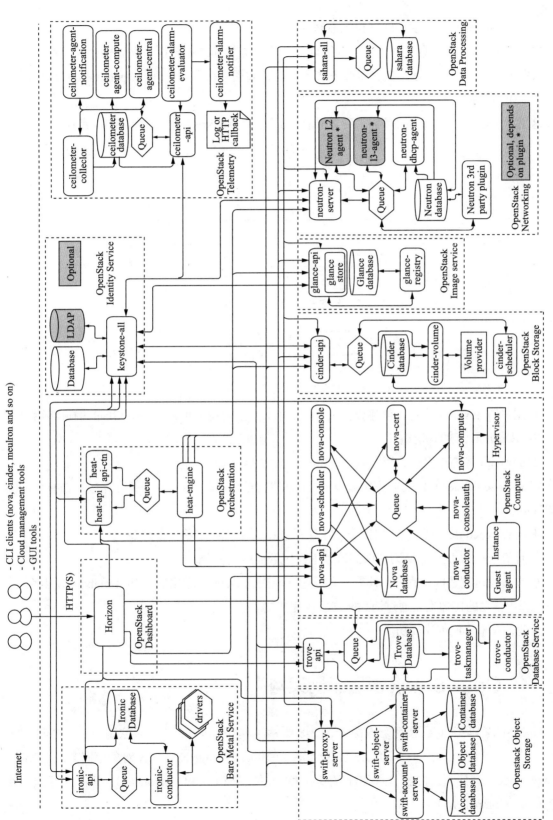

图 3-6　各服务之间的逻辑体系结构图

nova-consoleauth，提供访问虚拟机控制台的 Token 认证。

Message Queue，消息队列服务。为了降低各个服务之间的耦合性，提高平台灵活性，引入消息队列作为信息的中转站，负责将服务产生的消息数据进行队列管理。OpenStack 默认使用的是 RabbitMQ 服务。

这里，从创建一个实例的过程来看 Nova 各服务是如何协同工作的，如图 3-7 所示。

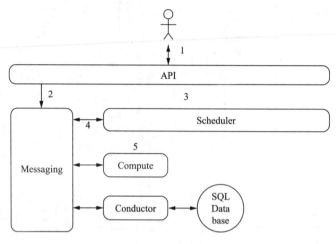

图 3-7　Nova 各服务运行关系

① 用户向 nova-api 发起"创建虚拟机"的请求。这里的用户可以是 OpenStack 的最终用户，也可以是 OpenStack 的其他组件。

② nova-api 将消息提交给消息队列（RabbitMQ）做处理。

③ nova-scheduler 从消息队列中获取 nova-api 发送的消息，然后执行调度算法，选择合适的计算节点 A。

④ nova-scheduler 向消息队列提交"在计算节点 A 上创建实例"的信息。

⑤ 计算节点 A 从消息队列获取到这条消息，开始启动实例。

⑥ 在虚拟机创建的过程中，计算节点通过 nova-conductor 与数据库 SQL Database 发生交互，这一过程中信息的传递也是通过消息队列服务完成的。

至此，一个实例创建成功。

3.3　OpenStack 部署示例架构

OpenStack 典型部署示例架构图如图 3-8 所示。典型部署架构包括控制节点（Controller Node）、计算节点（Compute Node）、块存储节点（Block Storage Node）及对象存储节点（Object Storage Node）。其中，控制节点和计算节点为核心组件（Core Component），其他为可选组件（Optional Component）。

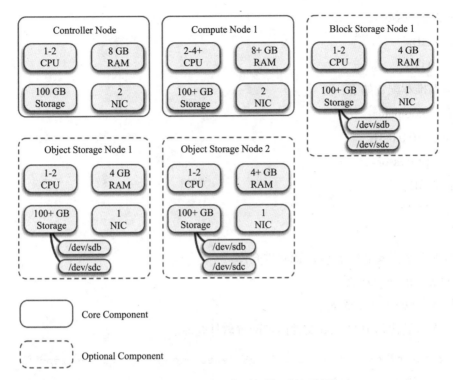

图 3-8　OpenStack 典型部署示例架构图

3.4　项目实战——搭建云计算 OpenStack 系统平台

项目背景

　　某公司为提高现有硬件资源的使用率，决定利用现有资源构建云计算平台，统一管理资源池中的计算资源、存储资源和网络资源，实现对资源的合理分配及灵活调度，并通过统一的身份验证机制 API 实现对资源的管理和配置。

项目实施

任务 1　基础环境搭建

1．硬件及网络配置

准备两台主机分别作为 OpenStack 集群中的控制节点和计算节点。

其中，控制节点的推荐配置如下：

主机名：controller。

IP 地址：10.0.0.11。

CPU：2 processor。

内存：4 GB。

磁盘：10 GB。

计算节点推荐配置如下：

主机名：compute。

IP 地址：10.0.0.31。

CPU：2 processor。

内存：2 GB。

磁盘：20 GB。

2. 软件仓库

在集群所有节点安装以下 yum 源软件仓库。

① 阿里云 centos yum 源。

② 阿里云 openstack yum 源。

按照以下示例代码中的 repo 文件内容配置软件仓库。

```
[root@controller yum.repos.d]# cat CentOS-OpenStack-queens.repo
[centos-openstack-queens]
name=openstack-queens
baseurl=https://mirrors.aliyun.com/centos-vault/7.5.1804/cloud/x86_64/openstack-queens/
enabled=1
gpgcheck=0
[qume-kvm]
name=qemu-kvm
baseurl= https://mirrors.aliyun.com/centos-vault/7.5.1804/virt/x86_64/kvm-common/
enabled=1
gpgcheck=0
[root@controller yum.repos.d]# yum repolist
......
```

repo id	repo name	status
base/7/x86_64	CentOS-7-Base-mirrors.aliyun.com	10,072
centos-openstack-queens	openstack-queens	2,449
extras/7/x86_64	CentOS-7 - Extras - mirrors.aliyun.com	500
qume-kvm	qemu-kvm	63
updates/7/x86_64	CentOS-7 - Updates - mirrors.aliyun.com	3,252
repolist: 16,336		

3. 时间服务

集群中节点之间时间要求严格同步,分别在控制节点和计算节点上安装 chrony 服务并进行配置,实现节点间的时间同步。

```
controller:
[root@controller yum.repos.d]# yum -y install chrony
Loaded plugins: fastestmirror
Loading mirror speeds from cached hostfile
 * base: mirrors.aliyun.com
 * extras: mirrors.aliyun.com
 * updates: mirrors.aliyun.com
centos-openstack-queens | 2.9 kB        00:00
qume-kvm                 | 2.9 kB        00:00
Resolving Dependencies
--> Running transaction check
---> Package chrony.x86_64 0:3.4-1.el7 will be installed
--> Processing Dependency: libseccomp.so.2()(64 bit) for package:
chrony-3.4-1.el7.x86_64
--> Running transaction check
---> Package libseccomp.x86_64 0:2.3.1-4.el7 will be installed
--> Finished Dependency Resolution


Dependencies Resolved


================================================
 Package     Arch     Version      Repository
                                        Size
================================================
Installing:
 chrony      x86_64   3.4-1.el7     base   251 k
Installing for dependencies:
 libseccomp  x86_64   2.3.1-4.el7   base   56 k

Transaction Summary
================================================
Install  1 Package (+1 Dependent package)

Total download size: 307 k
Installed size: 788 k
Downloading packages:
(1/2): libseccomp-2.3.1-4 |  56 kB    00:04
```

```
(2/2): chrony-3.4-1.el7.x | 251 kB   00:04
--------------------------------------------------
Total              64 kB/s | 307 kB  00:04
Running transaction check
Running transaction test
Transaction test succeeded
Running transaction
  Installing : libseccomp-2.3.1-4.el7.x8   1/2
  Installing : chrony-3.4-1.el7.x86_64     2/2
  Verifying  : libseccomp-2.3.1-4.el7.x8   1/2
  Verifying  : chrony-3.4-1.el7.x86_64     2/2

Installed:
  chrony.x86_64 0:3.4-1.el7

Dependency Installed:
  libseccomp.x86_64 0:2.3.1-4.el7

Complete!
[root@controller yum.repos.d]# cat -n /etc/chrony.conf
  1 # Use public servers from the pool.ntp.org project.
  2 # Please consider joining the pool (http://www.pool.ntp.org/join.html).
  3 server ntp1.aliyun.com iburst
  4
  5 # Record the rate at which the system clock gains/losses time.
  6 driftfile /var/lib/chrony/drift
  7
  8 # Allow the system clock to be stepped in the first three updates
  9 # if its offset is larger than 1 second.
 10 makestep 1.0 3
 11
 12 # Enable kernel synchronization of the real-time clock (RTC).
 13 rtcsync
 14
 15 # Enable hardware timestamping on all interfaces that support it.
 16 # hwtimestamp *
 17
 18 # Increase the minimum number of selectable sources required to adjust
 19 # the system clock.
 20 # minsources 2
 21
```

```
22 # Allow NTP client access from local network.
23 allow 10.0.0.0/24
24
25 # Serve time even if not synchronized to a time source.
26 # local stratum 10
27
28 # Specify file containing keys for NTP authentication.
29 # keyfile /etc/chrony.keys
30
31 # Specify directory for log files.
32 logdir /var/log/chrony
33
34 # Select which information is logged.
35 # log measurements statistics tracking
```

```
[root@compute yum.repos.d]# systemctl restart chronyd
[root@compute yum.repos.d]# systemctl enable chronyd
compute:
[root@controller yum.repos.d]# yum -y install chrony
[root@compute yum.repos.d]# cat -n /etc/chrony.conf
 1 # Use public servers from the pool.ntp.org project.
 2 # Please consider joining the pool (http://www.pool.ntp.org/join.html).
 3 server controller iburst
 4 # Record the rate at which the system clock gains/losses time.
 5 driftfile /var/lib/chrony/drift
[root@compute yum.repos.d]# systemctl restart chronyd
[root@compute yum.repos.d]# systemctl enable chronyd
```

与时间源同步，并验证时间是否已同步。

```
[root@controller yum.repos.d]# chronyc sources
210 Number of sources = 1
MS Name/IP address         Stratum Poll Reach LastRx Last sample
===============================================================================
^* 120.25.115.20              2   6   377    11    -79us[ -326us] +/-   17ms
[root@controller yum.repos.d]# date
Fri Dec 24 22:30:35 CST 2021
[root@compute yum.repos.d]# chronyc sources
210 Number of sources = 1
MS Name/IP address         Stratum Poll Reach LastRx Last sample
===============================================================================
^* controller                 3   6   77     49    +89us[ +956us] +/-   22ms
```

```
[root@compute yum.repos.d]# date
Fri Dec 24 22:30:35 CST 2021
```

4. 安装 OpenStack 软件

在集群中所有节点上安装 python-openstackclient 及 openstack-selinux。

```
[root@controller yum.repos.d]# yum-y install python-openstackclient
[root@controller yum.repos.d]# yum-y install openstack-selinux
```

5. 数据库服务

在控制节点上安装数据库服务，并进行初始化及权限设置。

```
[root@controller yum.repos.d]# yum -y install mariadb mariadb-server
[root@controller yum.repos.d]# vi /etc/my.cnf.d/openstack.cnf
[mysqld]
bind-address = 10.0.0.11

default-storage-engine = innodb
innodb_file_per_table = on
max_connections = 4096
collation-server = utf8_general_ci
character-set-server = utf8
[root@controller yum.repos.d]# systemctl restart mariadb
[root@controller yum.repos.d]# systemctl enable mariadb
Created symlink from /etc/systemd/system/multi-user.target.wants/mariadb.
service to /usr/lib/systemd/system/mariadb.service.
[root@controller yum.repos.d]# mysql_secure_installation


NOTE: RUNNING ALL PARTS OF THIS SCRIPT IS RECOMMENDED FOR ALL MariaDB
      SERVERS IN PRODUCTION USE!  PLEASE READ EACH STEP CAREFULLY!


In order to log into MariaDB to secure it, we'll need the current
password for the root user.  If you've just installed MariaDB, and
you haven't set the root password yet, the password will be blank,
so you should just press enter here.


Enter current password for root (enter for none):
OK, successfully used password, moving on...


Setting the root password ensures that nobody can log into the MariaDB root
user without the proper authorisation.
```

```
Set root password? [y/n] y
New password:
Re-enter new password:
Password updated successfully!
Reloading privilege tables.
 ... Success!
```

By default, a MariaDB installation has an anonymous user, allowing anyone to log into MariaDB without having to have a user account created for them. This is intended only for testing, and to make the installation go a bit smoother. You should remove them before moving into a production environment.

```
Remove anonymous users? [y/n] y
 ... Success!
```

Normally, root should only be allowed to connect from 'localhost'. This ensures that someone cannot guess at the root password from the network.

```
Disallow root login remotely? [y/n] y
 ... Success!
```

By default, MariaDB comes with a database named 'test' that anyone can access. This is also intended only for testing, and should be removed before moving into a production environment.

```
Remove test database and access to it? [y/n] y
 - Dropping test database...
 ... Success!
 - Removing privileges on test database...
 ... Success!
```

Reloading the privilege tables will ensure that all changes made so far will take effect immediately.

```
Reload privilege tables now? [y/n] y
 ... Success!

Cleaning up...
```

```
    All  done!   If  you've  completed  all  of  the  above  steps,  your  MariaDB
installation should now be secure.

    Thanks for using MariaDB!
```

6. 消息队列服务

在控制节点上安装 rabbitmq-server，并向数据库中添加相应用户并赋予权限。

```
[root@controller yum.repos.d]# yum -y install rabbitmq-server
[root@controller yum.repos.d]# systemctl start rabbitmq-server
[root@controller yum.repos.d]# systemctl enable rabbitmq-server
Created symlink from /etc/systemd/system/multi-user.target.wants/rabbitmq-
server.service to /usr/lib/systemd/system/rabbitmq-server.service.
[root@controller yum.repos.d]# rabbitmqctl add_user openstack RABBIT_PASS
Creating user "openstack" ...
[root@controller yum.repos.d]# rabbitmqctl set_permissions openstack ".*"
".*" ".*"
Setting permissions for user "openstack" in vhost "/" ...
```

可安装监控插件（web 界面监控工具）。

```
[root@controller yum.repos.d]# rabbitmq-plugins enable rabbitmq_management
The following plugins have been enabled:
 mochiweb
 webmachine
 rabbitmq_web_dispatch
 amqp_client
 rabbitmq_management_agent
 rabbitmq_management

Applying plugin configuration to rabbit@controller... started 6 plugins.
[root@controller yum.repos.d]# netstat -ntpl
Active Internet connections (only servers)
Proto Recv-Q Send-Q Local Address     Foreign Address State      PID/Program name
tcp   0      0 0.0.0.0:25672          0.0.0.0:*       LISTEN     2621/beam.smp
tcp   0      0 10.0.0.11:3306         0.0.0.0:*       LISTEN     2420/mysqld
tcp   0      0 0.0.0.0:4369           0.0.0.0:*       LISTEN     1/systemd
tcp   0      0 10.0.0.11:53           0.0.0.0:*       LISTEN     794/named
tcp   0      0 127.0.0.1:53           0.0.0.0:*       LISTEN     794/named
tcp   0      0 0.0.0.0:22             0.0.0.0:*       LISTEN     774/sshd
tcp   0      0 0.0.0.0:15672          0.0.0.0:*       LISTEN     2621/beam.smp
tcp   0      0 127.0.0.1:25           0.0.0.0:*       LISTEN     905/master
```

```
tcp      0      0 127.0.0.1:953         0.0.0.0:*        LISTEN       794/named
tcp      0      0 :::5672               :::*             LISTEN       2621/beam.smp
tcp6     0      0 ::1:53                :::*             LISTEN       794/named
tcp6     0      0 :::22                 :::*             LISTEN       774/sshd
tcp6     0      0 ::1:25                :::*             LISTEN       905/master
tcp6     0      0 ::1:953               :::*             LISTEN       794/named
```

监听 5672（客户端使用）、25672（集群节点间通信用）、15672（监控 web 界面）。

使用浏览器访问 http://10.0.0.11:15672，如图 3-9、图 3-10 所示。

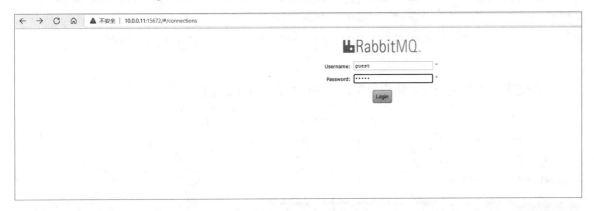

图 3-9　RabbitMQ 登录界面

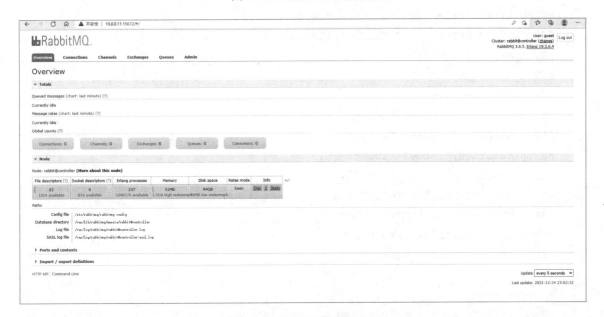

图 3-10　RabbitMQ 管理界面

7. 缓存服务

在控制节点上安装 memcached 服务。

```
[root@controller yum.repos.d]# yum -y install memcached python-memcached
[root@controller yum.repos.d]# cat /etc/sysconfig/memcached
PORT="11211"
USER="memcached"
MAXCONN="1024"
CACHESIZE="64"
OPTIONS="-l 127.0.0.1,::1,controller"
[root@controller yum.repos.d]# systemctl enable memcached.service
Created symlink from /etc/systemd/system/multi-user.target.wants/memcached.
service to /usr/lib/systemd/system/memcached.service.
[root@controller yum.repos.d]# systemctl start memcached.service
```

8. etcd

在控制节点上安装 etcd，并修改配置文件设置客户端 url 地址。

```
[root@controller yum.repos.d]# yum -y install etcd
[root@controller yum.repos.d]# vi /etc/etcd/etcd.conf
#[Member]
#ETCD_CORS=""
ETCD_DATA_DIR="/var/lib/etcd/default.etcd"
#ETCD_WAL_DIR=""
#ETCD_LISTEN_PEER_URLS="http://10.0.0.11:2380"
ETCD_LISTEN_CLIENT_URLS="http://10.0.0.11:2379"
#ETCD_MAX_SNAPSHOTS="5"
#ETCD_MAX_WALS="5"
ETCD_NAME="default"
#ETCD_SNAPSHOT_COUNT="100000"
#ETCD_HEARTBEAT_INTERVAL="100"
#ETCD_ELECTION_TIMEOUT="1000"
#ETCD_QUOTA_BACKEND_BYTES="0"
#ETCD_MAX_REQUEST_BYTES="1572864"
#ETCD_GRPC_KEEPALIVE_MIN_TIME="5s"
#ETCD_GRPC_KEEPALIVE_INTERVAL="2h0m0s"
#ETCD_GRPC_KEEPALIVE_TIMEOUT="20s"
#
#[Clustering]
#ETCD_INITIAL_ADVERTISE_PEER_URLS="http://10.0.0.11:2380"
ETCD_ADVERTISE_CLIENT_URLS="http://10.0.0.11:2379"
#ETCD_DISCOVERY=""
#ETCD_DISCOVERY_FALLBACK="proxy"
#ETCD_DISCOVERY_PROXY=""
```

```
#ETCD_DISCOVERY_SRV=""
#ETCD_INITIAL_CLUSTER="default=http://10.0.0.11:2380"
#ETCD_INITIAL_CLUSTER_TOKEN="etcd-cluster"
#ETCD_INITIAL_CLUSTER_STATE="new"
#ETCD_STRICT_RECONFIG_CHECK="true"
#ETCD_ENABLE_V2="true"
#
#[Proxy]
#ETCD_PROXY="off"
#ETCD_PROXY_FAILURE_WAIT="5000"
#ETCD_PROXY_REFRESH_INTERVAL="30000"
#ETCD_PROXY_DIAL_TIMEOUT="1000"
#ETCD_PROXY_WRITE_TIMEOUT="5000"
#ETCD_PROXY_READ_TIMEOUT="0"
#
#[Security]
#ETCD_CERT_FILE=""
#ETCD_KEY_FILE=""
#ETCD_CLIENT_CERT_AUTH="false"
:1,$s/localhost/10.0.0.11/g


[root@controller yum.repos.d]# systemctl restart etcd
[root@controller yum.repos.d]# systemctl enable etcd
Created symlink from /etc/systemd/system/multi-user.target.wants/etcd.
service to /usr/lib/systemd/system/etcd.service.
```

任务 2   部署 OpenStack 核心组件

1. keystone 组件部署

编辑脚本文件，并通过运行脚本 openstack-keystone-install.sh 安装 keystone。

```
[root@controller ~]# cat openstack-keystone-install.sh
#!/bin/bash
##############################################
# install keystone for Openstack on controller
# Author:Ann   Date:2022-1-5
##############################################


mysql -uroot -p000000 -e "create database IF NOT EXISTS keystone ;"
mysql -uroot -p000000 -e "GRANT ALL PRIVILEGES ON keystone.* TO 'keystone'
```

```
@'localhost' IDENTIFIED BY 'KEYSTONE_DBPASS';"
    mysql -uroot -p000000 -e "GRANT ALL PRIVILEGES ON keystone.* TO 'keystone'
@'%' IDENTIFIED BY 'KEYSTONE_DBPASS';"



    yum -y install openstack-keystone httpd mod_wsgi
    yum -y install openstack-utils


    # 配置数据库连接
    openstack-config --set /etc/keystone/keystone.conf database connection
mysql+pymysql://keystone:KEYSTONE_DBPASS@controller/keystone
    # 使用 fernet 生产 token
    openstack-config --set /etc/keystone/keystone.conf token provider  fernet
    # 同步数据库
    su -s /bin/sh -c "keystone-manage db_sync" keystone
    # 初始化 fernet
    keystone-manage fernet_setup --keystone-user keystone --keystone-group
keystone
    keystone-manage credential_setup --keystone-user keystone --keystone-group
keystone
    # 注册 keystone
    keystone-manage bootstrap --bootstrap-password ADMIN_PASS \
    --bootstrap-admin-url http://controller:5000/v3/ \
    --bootstrap-internal-url http://controller:5000/v3/ \
    --bootstrap-public-url http://controller:5000/v3/ \
    --bootstrap-region-id RegionOne


    # 修改 apache 配置
    sed -i "s/#ServerName www.example.com:80/ServerName controller/g" /etc/
httpd/conf/httpd.conf
    ln -s /usr/share/keystone/wsgi-keystone.conf /etc/httpd/conf.d/
    systemctl restart httpd
    systemctl enable httpd


    # 配置临时管理账户
    export OS_USERNAME=admin
    export OS_PASSWORD=ADMIN_PASS
    export OS_PROJECT_NAME=admin
    export OS_USER_DOMAIN_NAME=Default
    export OS_PROJECT_DOMAIN_NAME=Default
```

```
export OS_AUTH_URL=http://controller:5000/v3
export OS_IDENTITY_API_VERSION=3


# 创建域、单元、用户和角色
openstack domain create --description "An Example Domain" example
openstack project create --domain default --description "Service Project"
service
openstack project create --domain default --description "Demo Project" demo
openstack user create --domain default --password DEMO_PASS demo
openstack role create user
openstack role add --project demo --user demo user



unset OS_AUTH_URL OS_PASSWORD

cat > ~/admin-openrc.sh <<-EOF
export OS_PROJECT_DOMAIN_NAME=Default
export OS_USER_DOMAIN_NAME=Default
export OS_PROJECT_NAME=admin
export OS_USERNAME=admin
export OS_PASSWORD=ADMIN_PASS
export OS_AUTH_URL=http://controller:5000/v3
export OS_IDENTITY_API_VERSION=3
export OS_IMAGE_API_VERSION=2
EOF

cat > ~/demo-openrc.sh <<-EOF
export OS_PROJECT_DOMAIN_NAME=Default
export OS_USER_DOMAIN_NAME=Default
export OS_PROJECT_NAME=demo
export OS_USERNAME=demo
export OS_PASSWORD=DEMO_PASS
export OS_AUTH_URL=http://controller:5000/v3
export OS_IDENTITY_API_VERSION=3
export OS_IMAGE_API_VERSION=2
EOF

echo "source ~/admin-openrc.sh" >> ~/.bashrc
bash
```

```
[root@controller ~]# ./openstack-keystone-install.sh
```

引用 admin-openrc.sh 中的环境变量，使用 openstack token issue 命令获取 token，以验证 keystone 服务是否正常工作。

```
[root@controller ~]# source admin-openrc.sh
[root@controller ~]# openstack token issue

+------------+------------------------------------------------------------------------------------------------------------+
| Field | Value                                                                                                              |
+------------+------------------------------------------------------------------------------------------------------------+
| expires | 2021-12-25T08:21:49+0000                                                                                    |
| id | gAAAAABhxsaN1kzGK_ggpWGbM1W4OW87Krf3HPO1DQ26AD8Xy5A99uOiVnp2je5Admx
ih-HJvTVcZQuEEXuEuIrBpRT9oJReC-Tfvlchw3IStrR_Mhfdjx0hVxRSY_79JBDuoQxETQ7SCkKSI44
8GDKp7Xboyc06r8ECvf3qM7mPVIlzRzyd7NI |
| project_id | e3215238c9ac4193a42dbc88cab07150                                                                |
| user_id    | a431c36941aa4b058cb254fdc52a8a5e                                                                |
+------------+------------------------------------------------------------------------------------------------------------+
```

2. glance 组件部署

编辑脚本文件，并通过运行脚本 openstack-glance-install.sh 的方式安装 glance。

```
[root@controller ~]# cat openstack-glance-install.sh
#!/bin/bash
###############################################
# install glance for Openstack on controller
# Author:Ann   Date:2022-1-5
###############################################


source  ~/admin-openrc.sh
mysql -uroot -p000000 -e "create database IF NOT EXISTS glance;"
mysql -uroot -p000000 -e "GRANT ALL PRIVILEGES ON glance.* TO 'glance'@'localhost'
 IDENTIFIED BY 'GLANCE_DBPASS' ;"
mysql -uroot -p000000 -e "GRANT ALL PRIVILEGES ON glance.* TO 'glance'@'%'
IDENTIFIED BY 'GLANCE_DBPASS' ;"
```

```
openstack user create --domain default --password GLANCE_PASS glance
openstack role add --project service --user glance admin
openstack service create --name glance --description "OpenStack Image" image

openstack endpoint create --region RegionOne image public http://
controller:9292
openstack endpoint create --region RegionOne image internal http://
controller:9292
openstack endpoint create --region RegionOne image admin http://
controller:9292

yum install -y openstack-glance

openstack-config --set /etc/glance/glance-api.conf database connection
mysql+pymysql://glance:GLANCE_DBPASS@controller/glance
openstack-config --set /etc/glance/glance-api.conf keystone_authtoken auth_
uri http://controller:5000
openstack-config --set /etc/glance/glance-api.conf keystone_authtoken auth_
url http://controller:5000
openstack-config --set /etc/glance/glance-api.conf keystone_authtoken
memcached_servers controller:11211
openstack-config --set /etc/glance/glance-api.conf keystone_authtoken auth_
type password
openstack-config --set /etc/glance/glance-api.conf keystone_authtoken
project_domain_name default
openstack-config --set /etc/glance/glance-api.conf keystone_authtoken user_
domain_name default
openstack-config --set /etc/glance/glance-api.conf keystone_authtoken
project_name service
openstack-config --set /etc/glance/glance-api.conf keystone_authtoken
username glance
openstack-config --set /etc/glance/glance-api.conf keystone_authtoken
password GLANCE_PASS
openstack-config --set /etc/glance/glance-api.conf paste_deploy flavor
keystone
openstack-config --set /etc/glance/glance-api.conf glance_store stores
file,http
openstack-config --set /etc/glance/glance-api.conf glance_store default_store file
openstack-config --set /etc/glance/glance-api.conf glance_store filesystem_
```

```
store_datadir /var/lib/glance/images/

    openstack-config --set /etc/glance/glance-registry.conf database connection
mysql+pymysql://glance:GLANCE_DBPASS@controller/glance
    openstack-config --set /etc/glance/glance-registry.conf keystone_authtoken
auth_uri http://controller:5000
    openstack-config --set /etc/glance/glance-registry.conf keystone_authtoken
auth_url http://controller:35357
    openstack-config --set /etc/glance/glance-registry.conf keystone_authtoken
memcached_servers controller:11211
    openstack-config --set /etc/glance/glance-registry.conf keystone_authtoken
auth_type password
    openstack-config --set /etc/glance/glance-registry.conf keystone_authtoken
project_domain_name default
    openstack-config --set /etc/glance/glance-registry.conf keystone_authtoken
user_domain_name default
    openstack-config --set /etc/glance/glance-registry.conf keystone_authtoken
project_name service
    openstack-config --set /etc/glance/glance-registry.conf keystone_authtoken
username glance
    openstack-config --set /etc/glance/glance-registry.conf keystone_authtoken
password GLANCE_PASS
    openstack-config --set /etc/glance/glance-registry.conf paste_deploy flavor
keystone


    su -s /bin/sh -c "glance-manage db_sync" glance


    systemctl enable openstack-glance-api.service openstack-glance-registry.
service
    systemctl restart openstack-glance-api.service openstack-glance-registry.
service
    [root@controller ~]# ./openstack-glance-install.sh
```

下载 cirros 镜像文件，制作 OpenStack 镜像验证 glance 服务是否正常提供服务。

```
    [root@controller ~]# wget http://download.cirros-cloud.net/0.4.0/cirros-
0.4.0-i386-disk.img
    ……
    Saving to: 'cirros-0.4.0-i386-disk.img'
```

```
    100%[========================================================================
===============================>] 12,094,976  8.76MB/s   in 1.3s

    2021-12-25 16:08:00 (8.76 MB/s) - 'cirros-0.4.0-i386-disk.img' saved [12094976/12094976]

    [root@controller ~]# openstack image create "cirros" --file cirros-0.4.0-
i386-disk.img --disk-format qcow2 --container-format bare --public
    +------------------+------------------------------------------------------+
    | Field | Value |
    +------------------+------------------------------------------------------+
    | checksum | b7d8ac291c698c3f1dc0705ce52a3b64 |
    | container_format | bare |
    | created_at | 2021-12-25T08:09:31Z |
    | disk_format | qcow2 |
    | file | /v2/images/1c27bb51-9d4a-47dc-be9d-13445f29580d/file |
    | id | 1c27bb51-9d4a-47dc-be9d-13445f29580d |
    | min_disk | 0 |
    | min_ram | 0 |
    | name | cirros |
    | owner | 50467f099c214d8abba4e3f9cf8475bc |
    | protected | False |
    | schema | /v2/schemas/image |
    | size | 12094976 |
    | status | active |
    | tags | |
    | updated_at | 2021-12-25T08:09:31Z |
    | virtual_size | None |
    | visibility | public |
    +------------------+------------------------------------------------------+
    [root@controller ~]# openstack image list
    +--------------------------------------+--------+--------+
    | ID | Name   | Status |
    +--------------------------------------+--------+--------+
    | 1c27bb51-9d4a-47dc-be9d-13445f29580d | cirros | active |
    +--------------------------------------+--------+--------+
```

3. nova 组件部署

在控制节点编辑脚本文件，并通过运行脚本的方式安装 nova。

```
[root@controller ~]# cat openstack-nova-install.sh
#!/bin/bash
```

```
#################################################
# install nova for Openstack on controller
# Author:Ann   Date:2022-1-5
#################################################


source ~/admin-openrc.sh


mysql -uroot -p000000 -e "create database IF NOT EXISTS nova;"
mysql -uroot -p000000 -e "create database IF NOT EXISTS nova_api;"
mysql -uroot -p000000 -e "create database IF NOT EXISTS nova_cell0 ;"


mysql -uroot -p000000 -e " GRANT ALL PRIVILEGES ON nova_api.* TO 'nova'
@'localhost' IDENTIFIED BY 'NOVA_DBPASS';"
mysql -uroot -p000000 -e " GRANT ALL PRIVILEGES ON nova_api.* TO 'nova'@'%'
IDENTIFIED BY 'NOVA_DBPASS';"
mysql -uroot -p000000 -e " GRANT ALL PRIVILEGES ON nova.* TO 'nova'@'localhost'
 IDENTIFIED BY 'NOVA_DBPASS';"
mysql -uroot -p000000 -e " GRANT ALL PRIVILEGES ON nova.* TO 'nova'@'%'
IDENTIFIED BY 'NOVA_DBPASS';"
mysql -uroot -p000000 -e " GRANT ALL PRIVILEGES ON nova_cell0.* TO 'nova'@'
localhost' IDENTIFIED BY 'NOVA_DBPASS';"
mysql -uroot -p000000 -e " GRANT ALL PRIVILEGES ON nova_cell0.* TO 'nova'@'%'
 IDENTIFIED BY 'NOVA_DBPASS';"


openstack user create --domain default --password NOVA_PASS nova
openstack role add --project service --user nova admin
openstack service create --name nova --description "OpenStack Compute"
compute
openstack endpoint create --region RegionOne compute public http://
controller:8774/v2.1
openstack endpoint create --region RegionOne compute internal http://
controller:8774/v2.1
openstack endpoint create --region RegionOne compute admin http://
controller:8774/v2.1


openstack user create --domain default --password PLACEMENT_PASS placement
openstack role add --project service --user placement admin
openstack service create --name placement --description "Placement API"
placement
```

```
openstack endpoint create --region RegionOne placement public http://
controller:8778
openstack endpoint create --region RegionOne placement internal http://
controller:8778
openstack endpoint create --region RegionOne placement admin http://
controller:8778

yum install -y openstack-nova-api openstack-nova-conductor openstack-nova-
console openstack-nova-novncproxy openstack-nova-scheduler openstack-nova-
placement-api

openstack-config --set /etc/nova/nova.conf database connection
mysql+pymysql://nova:NOVA_DBPASS@controller/nova
openstack-config --set /etc/nova/nova.conf DEFAULT transport_url rabbit://
openstack:RABBIT_PASS@controller
openstack-config --set /etc/nova/nova.conf DEFAULT enabled_apis osapi_
compute,metadata
openstack-config --set /etc/nova/nova.conf DEFAULT my_ip 10.0.0.11
openstack-config --set /etc/nova/nova.conf DEFAULT use_neutron True
openstack-config --set /etc/nova/nova.conf DEFAULT firewall_driver nova.
virt.firewall.NoopFirewallDriver
openstack-config --set /etc/nova/nova.conf api auth_strategy keystone
openstack-config --set /etc/nova/nova.conf api_database connection
mysql+pymysql://nova:NOVA_DBPASS@controller/nova_api
openstack-config --set /etc/nova/nova.conf keystone_authtoken auth_url
http://controller:5000/v3
openstack-config --set /etc/nova/nova.conf keystone_authtoken memcached_
servers controller:11211
openstack-config --set /etc/nova/nova.conf keystone_authtoken auth_type
password
openstack-config --set /etc/nova/nova.conf keystone_authtoken project_domain_
name default
openstack-config --set /etc/nova/nova.conf keystone_authtoken user_domain_
name default
openstack-config --set /etc/nova/nova.conf keystone_authtoken project_name
service
openstack-config --set /etc/nova/nova.conf keystone_authtoken username nova
openstack-config --set /etc/nova/nova.conf keystone_authtoken password NOVA_
PASS
openstack-config --set /etc/nova/nova.conf vnc enabled true
```

```
    openstack-config --set /etc/nova/nova.conf vnc server_listen '$my_ip'
    openstack-config --set /etc/nova/nova.conf vnc server_proxyclient_address
'$my_ip'
    openstack-config --set /etc/nova/nova.conf glance api_servers http://
controller:9292
    openstack-config --set /etc/nova/nova.conf oslo_concurrency lock_path  /var/
lib/nova/tmp
    openstack-config --set /etc/nova/nova.conf placement os_region_name
RegionOne
    openstack-config --set /etc/nova/nova.conf placement  project_domain_name
Default
    openstack-config --set /etc/nova/nova.conf placement  project_name  service
    openstack-config --set /etc/nova/nova.conf placement  auth_type  password
    openstack-config --set /etc/nova/nova.conf placement  user_domain_name
Default
    openstack-config --set /etc/nova/nova.conf placement  auth_url  http://
controller:5000/v3
    openstack-config --set /etc/nova/nova.conf placement  username  placement
    openstack-config --set /etc/nova/nova.conf placement  password  PLACEMENT_
PASS

    cat > /etc/httpd/conf.d/00-nova-placement-api.conf <<-EOF
    Listen 8778

    <VirtualHost *:8778>
      WSGIProcessGroup nova-placement-api
      WSGIApplicationGroup %{GLOBAL}
      WSGIPassAuthorization On
      WSGIDaemonProcess nova-placement-api processes=3 threads=1 user=nova
group=nova
      WSGIScriptAlias / /usr/bin/nova-placement-api
      <IfVersion >= 2.4>
        ErrorLogFormat "%M"
      </IfVersion>
      ErrorLog /var/log/nova/nova-placement-api.log
      <Directory /usr/bin>
      <IfVersion >= 2.4>
         Require all granted
      </IfVersion>
      <IfVersion < 2.4>
```

```
      Order allow,deny
      Allow from all
    </IfVersion>
  </Directory>


  #SSLEngine On
  #SSLCertificateFile ...
  #SSLCertificateKeyFile ...
</VirtualHost>


Alias /nova-placement-api /usr/bin/nova-placement-api
<Location /nova-placement-api>
  SetHandler wsgi-script
  Options +ExecCGI
  WSGIProcessGroup nova-placement-api
  WSGIApplicationGroup %{GLOBAL}
  WSGIPassAuthorization On
</Location>


<Directory /usr/bin>
   <IfVersion >= 2.4>
      Require all granted
   </IfVersion>
   <IfVersion < 2.4>
      Order allow,deny
      Allow from all
   </IfVersion>
</Directory>
EOF
systemctl restart httpd


su -s /bin/sh -c "nova-manage api_db sync" nova
su -s /bin/sh -c "nova-manage cell_v2 map_cell0" nova
su -s /bin/sh -c "nova-manage cell_v2 create_cell --name=cell1" nova
su -s /bin/sh -c "nova-manage db sync" nova


systemctl enable openstack-nova-api.service openstack-nova-consoleauth.
service openstack-nova-scheduler.service openstack-nova-conductor.service
openstack-nova-novncproxy.service
   systemctl start openstack-nova-api.service openstack-nova-consoleauth.
```

```
service openstack-nova-scheduler.service openstack-nova-conductor.service
openstack-nova-novncproxy.service
    nova service-list
    [root@controller ~]# ./openstack-nova-install-controller.sh
```

在计算节点编辑脚本文件 openstack-nova-install-compute.sh，并运行脚本文件安装 nova 在计算节点上的服务。

```
    [root@compute ~]# vi openstack-nova-install-compute.sh
    #!/bin/bash
    ################################################
    # install nova for Openstack on compute
    # Author:Ann   Date:2022-1-6
    ################################################


    yum install -y openstack-utils
    yum install -y openstack-nova-compute


    openstack-config --set /etc/nova/nova.conf DEFAULT enabled_apis   osapi_
compute,metadata
    openstack-config --set /etc/nova/nova.conf DEFAULT transport_url   rabbit://
openstack:RABBIT_PASS@controller
    openstack-config --set /etc/nova/nova.conf DEFAULT my_ip  10.0.0.31
    openstack-config --set /etc/nova/nova.conf DEFAULT use_neutron  True
    openstack-config --set /etc/nova/nova.conf DEFAULT firewall_driver  nova.virt.
firewall.NoopFirewallDriver
    openstack-config --set /etc/nova/nova.conf api auth_strategy  keystone
    openstack-config --set /etc/nova/nova.conf keystone_authtoken auth_url
http://controller:5000/v3
    openstack-config --set /etc/nova/nova.conf keystone_authtoken memcached_
servers  controller:11211
    openstack-config --set /etc/nova/nova.conf keystone_authtoken auth_type
password
    openstack-config --set /etc/nova/nova.conf keystone_authtoken project_domain_
name  default
    openstack-config --set /etc/nova/nova.conf keystone_authtoken user_domain_
name  default
    openstack-config --set /etc/nova/nova.conf keystone_authtoken project_name
service
    openstack-config --set /etc/nova/nova.conf keystone_authtoken username  nova
```

```
openstack-config --set /etc/nova/nova.conf keystone_authtoken password NOVA_
PASS
    openstack-config --set /etc/nova/nova.conf vnc enabled True
    openstack-config --set /etc/nova/nova.conf vnc server_listen 0.0.0.0
    openstack-config --set /etc/nova/nova.conf vnc server_proxyclient_address
'$my_ip'
    openstack-config --set /etc/nova/nova.conf vnc novncproxy_base_url http://
controller:6080/vnc_auto.html
    openstack-config --set /etc/nova/nova.conf glance api_servers http://
controller:9292
    openstack-config --set /etc/nova/nova.conf oslo_concurrency lock_path /var/
lib/nova/tmp
    openstack-config --set /etc/nova/nova.conf placement os_region_name
RegionOne
    openstack-config --set /etc/nova/nova.conf placement project_domain_name
Default
    openstack-config --set /etc/nova/nova.conf placement project_name service
    openstack-config --set /etc/nova/nova.conf placement auth_type password
    openstack-config --set /etc/nova/nova.conf placement user_domain_name
Default
    openstack-config --set /etc/nova/nova.conf placement auth_url http://
controller:5000/v3
    openstack-config --set /etc/nova/nova.conf placement username placement
    openstack-config --set /etc/nova/nova.conf placement password PLACEMENT_PASS
    systemctl enable libvirtd.service openstack-nova-compute.service
    systemctl start libvirtd.service openstack-nova-compute.service

[root@compute ~]# ./openstack-nova-install-compute.sh
```

在控制节点上进行操作，将计算节点添加到数据库。

```
[root@controller ~]# openstack compute service list --service nova-compute
    +----+--------------+---------+------+---------+-------+--------------------
---------+
    | ID | Binary       | Host    | Zone | Status  | State | Updated At |
    +----+--------------+---------+------+---------+-------+--------------------
---------+
    | 8  | nova-compute | compute | nova | enabled | up    | 2021-12-
27T01:17:26.000000 |
    +----+--------------+---------+------+---------+-------+--------------------
---------+
```

```
[root@controller ~]# su -s /bin/sh -c "nova-manage cell_v2 discover_hosts
--verbose" nova
    /usr/lib/python2.7/site-packages/oslo_db/sqlalchemy/enginefacade.py:332:
NotSupportedWarning: Configuration option(s) ['use_tpool'] not supported
    exception.NotSupportedWarning
Found 2 cell mappings.
Skipping cell0 since it does not contain hosts.
Getting computes from cell 'cell1': a4a04e93-ca56-44df-8262-f5ac902d0ca6
Checking host mapping for compute host 'compute': 9a4ac031-6276-48a6-b02e-
589b9005c284
Creating host mapping for compute host 'compute': 9a4ac031-6276-48a6-b02e-
589b9005c284
Found 1 unmapped computes in cell: a4a04e93-ca56-44df-8262-f5ac902d0ca6
[root@controller ~]# openstack compute service list
    +------------------------------------+------------------+------------+----
------+---------+-------+------------------------------+------------------+--------
-----+
    | Id | Binary | Host | Zone | Status | State | Updated_at | Disabled Reason
| Forced down |
    +------------------------------------+------------------+------------+----
------+---------+-------+------------------------------+------------------+--------
-----+
    | b1984dce-b9d0-40ad-9c34-d9f28380fdc1 | nova-consoleauth | controller |
internal | enabled | up | 2021-12-27T01:19:36.000000 | - | False |
    | dc112131-d714-4464-8208-d7357fbcd4f0 | nova-conductor | controller |
internal | enabled | up | 2021-12-27T01:19:39.000000 | - | False |
    | ad83513a-0c2a-45de-a269-c7abb62a2a7d | nova-scheduler | controller |
internal | enabled | up | 2021-12-27T01:19:31.000000 | - | False |
    | 7a6c8b70-0990-4427-8076-a8ce55bc8e72 | nova-compute | compute | nova |
enabled | up | 2021-12-27T01:19:36.000000 | - | False |
    +------------------------------------+------------------+------------+----
------+---------+-------+------------------------------+------------------+--------
-----+
```

4. neutron 组件部署

在控制节点编辑脚本文件，并通过运行脚本的方式安装 neutron。

```
[root@controller ~]# cat openstack-neutron-install.sh
#!/bin/bash
##############################################
# install neutron for Openstack on controller
```

```
# Author:Ann   Date:2022-1-6
###############################################

source  ~/admin-openrc.sh

mysql -uroot -p000000 -e "create database IF NOT EXISTS neutron;"
mysql -uroot -p000000 -e "GRANT ALL PRIVILEGES ON neutron.* TO 'neutron'
@'localhost' IDENTIFIED BY 'NEUTRON_DBPASS';"
mysql -uroot -p000000 -e "GRANT ALL PRIVILEGES ON neutron.* TO 'neutron'@'%'
IDENTIFIED BY 'NEUTRON_DBPASS';"

openstack user create --domain default --password NEUTRON_PASS neutron
openstack role add --project service --user neutron admin
openstack service create --name neutron --description "OpenStack Networking"
network
openstack endpoint create --region RegionOne network public http://control-
ler:9696
openstack endpoint create --region RegionOne network internal http://con-
troller:9696
openstack endpoint create --region RegionOne network admin http://control-
ler:9696

yum -y install openstack-neutron openstack-neutron-ml2 openstack-neutron-li-
nuxbridge ebtables

openstack-config --set /etc/neutron/neutron.conf database connection
mysql://neutron:NEUTRON_DBPASS@controller/neutron
openstack-config --set /etc/neutron/neutron.conf DEFAULT core_plugin  ml2
openstack-config --set /etc/neutron/neutron.conf DEFAULT service_plugins
openstack-config --set /etc/neutron/neutron.conf DEFAULT transport_url  rab-
bit://openstack:RABBIT_PASS@controller
openstack-config --set /etc/neutron/neutron.conf DEFAULT auth_strategy  key-
stone
openstack-config --set /etc/neutron/neutron.conf keystone_authtoken auth_uri
http://controller:5000
openstack-config --set /etc/neutron/neutron.conf keystone_authtoken auth_url
http://controller:35357
openstack-config --set /etc/neutron/neutron.conf keystone_authtoken mem-
cached_servers  controller:11211
```

```
    openstack-config --set /etc/neutron/neutron.conf keystone_authtoken auth_type
password
    openstack-config --set /etc/neutron/neutron.conf keystone_authtoken project_
domain_name  default
    openstack-config --set /etc/neutron/neutron.conf keystone_authtoken user_do-
main_name default
    openstack-config --set /etc/neutron/neutron.conf keystone_authtoken project_
name  service
    openstack-config --set /etc/neutron/neutron.conf keystone_authtoken username
neutron
    openstack-config --set /etc/neutron/neutron.conf keystone_authtoken password
NEUTRON_PASS
    openstack-config --set /etc/neutron/neutron.conf DEFAULT notify_nova_on_port_
status_changes  true
    openstack-config --set /etc/neutron/neutron.conf DEFAULT notify_nova_on_port_
data_changes  true
    openstack-config --set /etc/neutron/neutron.conf  nova auth_url  http://con-
troller:35357
    openstack-config --set /etc/neutron/neutron.conf  nova auth_type  password
    openstack-config --set /etc/neutron/neutron.conf  nova project_domain_name
default
    openstack-config --set /etc/neutron/neutron.conf  nova user_domain_name  de-
fault
    openstack-config --set /etc/neutron/neutron.conf  nova region_name  RegionOne
    openstack-config --set /etc/neutron/neutron.conf  nova project_name  service
    openstack-config --set /etc/neutron/neutron.conf  nova username  nova
    openstack-config --set /etc/neutron/neutron.conf  nova password  NOVA_PASS
    openstack-config --set /etc/neutron/neutron.conf oslo_concurrency lock_path
/var/lib/neutron/tmp

    openstack-config --set /etc/neutron/plugins/ml2/ml2_conf.ini ml2 type_drivers
flat,vlan
    openstack-config --set /etc/neutron/plugins/ml2/ml2_conf.ini ml2 tenant_net-
work_types
    openstack-config --set /etc/neutron/plugins/ml2/ml2_conf.ini ml2 mechanism_
drivers  linuxbridge
    openstack-config --set /etc/neutron/plugins/ml2/ml2_conf.ini ml2 extension_
drivers  port_security
    openstack-config --set  /etc/neutron/plugins/ml2/ml2_conf.ini ml2_type_flat
```

```
flat_networks  provider
    openstack-config --set  /etc/neutron/plugins/ml2/ml2_conf.ini securitygroup
enable_ipset  true


    openstack-config --set /etc/neutron/plugins/ml2/linuxbridge_agent.ini linux_
bridge physical_interface_mappings provider:ens34
    openstack-config --set /etc/neutron/plugins/ml2/linuxbridge_agent.ini vxlan
enable_vxlan false
    openstack-config --set /etc/neutron/plugins/ml2/linuxbridge_agent.ini securi-
tygroup enable_security_group true
    openstack-config --set /etc/neutron/plugins/ml2/linuxbridge_agent.ini secu-
ritygroup firewall_driver neutron.agent.linux.iptables_firewall.IptablesFirewall-
Driver


    openstack-config --set /etc/neutron/dhcp_agent.ini DEFAULT interface_driver
linuxbridge
    openstack-config --set /etc/neutron/dhcp_agent.ini DEFAULT dhcp_driver neu-
tron.agent.linux.dhcp.Dnsmasq
    openstack-config --set /etc/neutron/dhcp_agent.ini DEFAULT enable_isolated_
metadata true


    openstack-config --set /etc/neutron/metadata_agent.ini DEFAULT nova_metadata_
host  controller
    openstack-config --set /etc/neutron/metadata_agent.ini DEFAULT metadata_
proxy_shared_secret METADATA_SECRET


    openstack-config --set /etc/nova/nova.conf  neutron url  http://control-
ler:9696
    openstack-config --set /etc/nova/nova.conf  neutron auth_url  http://control-
ler:35357
    openstack-config --set /etc/nova/nova.conf  neutron auth_type  password
    openstack-config --set /etc/nova/nova.conf  neutron project_domain_name  de-
fault
    openstack-config --set /etc/nova/nova.conf  neutron user_domain_name default
    openstack-config --set /etc/nova/nova.conf  neutron region_name  RegionOne
    openstack-config --set /etc/nova/nova.conf  neutron project_name  service
    openstack-config --set /etc/nova/nova.conf  neutron username  neutron
    openstack-config --set /etc/nova/nova.conf  neutron password  NEUTRON_PASS
    openstack-config --set /etc/nova/nova.conf  neutron service_metadata_proxy
```

```
true
    openstack-config --set /etc/nova/nova.conf  neutron metadata_proxy_shared_se-
cret  METADATA_SECRET

    ln -s /etc/neutron/plugins/ml2/ml2_conf.ini /etc/neutron/plugin.ini
    su -s /bin/sh -c "neutron-db-manage --config-file /etc/neutron/neutron.conf
--config-file /etc/neutron/plugins/ml2/ml2_conf.ini upgrade head" neutron
    systemctl restart openstack-nova-api.service
    systemctl enable neutron-server.service neutron-linuxbridge-agent.service
neutron-dhcp-agent.service neutron-metadata-agent.service
    systemctl start neutron-server.service neutron-linuxbridge-agent.service
neutron-dhcp-agent.service neutron-metadata-agent.service
    openstack network agent list
    [root@controller ~]# ./openstack-neutron-install-controller.sh
```

在计算节点编辑脚本文件 openstack-neutron-install-compute.sh，并运行脚本文件安装 neutron 在计算节点上的服务。

```
    [root@compute ~]# vi openstack-nova-install-compute.sh
    #!/bin/bash
    ###############################################
    # install nova for Openstack on controller
    # Author:Ann  Date:2022-1-6
    ###############################################

    source  ~/admin-openrc.sh

    yum install -y openstack-neutron-linuxbridge ebtables ipset

    openstack-config --set /etc/neutron/neutron.conf DEFAULT transport_url rab-
bit://openstack:RABBIT_PASS@controller
    openstack-config --set /etc/neutron/neutron.conf DEFAULT auth_strategy key-
stone
    openstack-config --set /etc/neutron/neutron.conf keystone_authtoken auth_uri
http://controller:5000
    openstack-config --set /etc/neutron/neutron.conf keystone_authtoken auth_url
http://controller:35357
    openstack-config --set /etc/neutron/neutron.conf keystone_authtoken mem-
cached_servers  controller:11211
    openstack-config --set /etc/neutron/neutron.conf keystone_authtoken auth_type
```

```
password
    openstack-config --set /etc/neutron/neutron.conf keystone_authtoken project_
domain_name  default
    openstack-config --set /etc/neutron/neutron.conf keystone_authtoken user_do-
main_name  default
    openstack-config --set /etc/neutron/neutron.conf keystone_authtoken project_
name  service
    openstack-config --set /etc/neutron/neutron.conf keystone_authtoken username
neutron
    openstack-config --set /etc/neutron/neutron.conf keystone_authtoken password
NEUTRON_PASS
    openstack-config --set /etc/neutron/neutron.conf oslo_concurrency lock_path
/var/lib/neutron/tmp
    openstack-config --set /etc/neutron/neutron.conf linux_bridge physical_inter-
face_mappings  provider:ens34
    openstack-config --set /etc/neutron/neutron.conf vxlan enable_vxlan  false
    openstack-config --set /etc/neutron/neutron.conf securitygroup enable_securi-
ty_group  true
    openstack-config --set /etc/neutron/neutron.conf securitygroup firewall_driver
neutron.agent.linux.iptables_firewall.IptablesFirewallDriver


    openstack-config --set /etc/nova/nova.conf neutron url  http://control-
ler:9696
    openstack-config --set /etc/nova/nova.conf neutron auth_url  http://control-
ler:35357
    openstack-config --set /etc/nova/nova.conf neutron auth_type  password
    openstack-config --set /etc/nova/nova.conf neutron project_domain_name  de-
fault
    openstack-config --set /etc/nova/nova.conf neutron user_domain_name  default
    openstack-config --set /etc/nova/nova.conf neutron region_name  RegionOne
    openstack-config --set /etc/nova/nova.conf neutron project_name  service
    openstack-config --set /etc/nova/nova.conf neutron username  neutron
    openstack-config --set /etc/nova/nova.conf neutron password  NEUTRON_PASS


    systemctl restart openstack-nova-compute.service
    systemctl enable neutron-linuxbridge-agent.service
    systemctl start neutron-linuxbridge-agent.service


    [root@compute ~]# ./openstack-neutron-install-compute.sh
    neutron
```

任务 3　安装 dashboard 管理界面

编辑脚本文件，并通过运行脚本的方式安装 dashboard。在控制节点编辑脚本文件 openstack-dashboard-install.sh。

```
[root@controller ~]# cat openstack-dashboard-install.sh
#!/bin/bash
###############################################
# install dashboard for Openstack on controller
# Author:Ann   Date:2022-1-6
###############################################

source  ~/admin-openrc.sh
yum -y install openstack-dashboard
cat > /etc/openstack-dashboard/local_settings <<-EOF
import os
from django.utils.translation import ugettext_lazy as _
from openstack_dashboard.settings import HORIZON_CONFIG
DEBUG = False
WEBROOT = '/dashboard/'
ALLOWED_HOSTS = ['*',]
SESSION_ENGINE = 'django.contrib.sessions.backends.cache'
OPENSTACK_API_VERSIONS = {
    "identity": 3,
    "image": 2,
    "volume": 2,
}
OPENSTACK_KEYSTONE_MULTIDOMAIN_SUPPORT = True
OPENSTACK_KEYSTONE_DEFAULT_DOMAIN = 'Default'
LOCAL_PATH = '/tmp'
SECRET_KEY='a98fb726ae49aeefb5ab'
CACHES = {
  'default': {
    'BACKEND': 'django.core.cache.backends.memcached.MemcachedCache',
    'LOCATION': 'controller:11211',
  },
}
EMAIL_BACKEND = 'django.core.mail.backends.console.EmailBackend'
OPENSTACK_HOST = "controller"
OPENSTACK_KEYSTONE_URL = "http://%s:5000/v3" % OPENSTACK_HOST
OPENSTACK_KEYSTONE_DEFAULT_ROLE = "user"
```

```
OPENSTACK_KEYSTONE_BACKEND = {
  'name': 'native',
  'can_edit_user': True,
  'can_edit_group': True,
  'can_edit_project': True,
  'can_edit_domain': True,
  'can_edit_role': True,
}
OPENSTACK_HYPERVISOR_FEATURES = {
  'can_set_mount_point': False,
  'can_set_password': False,
  'requires_keypair': False,
  'enable_quotas': True
}
OPENSTACK_CINDER_FEATURES = {
  'enable_backup': False,
}
OPENSTACK_NEUTRON_NETWORK = {
  'enable_router': False,
  'enable_quotas': False,
  'enable_distributed_router': False,
  'enable_ha_router': False,
  'enable_lb': False,
  'enable_firewall': False,
  'enable_vpn': False,
  'enable_fip_topology_check': False,
  'supported_vnic_types': ['*'],
  'physical_networks': [],
}
OPENSTACK_HEAT_STACK = {
  'enable_user_pass': True,
}
IMAGE_CUSTOM_PROPERTY_TITLES = {
  "architecture": _("Architecture"),
  "kernel_id": _("Kernel ID"),
  "ramdisk_id": _("Ramdisk ID"),
  "image_state": _("Euca2ools state"),
  "project_id": _("Project ID"),
  "image_type": _("Image Type"),
}
IMAGE_RESERVED_CUSTOM_PROPERTIES = []
```

```
API_RESULT_LIMIT = 1000
API_RESULT_PAGE_SIZE = 20
SWIFT_FILE_TRANSFER_CHUNK_SIZE = 512 * 1024
INSTANCE_LOG_LENGTH = 35
DROPDOWN_MAX_ITEMS = 30
TIME_ZONE = "Asia/Shanghai"
POLICY_FILES_PATH = '/etc/openstack-dashboard'
LOGGING = {
  'version': 1,
  'disable_existing_loggers': False,
  'formatters': {
    'console': {
      'format': '%(levelname)s %(name)s %(message)s'
    },
    'operation': {
      'format': '%(message)s'
    },
  },
  'handlers': {
    'null': {
      'level': 'DEBUG',
      'class': 'logging.NullHandler',
    },
    'console': {
      'level': 'INFO',
      'class': 'logging.StreamHandler',
      'formatter': 'console',
    },
    'operation': {
      'level': 'INFO',
      'class': 'logging.StreamHandler',
      'formatter': 'operation',
    },
  },
  'loggers': {
    'horizon': {
      'handlers': ['console'],
      'level': 'DEBUG',
      'propagate': False,
    },
    'horizon.operation_log': {
```

```
          'handlers': ['operation'],
          'level': 'INFO',
          'propagate': False,
      },
      'openstack_dashboard': {
          'handlers': ['console'],
          'level': 'DEBUG',
          'propagate': False,
      },
      'novaclient': {
          'handlers': ['console'],
          'level': 'DEBUG',
          'propagate': False,
      },
      'cinderclient': {
          'handlers': ['console'],
          'level': 'DEBUG',
          'propagate': False,
      },
      'keystoneauth': {
          'handlers': ['console'],
          'level': 'DEBUG',
          'propagate': False,
      },
      'keystoneclient': {
          'handlers': ['console'],
          'level': 'DEBUG',
          'propagate': False,
      },
      'glanceclient': {
          'handlers': ['console'],
          'level': 'DEBUG',
          'propagate': False,
      },
      'neutronclient': {
          'handlers': ['console'],
          'level': 'DEBUG',
          'propagate': False,
      },
      'swiftclient': {
          'handlers': ['console'],
```

```
            'level': 'DEBUG',
            'propagate': False,
        },
        'oslo_policy': {
            'handlers': ['console'],
            'level': 'DEBUG',
            'propagate': False,
        },
        'openstack_auth': {
            'handlers': ['console'],
            'level': 'DEBUG',
            'propagate': False,
        },
        'nose.plugins.manager': {
            'handlers': ['console'],
            'level': 'DEBUG',
            'propagate': False,
        },
        'django': {
            'handlers': ['console'],
            'level': 'DEBUG',
            'propagate': False,
        },
        'django.db.backends': {
            'handlers': ['null'],
            'propagate': False,
        },
        'requests': {
            'handlers': ['null'],
            'propagate': False,
        },
        'urllib3': {
            'handlers': ['null'],
            'propagate': False,
        },
        'chardet.charsetprober': {
            'handlers': ['null'],
            'propagate': False,
        },
        'iso8601': {
            'handlers': ['null'],
```

```
            'propagate': False,
        },
        'scss': {
            'handlers': ['null'],
            'propagate': False,
        },
    },
}
SECURITY_GROUP_RULES = {
    'all_tcp': {
        'name': _('All TCP'),
        'ip_protocol': 'tcp',
        'from_port': '1',
        'to_port': '65535',
    },
    'all_udp': {
        'name': _('All UDP'),
        "ip_protocol': 'udp',
        'from_port': '1',
        'to_port': '65535',
    },
    'all_icmp': {
        'name': _('All ICMP'),
        'ip_protocol': 'icmp',
        'from_port': '-1',
        'to_port': '-1',
    },
    'ssh': {
        'name': 'SSH',
        'ip_protocol': 'tcp',
        'from_port': '22',
        'to_port': '22',
    },
    'smtp': {
        'name': 'SMTP',
        'ip_protocol': 'tcp',
        'from_port': '25',
        'to_port': '25',
    },
    'dns': {
        'name': 'DNS',
```

```
        'ip_protocol': 'tcp',
        'from_port': '53',
        'to_port': '53',
    },
    'http': {
        'name': 'HTTP',
        'ip_protocol': 'tcp',
        'from_port': '80',
        'to_port': '80',
    },
    'pop3': {
        'name': 'POP3',
        'ip_protocol': 'tcp',
        'from_port': '110',
        'to_port': '110',
    },
    'imap': {
        'name': 'IMAP',
        'ip_protocol': 'tcp',
        'from_port': '143',
        'to_port': '143',
    },
    'ldap': {
        'name': 'LDAP',
        'ip_protocol': 'tcp',
        'from_port': '389',
        'to_port': '389',
    },
    'https': {
        'name': 'HTTPS',
        'ip_protocol': 'tcp',
        'from_port': '443',
        'to_port': '443',
    },
    'smtps': {
        'name': 'SMTPS',
        'ip_protocol': 'tcp',
        'from_port': '465',
        'to_port': '465',
    },
    'imaps': {
```

```
        'name': 'IMAPS',
        'ip_protocol': 'tcp',
        'from_port': '993',
        'to_port': '993',
    },
    'pop3s': {
        'name': 'POP3S',
        'ip_protocol': 'tcp',
        'from_port': '995',
        'to_port': '995',
    },
    'ms_sql': {
        'name': 'MS SQL',
        'ip_protocol': 'tcp',
        'from_port': '1433',
        'to_port': '1433',
    },
    'mysql': {
        'name': 'MYSQL',
        'ip_protocol': 'tcp',
        'from_port': '3306',
        'to_port': '3306',
    },
    'rdp': {
        'name': 'RDP',
        'ip_protocol': 'tcp',
        'from_port': '3389',
        'to_port': '3389',
    },
}
REST_API_REQUIRED_SETTINGS = ['OPENSTACK_HYPERVISOR_FEATURES',
                              'LAUNCH_INSTANCE_DEFAULTS',
                              'OPENSTACK_IMAGE_FORMATS',
                              'OPENSTACK_KEYSTONE_DEFAULT_DOMAIN',
                              'CREATE_IMAGE_DEFAULTS',
                              'ENFORCE_PASSWORD_CHECK']
ALLOWED_PRIVATE_SUBNET_CIDR = {'ipv4': [], 'ipv6': []}
EOF


sed -i '1i\WSGIApplicationGroup %{GLOBAL}' /etc/httpd/conf.d/openstack-dashboard.conf
```

```
systemctl restart httpd.service memcached.service
[root@controller ~]# ./openstack-keystone-install.sh
```

验证 dashboard 访问情况。

浏览器访问 10.0.0.11/dashboard 验证访问，其中域为"Default"，用户为"admin"，密码为"ADMIN_PASS"。登录成功后，可以在 dashboard 界面看到整个云平台的相关信息，如图 3-11、图 3-12 所示。

图 3-11　openstack dashboard 登录界面

图 3-12　openstack dashboard 管理界面

3.5 项目实战——使用仪表盘 web 界面管理云主机

 项目背景

公司为了方便对 OpenStack 平台进行日常运维管理工作，提供了仪表盘 web 图形管理界面，通过浏览器更加直观地对云主机的生命周期、镜像、虚拟网络等进行运维管理。

项目实施

任务 1　创建镜像

准备本地 CentOS7.5 的 qcow2 镜像文件，并上传到控制节点，创建 glance 镜像文件 CentOS7.5 以供云服务器使用。

```
[root@controller ~]# glance image-create --name "CentOS7.5" --disk-format
qcow2  --container-format bare --progress < /opt/CentOS_7.5_x86_64_XD.qcow2
[=============================>] 100%

+------------------+--------------------------------------+
| Property         | Value                                |
+------------------+--------------------------------------+
| checksum         | 3d3e9c954351a4b6953fd156f0c29f5c     |
| container_format | bare                                 |
| created_at       | 2022-03-19T15:22:26Z                 |
| disk_format      | qcow2                                |
| id               | 91af8dd4-581e-4c8a-8bb5-8a4d4f09a4ab |
| min_disk         | 0                                    |
| min_ram          | 0                                    |
| name             | CentOS7.5                            |
| owner            | 2b2476cd9c1a4429b3d1f6f5ad29e686     |
| protected        | False                                |
| size             | 510459904                            |
| status           | active                               |
| tags             | []                                   |
| updated_at       | 2022-03-19T15:22:38Z                 |
| virtual_size     | None                                 |
| visibility       | shared                               |
+------------------+--------------------------------------+
```

也可以在 dashboard 中，在 "管理员" → "计算" → "镜像" 界面进行镜像的上传，如图 3-13、图 3-14 所示。

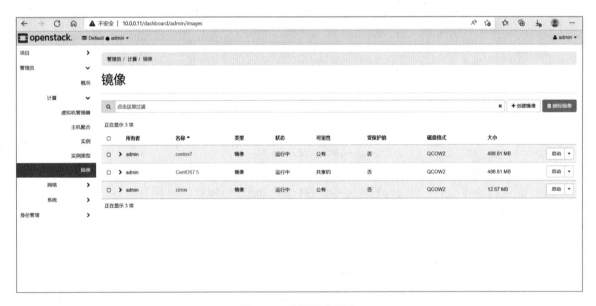

图 3-13   上传镜像

图 3-14   查看镜像列表

任务 2   创建虚拟网络

创建一个 flat 类型的虚拟网络 network，该网络所对应的物理网卡为 provider。

```
[root@controller ~]# openstack network create --share --external \
>   --provider-physical-network provider \
>   --provider-network-type flat network
+-------------------------+-------------------------------------+
| Field                   | Value                               |
```

```
+----------------------------+------------------------------------------+
| admin_state_up             | UP                                       |
| availability_zone_hints    |                                          |
| availability_zones         |                                          |
| created_at                 | 2022-01-06T05:56:35Z                     |
| description                |                                          |
| dns_domain                 | None                                     |
| id                         | c08d7e52-aca3-4946-8bbf-808787b36fe1     |
| ipv4_address_scope         | None                                     |
| ipv6_address_scope         | None                                     |
| is_default                 | None                                     |
| is_vlan_transparent        | None                                     |
| mtu                        | 1500                                     |
| name                       | network                                  |
| port_security_enabled      | True                                     |
| project_id                 | 928b3512a31c47e59b4fc1287db1ad7b         |
| provider:network_type      | flat                                     |
| provider:physical_network  | provider                                 |
| provider:segmentation_id   | None                                     |
| qos_policy_id              | None                                     |
| revision_number            | 4                                        |
| router:external            | External                                 |
| segments                   | None                                     |
| shared                     | True                                     |
| status                     | ACTIVE                                   |
| subnets                    |                                          |
| tags                       |                                          |
| updated_at                 | 2022-01-06T05:56:35Z                     |
+----------------------------+------------------------------------------+
```

在该虚拟网络中创建子网 subnet。在 flat 类型的网络中，虚拟网络与相对应物理网卡 provide 桥接在一起，设置子网网段为 192.168.0.0/24，地址池的范围定义为 192.168.0.2 ～ 192.168.0.254，网关通常为子网当中第一个可用地址，即 192.168.0.1。

```
[root@controller ~]# openstack subnet create --network provider
--allocation-pool start=192.168.0.2,end=192.168.0.254  --dns-nameserver
192.168.0.1 --gateway 192.168.0.1 --subnet-range 192.168.0.0/24 provider
+-------------------+---------------------------------------------+
| Field             | Value                                       |
+-------------------+---------------------------------------------+
| allocation_pools  | 192.168.0.2-192.168.0.254                   |
| cidr              | 192.168.0.0/24                              |
```

```
| created_at         | 2022-01-06T06:00:15Z                 |
| description        |                                      |
| dns_nameservers    | 192.168.0.1                          |
| enable_dhcp        | True                                 |
| gateway_ip         | 192.168.0.1                          |
| host_routes        |                                      |
| id                 | ecf65ebe-91f0-4ea5-876b-dbba0ad870f8 |
| ip_version         | 4                                    |
| ipv6_address_mode  | None                                 |
| ipv6_ra_mode       | None                                 |
| name               | provider                             |
| network_id         | c08d7e52-aca3-4946-8bbf-808787b36fe1 |
| project_id         | 928b3512a31c47e59b4fc1287db1ad7b     |
| revision_number    | 0                                    |
| segment_id         | None                                 |
| service_types      |                                      |
| subnetpool_id      | None                                 |
| tags               |                                      |
| updated_at         | 2022-01-06T06:00:15Z                 |
+--------------------+--------------------------------------+
```

通过查看网络列表，可以获取虚拟网络的 ID，使用 ID 可以进一步查看网络的详细信息。

```
[root@controller ~]# openstack network list
+--------------------------------------+---------+--------------------------------------+
| ID                                   | Name    | Subnets                              |
+--------------------------------------+---------+--------------------------------------+
| 0d91e83b-3d1d-4ebe-9c37-7964bca3c2a3 | network | 65c7720c-ac26-4a02-b15e-59910c0b4fd7 |
+--------------------------------------+---------+--------------------------------------+
[root@controller ~]# openstack network show 0d91e83b-3d1d-4ebe-9c37-7964bca3c2a3
+---------------------------+--------------------------------------+
| Field                     | Value                                |
+---------------------------+--------------------------------------+
| admin_state_up            | UP                                   |
| availability_zone_hints   |                                      |
| availability_zones        | nova                                 |
| created_at                | 2022-03-16T07:18:22Z                 |
| description               |                                      |
| dns_domain                | None                                 |
| id                        | 0d91e83b-3d1d-4ebe-9c37-7964bca3c2a3 |
```

```
| ipv4_address_scope         | None                                   |
| ipv6_address_scope         | None                                   |
| is_default                 | None                                   |
| is_vlan_transparent        | None                                   |
| mtu                        | 1500                                   |
| name                       | network                                |
| port_security_enabled      | True                                   |
| project_id                 | 2b2476cd9c1a4429b3d1f6f5ad29e686       |
| provider:network_type      | flat                                   |
| provider:physical_network  | provider                               |
| provider:segmentation_id   | None                                   |
| qos_policy_id              | None                                   |
| revision_number            | 5                                      |
| router:external            | External                               |
| segments                   | None                                   |
| shared                     | True                                   |
| status                     | ACTIVE                                 |
| subnets                    | 65c7720c-ac26-4a02-b15e-59910c0b4fd7   |
| tags                       |                                        |
| updated_at                 | 2022-03-16T07:18:22Z                   |
+----------------------------+----------------------------------------+
```

在 dashboard 中，可以在"管理员"→"网络"→"网络"界面中创建虚拟网络及子网，如图 3-15 ～图 3-18 所示。

图 3-15   配置网络信息

图 3-16　配置子网信息

图 3-17　配置子网详情

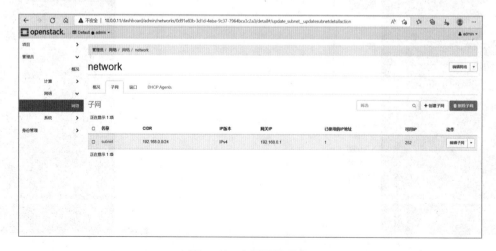

图 3-18　查看网络列表

在"单元"→"网络"→"安全组"界面中修改防火墙规则，删除原有规则，添加新的规则，放行 ICMP、TCP、UDP 协议所有方向、所有端口的流量，以便后续远程访问云主机或访问实例中的服务，如图 3-19、图 3-20 所示。

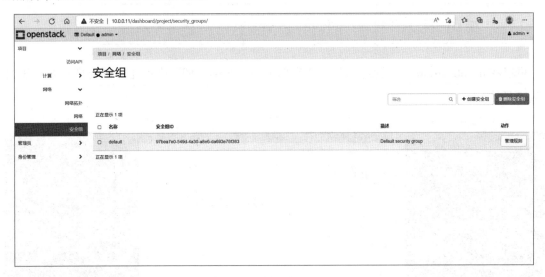

图 3-19　选择安全组

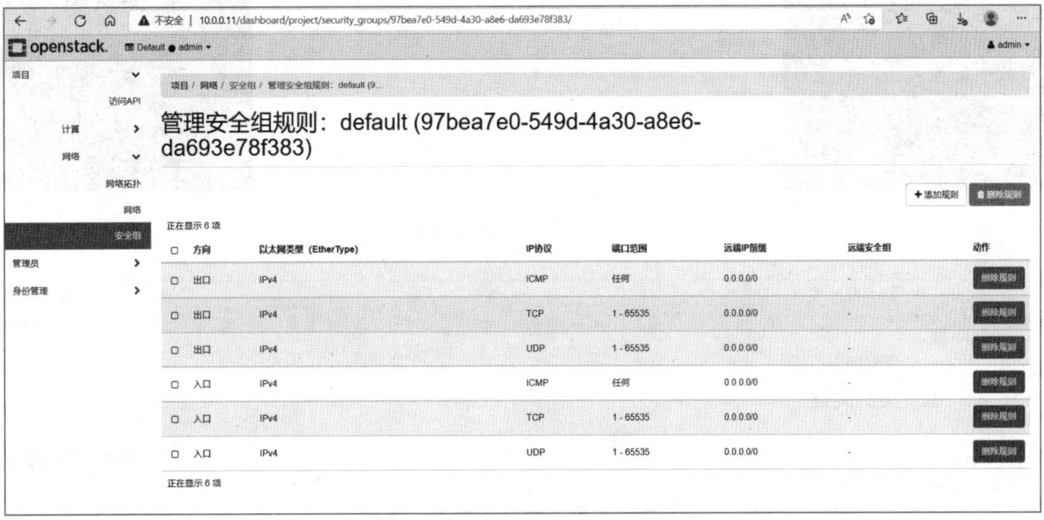

图 3-20　配置安全组规则

任务 3　启动云主机实例

首先创建一个名为 m1.small 的模板，虚拟 CPU 数量为 1，内存为 2 048 MB，磁盘容量为 20 GB，查看模板列表，可以看到创建成功的 m1.small 模板。

```
[root@controller ~]# openstack flavor create --id 0 --vcpus 1 --ram 2048
--disk 20 m1.small
```

```
[root@controller ~]# openstack flavor list
+----+----------+------+------+-----------+-------+-----------+
| ID | Name     | RAM  | Disk | Ephemeral | VCPUs | Is Public |
+----+----------+------+------+-----------+-------+-----------+
| 0  | m1.small | 2048 |  20  |        0  |    1  | True      |
+----+----------+------+------+-----------+-------+-----------+
```

可以在 dashboard 界面"管理员"→"计算"→"实例类型"中创建实例类型，如图 3-21、图 3-22 所示。

图 3-21　配置实例类型信息

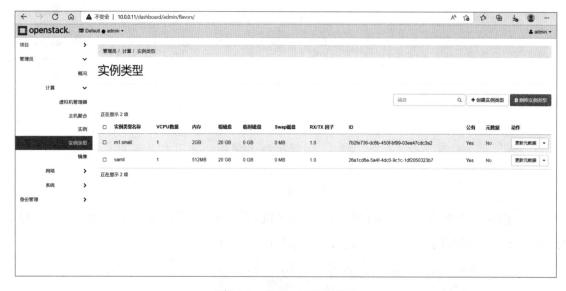

图 3-22　查看实例类型列表

　　启动实例 test，使用模板 m1.small，镜像使用 CentOS7.5，网络 id 为任务 2 中创建的虚拟网络，安全组选择默认安全组 default。

```
[root@controller ~]# openstack server create --flavor m1.small --image cirros
--nic net-id=0d91e83b-3d1d-4ebe-9c37-7964bca3c2a3 --security-group default test
+-----------------------------------+-----------------------------------+
| Field                             | Value                             |
+-----------------------------------+-----------------------------------+
| OS-DCF:diskConfig                 | MANUAL                            |
| OS-EXT-AZ:availability_zone       |                                   |
| OS-EXT-SRV-ATTR:host              | None                              |
| OS-EXT-SRV-ATTR:hypervisor_hostname | None                            |
| OS-EXT-SRV-ATTR:instance_name     |                                   |
| OS-EXT-STS:power_state            | NOSTATE                           |
| OS-EXT-STS:task_state             | scheduling                        |
| OS-EXT-STS:vm_state               | building                          |
| OS-SRV-USG:launched_at            | None                              |
| OS-SRV-USG:terminated_at          | None                              |
| accessIPv4                        |                                   |
| accessIPv6                        |                                   |
| addresses                         |                                   |
| adminPass                         | bzq4S9SgeExs                      |
| config_drive                      |                                   |
| created                           | 2021-12-29T02:08:20Z              |
| flavor                            | m1.small                          |
| hostId                            |                                   |
| id                                | e415d184-c0dc-4e14-bc3d-fb8a3f071ade|
| image                             | cirros (1c27bb51-9d4a-47dc-be9d-  |
|                                   | 13445f29580d)                     |
| key_name                          | none                              |
| name                              | test                              |
| progress                          | 0                                 |
| project_id                        | 50467f099c214d8abba4e3f9cf8475bc  |
| properties                        |                                   |
| security_groups                   | name='81b12d54-314f-4190-8d06-    |
|                                   | 3ac9db21492e'                     |
| status                            | BUILD                             |
| updated                           | 2021-12-29T02:08:20Z              |
| user_id                           | c346da06399c429f864eed7afff641e6  |
| volumes_attached                  |                                   |
```

```
+-----------------------------------+-------+---------------------------------+
[root@controller ~]# openstack server list
+-----------------------------------+-------+-----+------+---------+--------+
| ID        | Name | Status | Networks          | Image | Flavor  |
+-----------------------------------+-------+-----+------+---------+--------+
| e415d184-c0dc-4e14-bc3d-fb8a3f071ade | test | ACTIVE | network=192.168.0.3
| cirros | m1.small |
+-----------------------------------+-------+-----+------+---------+--------+
[root@controller ~]# openstack console url show test
+-------+---------------------------------------------------------------------+
| Field | Value          |
+-------+---------------------------------------------------------------------+
| type  | novnc         |
| url   | http://controller:6080/vnc_auto.html?token=4c3287c5-520b-44e0-
91e1-289743f19296 |
+-------+---------------------------------------------------------------------+
```

使用浏览器通过 VNC 访问实例，如图 3-23 所示。

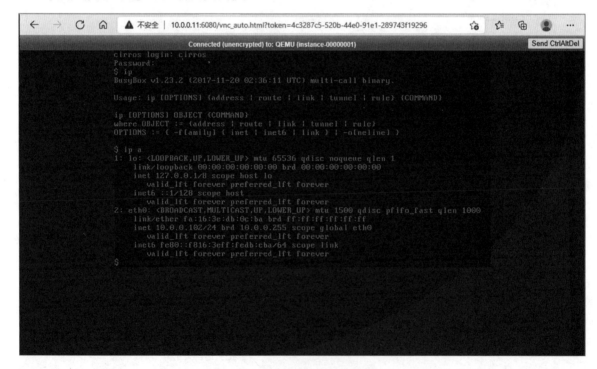

图 3-23　实例 VNC 访问

在 dashboard 的 "单元" → "计算" → "实例" 界面中，可以通过界面操作完成实例的创建，如图 3-24 ～图 3-30 所示。

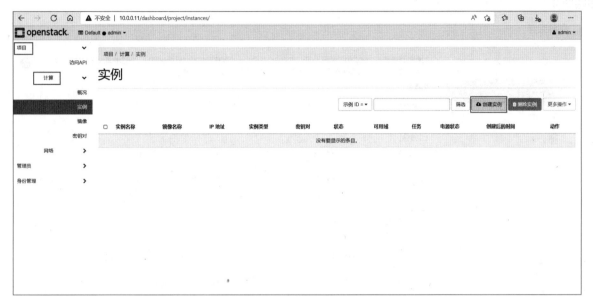

图 3-24 创建实例

图 3-25 配置实例信息

图 3-26　选择实例系统镜像

图 3-27　选择实例类型

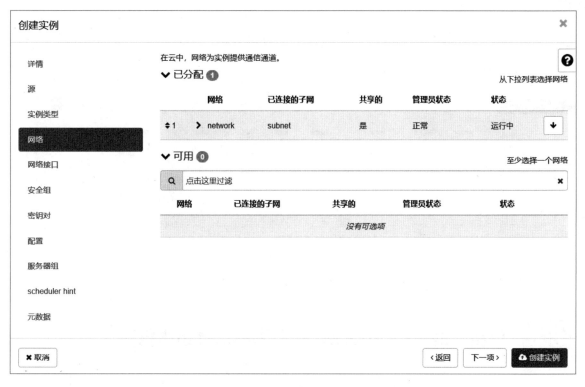

图 3-28　选择网络

图 3-29　选择安全组

图 3-30   查看实例列表

3.6   项目实战——OpenStack 平台组件运维管理

项目背景

在 OpenStack 平台日常运行中，运维人员需对平台各组件进行运维管理，包括身份认证、镜像、云主机、网络等组件。

项目实施

任务 1   Keystone 运维

Keystone 为 OpenStack 平台提供身份认证服务，对平台用户进行身份认证。首先创建一个名称为 asus 的账户，密码为 mypassword123，邮箱为 asus@example.com。

```
[root@container ~]# source/etc/keystone/admin-openrc.sh
[root@container ~]# openstack user create --password mypassword123 --email
asus@example.com --domain demo asus
+--------------------+----------------------------------+
| Field              | Value                            |
+--------------------+----------------------------------+
| domain_id          | b5fb5dbb3a0247f9b9d964457aab5d52 |
| email              | asus@example.com                 |
| enabled            | True                             |
| id                 | 0107f0547cd84c8db5fd8c7d98a6c430 |
```

```
| name                | asus                             |
| options             | {}                               |
| password_expires_at | None                             |
+---------------------+----------------------------------+
```

在创建用户时，还可以指定上述之外的其他参数，具体如下所示：

```
[root@controller ~]# openstack user create
usage: openstack user create [-h]
                             [-f {html,json,json,shell,table,value,yaml,yaml}]
                                    [-c COLUMN] [--max-width <integer>]
[--noindent]
                             [--prefix PREFIX] [--domain <domain>]
                             [--project <project>]
                             [--project-domain <project-domain>]
                             [--password <password>] [--password-prompt]
                             [--email <email-address>]
                             [--description <description>]
                             [--enable | --disable] [--or-show]
                             <name>
```

更改"asus"账户的密码为"000000"。

```
[root@container ~]# openstack user set --password 000000 asus
```

在 OpenStack 中，所有的资源以单元为单位进行划分，通常一个团队或一个单元可以定义在一个 project 中。创建一个名为 myproject 的单元。

```
[root@conytainer ~]# openstack project create --domain demo  myproject
+-------------+----------------------------------+
| Field       | Value                            |
+-------------+----------------------------------+
| description |                                  |
| domain_id   | b5fb5dbb3a0247f9b9d964457aab5d52 |
| enabled     | True                             |
| id          | c1bb4c90c7884e5c954300fcf009ade3 |
| is_domain   | False                            |
| name        | myproject                        |
| parent_id   | b5fb5dbb3a0247f9b9d964457aab5d52 |
| tags        | []                               |
+-------------+----------------------------------+
```

在创建单元时，还可以指定除上述之外的其他参数，如下所示：

```
[root@controller ~]# openstack project create
usage: openstack project create [-h]
                                [-f {html,json,json,shell,table,value,yaml,yaml}]
                                [-c COLUMN] [--max-width <integer>]
                                [--noindent] [--prefix PREFIX]
                                [--domain <domain>] [--parent <project>]
                                [--description <description>]
                                [--enable | --disable]
                                [--property <key=value>] [--or-show]
                                <project-name>
```

修改单元 myproject 的名字为 key_project。

```
[root@container ~]# openstack project set --name key_project myproject
```

用户对单元中资源的使用权限通过所关联的角色决定。为用户绑定不同的角色代表了用户具有不同权限。创建一个角色 compute-user。

```
[root@container ~]# openstack role create compute-user
+-----------+----------------------------------+
| Field     | Value                            |
+-----------+----------------------------------+
| domain_id | None                             |
| id        | 6c7bf68f8d1f439d8b420a49d8f725a5 |
| name      | compute-user                     |
+-----------+----------------------------------+
```

当需要为用户分配权限时，可将用户与相应角色进行绑定。例如，给用户 asus 分配 key_project 单元下的 compute-user 角色，命令如下：

```
[root@container ~]#  openstack role add --user alice --project key_project
compute-user
```

任务 2　Glance 运维

创建一个名称为 cirros，磁盘格式为 qcow2 的镜像，镜像文件使用 cirros-0.4.0-x86_64-disk.img。

```
[root@container ~]# glance image-create --name cirros --disk-format qcow2
--container-format bare --progress <cirros-0.4.0-x86_64-disk.img
[==============================>] 100%
```

```
+-------------------+------------------------------------------+
| Property          | Value                                    |
+-------------------+------------------------------------------+
| checksum          | 443b7623e27ecf03dc9e01ee93f67afe         |
| container_format  | bare                                     |
| created_at        | 2022-08-14T15:27:08Z                     |
| disk_format       | qcow2                                    |
| id                | c7f5e328-89a9-46ec-b5b9-c5a1c1040295     |
| min_disk          | 0                                        |
| min_ram           | 0                                        |
| name              | cirros                                   |
| owner             | b9cd2a3e9a2345128def2d8236cdd6a5         |
| protected         | False                                    |
| size              | 12716032                                 |
| status            | active                                   |
| tags              | []                                       |
| updated_at        | 2022-08-14T15:27:09Z                     |
| virtual_size      | None                                     |
| visibility        | shared                                   |
```

查询当前镜像的列表。

```
[root@container ~]# glance image-list
+----------------------------------------+--------+
| ID                                     | Name|
+----------------------------------------+--------+
| c7f5e328-89a9-46ec-b5b9-c5a1c1040295   | cirros |
+----------------------------------------+--------+
```

使用 list 命令只能看到关于镜像的列表信息，若要查看镜像的详细信息，可以使用 glance image-show 命令，其中参数可以是镜像 ID 或者镜像名称，如下所示：

```
[root@container ~]# glance image-show c7f5e328-89a9-46ec-b5b9-c5a1c1040295
```

查询结果如下所示：

```
+-------------------+------------------------------------------+
| Property          | Value                                    |
+-------------------+------------------------------------------+
| checksum          | 443b7623e27ecf03dc9e01ee93f67afe         |
```

```
| container_format | bare                                   |
| created_at       | 2022-08-14T15:27:08Z                   |
| disk_format      | qcow2                                  |
| id               | c7f5e328-89a9-46ec-b5b9-c5a1c1040295   |
| min_disk         | 0                                      |
| min_ram          | 0                                      |
| name             | cirros                                 |
| owner            | b9cd2a3e9a2345128def2d8236cdd6a5       |
| protected        | False                                  |
| size             | 12716032                               |
| status           | active                                 |
| tags             | []                                     |
| updated_at       | 2022-08-14T15:27:09Z                   |
| virtual_size     | None                                   |
| visibility       | shared                                 |
+------------------+----------------------------------------+
```

删除镜像 cirros 。

```
[root@container ~]# glance image-delete c7f5e328-89a9-46ec-b5b9-c5a1c1040295
[root@container ~]# glance image-list
+----+------+
| ID | Name |
+----+------+
+----+------+
```

任务 3　Nova 运维

Nova 负责管理云主机的整个生命周期。

在创建云主机前，通常需要先创建云主机类型，用来定义分配给云主机的资源情况。例如，创建一个名为 Fmin，ID 为 1，内存为 2 048 MB，磁盘容量为 20 GB，VCPU 数量为 2 的云主机类型。

```
[root@container ~]#  nova flavor-create Fmin 1 2048 20 2
+-------+-----------+-----------+------+-----------+------+-------+-------------+-----------+
| ID    | Name      | Memory_MB | Disk | Ephemeral | Swap | VCPUs | RXTX_Factor | Is_Public |
+-------+-----------+-----------+------+-----------+------+-------+-------------+-----------+
| 19999 | m1.flavor | 2048      | 20   | 0         |      | 2     | 1.0         | True      |
```

查看云主机类型。

```
[root@container ~]# nova flavor-show m1.flavor
+----------------------------+-------+
| Property                   | Value |
+----------------------------+-------+
| OS-FLV-DISABLED:disabled   | False |
| OS-FLV-EXT-DATA:ephemeral  | 0     |
| description                | -     |
| disk                       | 20    |
| extra_specs                | {}    |
| id                         | 1     |
| name                       | Fmin  |
| os-flavor-access:is_public | True  |
| ram                        | 2048  |
| rxtx_factor                | 1.0   |
| swap                       |       |
| vcpus                      | 2     |
+----------------------------+-------+
```

使用创建的云主机类型创建云主机，使用 cirros 镜像、public 的网络。云主机名称为 VM1。

```
# openstack server create VM1 -image cirros --flavor Fmin --network public
```

查看创建好的云主机。

```
[root@container ~]# nova list
+--------------------------------------+------+--------+------------+----------+-------------+
| ID                                   | Name | Status | Task State | Power
State | Networks            |
+--------------------------------------+------+--------+------------+----------+-------------+
| 6924f0ac-81da-42a5-9cb2-4cf3ea5bf503 | VM1  | ACTIVE | -| Running
|public=192.168.1.2 |
+--------------------------------------+------+--------+------------+----------+-------------+
```

创建云主机时，除了上述参数，还可以指定其他参数，如下所示：

```
[root@controller ~]# openstack server create
usage: openstack server create [-h]
                               [-f {html,json,json,shell,table,value,yaml,yaml}]
                               [-c COLUMN] [--max-width <integer>]
                               [--noindent] [--prefix PREFIX]
```

```
                              (--image <image> | --volume <volume>) --flavor
                              <flavor>
                              [--security-group <security-group-name>]
                              [--key-name <key-name>]
                              [--property <key=value>]
                              [--file <dest-filename=source-filename>]
                              [--user-data <user-data>]
                              [--availability-zone <zone-name>]
                              [--block-device-mapping <dev-name=mapping>]
                                      [--nic <net-id=net-uuid,v4-fixed-ip=ip-
addr,v6-fixed-ip=ip-addr,port-id=port-uuid>]
                              [--hint <key=value>]
                              [--config-drive <config-drive-volume>|True]
                              [--min <count>] [--max <count>] [--wait]
                              <server-name>
```

删除虚拟机 VM1。

```
[root@container ~]# openstack server delete VM1
```

任务 4　Neutron 运维

Neutron 为 OpenStack 平台提供网络服务，为云主机提供网络访问。

创建一个名为 public 的网络。

```
[root@container ~]# neutron net-create public
neutron CLI is deprecated and will be removed in the future. Use openstack
CLI instead.
Created a new network:
+----------------------------+----------------------------------------+
| Field                      | Value                                  |
+----------------------------+----------------------------------------+
| admin_state_up             | True                                   |
| availability_zone_hints    |                                        |
| availability_zones         |                                        |
| created_at                 | 2022-08-15T09:19:39Z                   |
| description                |                                        |
| id                         | 0bbb8a16-f04c-4b08-b964-52802fa33b60   |
| ipv4_address_scope         |                                        |
| ipv6_address_scope         |                                        |
```

```
| is_default                | False                              |
| mtu                       | 1450                               |
| name                      | public                             |
| port_security_enabled     | True                               |
| project_id                | b9cd2a3e9a2345128def2d8236cdd6a5   |
| provider:network_type     | vxlan                              |
| provider:physical_network |                                    |
| provider:segmentation_id  | 13                                 |
| revision_number           | 2                                  |
| router:external           | False                              |
| shared                    | False                              |
| status                    | ACTIVE                             |
| subnets                   |                                    |
| tags                      |                                    |
| tenant_id                 | b9cd2a3e9a2345128def2d8236cdd6a5   |
| updated_at                | 2022-08-15T09:19:39Z               |
+---------------------------+------------------------------------+
```

创建子网 newnet，设置 DNS 为 8.8.8.8，网关为 192.168.1.1，地址池范围为 192.168.1.2 ～ 192.168.1.254。

```
[root@container ~]# neutron subnet-create --name newnet --dns-nameserver
8.8.8.8 --gateway 192.168.1.1 public 192.168.1.0/24
neutron CLI is deprecated and will be removed in the future. Use openstack
CLI instead.
Created a new subnet:
+-------------------+----------------------------------------------------+
| Field             | Value                                              |
+-------------------+----------------------------------------------------+
| allocation_pools  | {"start": "192.168.1.2", "end": "192.168.1.254"}   |
| cidr              | 192.168.1.0/24                                     |
| created_at        | 2022-08-15T09:19:54Z                              |
| description       |                                                    |
| dns_nameservers   | 8.8.8.8                                            |
| enable_dhcp       | True                                               |
| gateway_ip        | 192.168.1.1                                        |
| host_routes       |                                                    |
| id                | e97d3ebd-6fec-43f3-82b2-bff97f3b71c2              |
| ip_version        | 4                                                  |
```

```
| ipv6_address_mode  |                                            |
| ipv6_ra_mode       |                                            |
| name               | newnet                                     |
| network_id         | 0bbb8a16-f04c-4b08-b964-52802fa33b60       |
| project_id         | b9cd2a3e9a2345128def2d8236cdd6a5           |
| revision_number    | 0                                          |
| service_types      |                                            |
| subnetpool_id      |                                            |
| tags               |                                            |
| tenant_id          | b9cd2a3e9a2345128def2d8236cdd6a5           |
| updated_at         | 2022-08-15T09:19:54Z                       |
+--------------------+--------------------------------------------+
```

查询网络详细信息。

```
[root@container ~]# neutron net-list
neutron CLI is deprecated and will be removed in the future. Use openstack
CLI instead.
+--------------------------+--------+--------------------------+--------------------+
| id                       | name   | tenant_id                | subnets            |--
+--------------------------+--------+--------------------------+--------------------+
  0bbb8a16-f04c-4b08-b964-52802fa33b60 | public | b9cd2a3e9a2345128def2d-
8236cdd6a5 |
  e97d3ebd-6fec-43f3-82b2-bff97f3b71c2 192.168.1.0/24 |
+--------------------------+--------+--------------------------+--------------------+
[root@container ~]# neutron subnet-list
+--------------------------+--------+--------------------------+----------------+------------------+
| id                       | name   | tenant_id                | cidr           | allocation_pools |
+--------------------------+--------+--------------------------+----------------+------------------+
| e97d3ebd-6fec-43f3-82b2-bff97f3b71c2 | newnet | b9cd2a3e9a2345128def2d-
8236cdd6a5 | 192.168.1.0/24 | {"start": "192.168.1.2", "end": "192.168.1.254"} |
+--------------------------+--------+--------------------------+----------------+------------------+
```

查询网络服务的相关信息。

```
[root@controller ~]# neutron agent-list
+----------+----------+----------+------------+-------+----------+----------+
| id       | agent_type | host     | availability_zone | alive | admin_
state_up | binary   |
+----------+----------+----------+------------+-------+----------+----------+
```

id	agent_type	host	availability_zone	alive	admin_state_up	binary
1ccc8d2d-b40a-44 06-b15b- ab4b2b8e 8398	Open vSwitch agent	controll er		:-)	True	neutron- openvswi tch- agent
245ae2f9 -6220-4c 00-a36d- 542a746a 3f9e	DHCP agent	controll er	nova	:-)	True	neutron- dhcp- agent
26f38d05 -3553-46 f6-8740- 847f9774 1f89	L3 agent	controll er	nova	:-)	True	neutron- l3-agent
5bd0c9e2 -a2ef-48 90-a4ac- a4cb6bcc	Metadata agent	controll er		:-)	True	neutron- metadata -agent

```
| f71b       |            |           |           |        |           |          |
|            |            |           |           |        |           |          |
| f1d1fc34   | Loadbalanc | controll  |           |        | :-)       | True     |
| neutron-  |            |           |           |        |           |          |
| -6e9d-46   | er agent   | er        |           |        |           |          |
| lbaas-    |            |           |           |        |           |          |
| b5-8370-   |            |           |           |        |           |          |
| agent     |            |           |           |        |           |          |
| 0579f3b4   |            |           |           |        |           |          |
|            |            |           |           |        |           |          |
| 6b6f       |            |           |           |        |           |          |
|            |            |           |           |        |           |          |
+-----------+------------+-----------+-----------+--------+-----------+----------+
```

单元 4

容器云技术——Docker

4.1 Docker 容器引擎

Docker 是一种开源的容器引擎技术，用于将软件打包成标准化单元，用于开发、装运和部署软件应用程序。与传统虚拟化技术相比，Docker 容器共享宿主机操作系统内核，因此容器更加轻量级、独立，大大提高了服务器使用效率。

容器也是一种虚拟化技术，但他和传统虚拟化技术有很大的不同。虚拟机是在物理宿主机的基础上，对计算资源进行虚拟化，构建虚拟主机。每台虚拟机都需要安装操作系统，部署应用程序，完成业务功能。由于需要引导操作系统，虚拟机通常开机启动时间较长。而容器是应用层的抽象，它将代码和依赖项打包在一起。多个容器可以在同一台计算机上运行，并与其他容器共享操作系统内核，每个容器在用户空间中作为隔离的进程运行。容器的大小与虚拟机相比要小得多，通常只有数十兆字节，而虚拟机可能需要几吉字节。因此，在相同的条件下，使用容器技术可以运行更多的应用程序，同时需要更少的虚拟机和操作系统资源。

容器和虚拟机的模型对比如图 4-1 所示。

容器的思想是"一次打包，随处运行。"这决定了 Docker 具有三大优势：

1. 标准化

服务器只要配置好标准的运行环境，就可以运行任何容器，这使得运维工作的效率大大提升。在传统产品生产的过程中，时常出现开发环境、测试环境、生产环境不一致的情况，这使得每个环节的对接难度增加，运维人员需要耗费大量的精力解决环境不一致问题。有了容器以后，只要将容器一次打包好，就可以重复性地在任意环境中使用，只需要配置好容器运行环境。

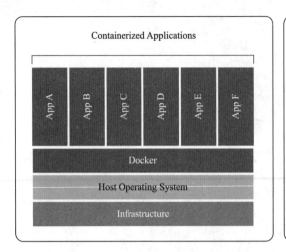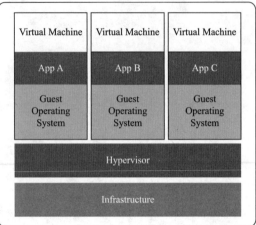

图 4-1   容器和虚拟机的模型对比

2. 轻量级

容器被称为"轻量级"的虚拟化技术。在传统的虚拟化当中，每当我们需要搭建业务、运行一些程序，都需要先从创建虚拟机开始，进一步安装操作系统并部署相应的环境。但是在某些使用场景中，可能只是要跑一个简单的程序或者仅仅是一个进程，如果还是用传统虚拟化方式就有点小题大做。这时就可以通过容器来实现。容器不需要创建虚拟机，也不需要安装额外的操作系统，只需要一条命令就可以完成容器的运行。

3. 安全

在传统计算机中，应用程序都在同一个操作系统环境中运行，而当应用被封装在容器中后，由于容器具有强大的隔离功能，这时应用程序在容器中运行更加安全。

Docker 开源单元的官方网站界面，如图 4-2 所示。在这里有最权威的文档、操作示例以及最佳实践。

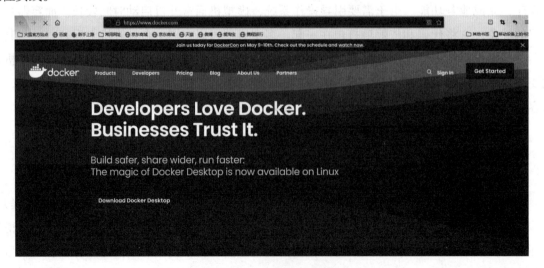

图 4-2   Docker 官方网站界面

4.2　Docker 架构

Docker 是 C/S（Client/Server，客户 / 服务器）结构的系统，客户端向 Docker 守护端，也就是其服务端发送请求，守护程序负责构建、运行和分发 Docker 容器。Docker 的核心组件包括 Docker 客户端——Client，Docker 服务端——Docker daemon，Docker 镜像——Images，镜像仓库——Registry，以及 Docker 容器——Containers。图 4-3 为 Docker 架构示意图。

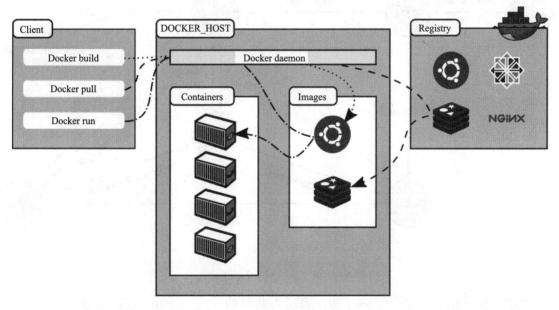

图 4-3　Docker 架构示意图

容器的整个生命周期是通过各组件相互协同工作完成的。运行一个容器，首先要从仓库 Registry 里面拉取镜像到本地，仓库分为公有仓库和私有仓库两种。Docker 默认使用的是 Docker Hub 仓库，这是由 Docker 公司维护的镜像仓库，其中维护了上万个容器镜像供用户使用。获取镜像后，将镜像实例化，启动并运行，就获得了一个容器。在这个过程中，对于镜像的拉取、容器的启动等操作都是通过客户端 Docker 命令工具实现的。客户端将这些操作请求发送到服务端 Docker daemon，由宿主机 Host 上的服务端程序完成容器的创建、运行、监控等。反过来，有时候构建好了一个容器，想要把当前的状态保留下来，打包并放在其他环境中来运行，这时候需要将容器构建提交为镜像，并推送到仓库当中，以便在其他环境中拉取并实例化。上述过程如图 4-4 所示。

容器的整个生命周期是从镜像开始的，首先要检查本地是否存在所需的镜像，如果不存在，要从仓库进

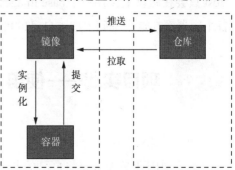

图 4-4　容器工作过程

行拉取；然后利用拉去下来的本地镜像启动容器，并分配一个文件系统，即在原始镜像的基础上在顶层加入一层容器可写层，这一层是用户可以写入新的数据的一层；接下来宿主机网桥分配一个虚拟接口到容器，并从地址池当中分配一个可用的 IP 地址给容器，以便用户通过网络访问容器；在容器中执行用户指定指令；业务执行完毕容器终止。整个容器的生命周期流程如图 4-5 所示。

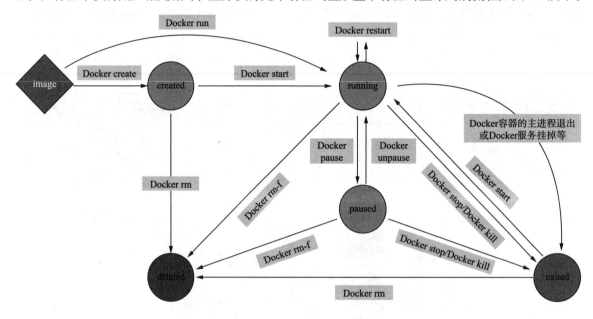

图 4-5　整个容器的生命周期流程

容器镜像采用了一种分层的结构，底层为共享的 kernel，在此基础上是第一层镜像 base image，顶层为可写的容器层。在容器的层次结构中，每一层都是在下层的基础上添加的软件，下层称为上层的父镜像。容器层以下都是只读 read-only 层。容器的分层结构如图 4-6 所示。容器的这种分层结构，可以充分实现资源共享，宿主机 Host 只会在磁盘上保存一个 base image，被所有容器共享，各个容器都在这个基础上加入自己的可写层。

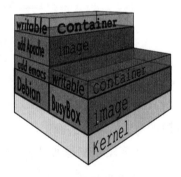

图 4-6　容器的分层结构

4.3　项目实战——使用 Docker 引擎技术管理容器

项目背景

某公司因业务需求要将自己的业务迁移上容器云，需要运维人员首先搭建容器运行环境，并掌握涉及的镜像、容器等简单的运维方法。

项目实施

任务 1　安装容器引擎

搭建 yum 源。首先搭建好 centos 的 base 网络源，使用网络源安装 yum-utils 工具，下载 docker-ce 软件包的 yum 仓库源配置文件 docker-ce.repo。

```
[root@localhost ~]# curl -o /etc/yum.repos.d/CentOS-Base.repo https://
mirrors.aliyun.com/repo/Centos-7.repo
[root@localhost ~]# yum -y install yum-utils
[root@localhost ~]# yum-config-manager \
--add-repo \
http://mirrors.aliyun.com/docker-ce/linux/centos/docker-ce.repo
[root@localhost ~]# yum repolist
Loaded plugins: fastestmirror
Loading mirror speeds from cached hostfile
* base: mirrors.aliyun.com
* epel: my.mirrors.thegigabit.com
* extras: mirrors.aliyun.com
* updates: mirrors.aliyun.com
docker-ce-stable                              | 3.5 kB  00:00:00
(1/2): docker-ce-stable/7/x86_64/primary_db   | 70 kB 00:00:00
(2/2): docker-ce-stable/7/x86_64/updateinfo   | 55 B 00:00:00
repo id                       repo name                              status
base/7/x86_64                        CentOS-7 - Base - mirrors.aliyun.com  10,072
docker-ce-stable/7/x86_64            Docker CE Stable - x86_64             139
*epel/x86_64         Extra Packages for Enterprise Linux 7 - x86_64 13,712
extras/7/x86_64      CentOS-7 - Extras - mirrors.aliyun.com         500
local                local                                          3,971
updates/7/x86_64     CentOS-7 - Updates - mirrors.aliyun.com        3,252
repolist: 31,646
```

安装 docker-ce 并开启服务。

```
[root@localhost ~]# yum -y install docker-ce
[root@localhost ~]# systemctl enable docker
[root@localhost ~]# systemctl start docker
```

安装成功后，使用 docker info 查看 docker 相关信息。

```
[root@localhost ~]# docker info
Client:
Context:    default
```

```
Debug Mode: false
Plugins:
app: Docker App (Docker Inc., v0.9.1-beta3)
buildx: Docker Buildx (Docker Inc., v0.7.1-docker)
scan: Docker Scan (Docker Inc., v0.12.0)

Server:
Containers: 0
Running: 0
Paused: 0
Stopped: 0
Images: 0
Server Version: 20.10.12
Storage Driver: overlay2
Backing Filesystem: xfs
Supports d_type: true
Native Overlay Diff: true
userxattr: false
Logging Driver: json-file
Cgroup Driver: cgroupfs
Cgroup Version: 1
Plugins:
Volume: local
Network: bridge host ipvlan macvlan null overlay
Log: awslogs fluentd gcplogs gelf journald json-file local logentries splunk syslog
Swarm: inactive
Runtimes: io.containerd.runc.v2 io.containerd.runtime.v1.linux runc
Default Runtime: runc
Init Binary: docker-init
containerd version: 7b11cfaabd73bb80907dd23182b9347b4245eb5d
runc version: v1.0.2-0-g52b36a2
init version: de40ad0
Security Options:
seccomp
Profile: default
Kernel Version: 3.10.0-862.el7.x86_64
Operating System: CentOS Linux 7 (Core)
OSType: linux
Architecture: x86_64
CPUs: 2
Total Memory: 1.936GiB
```

```
Name: localhost.localdomain
ID: HWX4:KNYS:IDY6:T3LF:N6B4:GTXW:UAE5:UET4:QVFH:34GE:FFER:AZCK
Docker Root Dir: /var/lib/docker
Debug Mode: false
Registry: https://index.docker.io/v1/
Labels:
Experimental: false
Insecure Registries:
127.0.0.0/8
Live Restore Enabled: false
```

至此 docker-ce 安装成功。接下来配置 docker 进程配置文件 daemon.json。在配置文件中可以设置数据存储路径 graph，存储驱动类型 storage-driver，添加信任区域 insecure-registries，容器网络网桥 IP 地址 bip 等参数。

```
[root@localhost ~]# vi /etc/docker/daemon.json
{
"graph": "/data/docker",
"storage-driver": "overlay",
"insecure-registries": ["registry.access.redhat.com","quay.io","harbor.od.com"],
"bip": "172.7.21.1/24",
"exec-opts": ["native.cgroupdriver=systemd"]
}
[root@master ~]# systemctl daemon-reload
[root@localhost ~]# systemctl restart docker
```

任务 2　启动一个最简单的容器 "hello world"

使用 docker run 命令运行一个 hello-world 镜像的容器，这是一个最简单的容器镜像，一般用来测试使用。容器启动后打印一条信息 "Hello from Docker!"。

```
[root@localhost ~]# docker run hello-world
Unable to find image 'hello-world:latest' locally
latest: Pulling from library/hello-world
2db29710123e: Pull complete
Digest: sha256:2498fce14358aa50ead0cc6c19990fc6ff866ce72aeb5546e1d59caac3d0d60f
Status: Downloaded newer image for hello-world:latest
Hello from Docker!
This message shows that your installation appears to be working correctly.

To generate this message, Docker took the following steps:
1. The Docker client contacted the Docker daemon.
```

2. The Docker daemon pulled the "hello-world" image from the Docker Hub. (amd64)

3. The Docker daemon created a new container from that image which runs the executable that produces the output you are currently reading.

4. The Docker daemon streamed that output to the Docker client, which sent it to your terminal.

To try something more ambitious, you can run an Ubuntu container with:
```
$ docker run -it ubuntu bash
```

Share images, automate workflows, and more with a free Docker ID:
https://hub.docker.com/

For more examples and ideas, visit:
https://docs.docker.com/get-started/

任务3 管理镜像

1. 连接镜像仓库

容器启动需要由镜像进行实例化，Docker 默认使用的镜像仓库是 Docker Hub。要使用 Docker Hub，先要进行注册。访问 Docker Hub 的官网，根据提示完成注册，如图 4-7 所示。

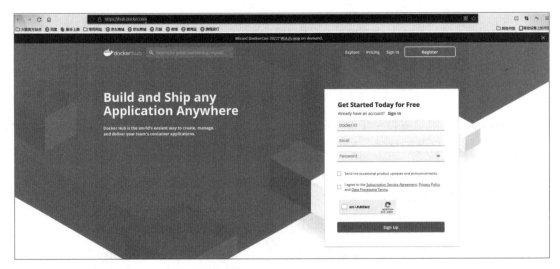

图 4-7 Docker Hub 注册界面

登录 Docker Hub。

```
[root@localhost ~]# docker login docker.io
Login with your Docker ID to push and pull images from Docker Hub. If you don't have a Docker ID, head over to https://hub.docker.com to create one.
```

```
Username: test123
Password:
WARNING! Your password will be stored unencrypted in /root/.docker/config.json.
Configure a credential helper to remove this warning. See
https://docs.docker.com/engine/reference/commandline/login/#credentials-store
Login Succeeded
```

由于从 Docker Hub 拉取镜像速度较慢，因此可以配置镜像加速，可以选择使用阿里云的加速器产品，即选择"产品"→"容器与中间件"→"容器镜像服务 ACR"命令，如图 4-8 所示。

图 4-8　使用阿里云的加速器产品

进入容器镜像服务管理控制台，如图 4-9 所示。

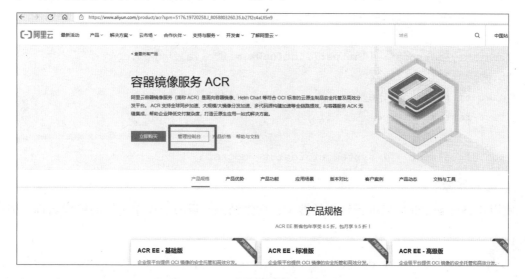

图 4-9　进入容器镜像服务管理控制台

复制加速器地址，在 daemon.json 里添加仓库镜像地址配置 registry-mirrors，如图 4-10 所示，获取镜像加速器配置信息。

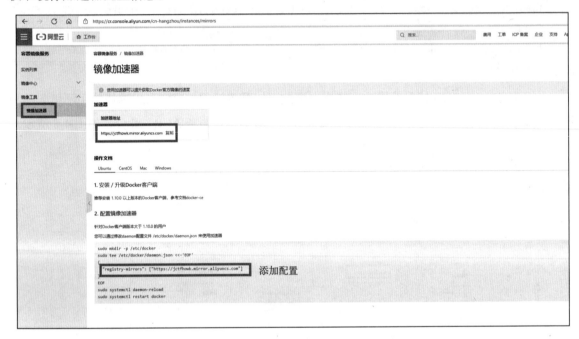

图 4-10　获取镜像加速器配置信息

根据页面显示的配置信息编写配置文件 daemon.json。

```
[root@localhost ~]# cat /etc/docker/daemon.json
{
"graph": "/data/docker",
"storage-driver": "overlay",
"insecure-registries": ["registry.access.redhat.com","quay.io","harbor.
od.com"],
"registry-mirrors": ["https://jctfhowk.mirror.aliyuncs.com"],
"bip": "172.7.21.1/24",
"exec-opts": ["native.cgroupdriver=systemd"],
"live-restore": true
}
[root@localhost ~]# systemctl restart docker
```

2. 搜索镜像

使用 docker search 可以进行镜像的搜索，搜索结果会返回仓库中可用的镜像名称、描述等信息。

```
[root@localhost ~]# docker search centos
NAME                            DESCRIPTION      STARS    OFFICIAL    AUTOMATED
```

centos	The official build of CentOS.	6962	[OK]
ansible/centos7-ansible	Ansible on Centos7	135	[OK]
consol/centos-xfce-vnc	Centos container with "headless" VNC session…		
		132	[OK]
jdeathe/centos-ssh	OpenSSH / Supervisor / EPEL/IUS/SCL Repos - …		
		121	[OK]
centos/systemd	systemd enabled base container.	105	[OK]
centos/mysql-57-centos7	MySQL 5.7 SQL database server	92	
imagine10255/centos6-lnmp-php56	centos6-lnmp-php56	58	[OK]
tutum/centos	Simple CentOS docker image with SSH access		
		48	
centos/postgresql-96-centos7	PostgreSQL is an advanced Object-Relational …		
		45	
centos/httpd-24-centos7	Platform for running Apache httpd 2.4 or bui…		
		41	
kinogmt/centos-ssh	CentOS with SSH	29	[OK]
guyton/centos6	From official centos6 container with full up…		
		10	[OK]
centos/tools	Docker image that has systems administration…		
		7	[OK]
drecom/centos-ruby	centos ruby	6	[OK]
centos/redis	Redis built for CentOS	6	[OK]
mamohr/centos-java	Oracle Java 8 Docker image based on Centos 7	4	[OK]
roboxes/centos8	A generic CentOS 8 base image.	4	
darksheer/centos	Base Centos Image -- Updated hourly	3	[OK]
dokken/centos-7	CentOS 7 image for kitchen-dokken	2	
miko2u/centos6	CentOS6 日本语环境	2	[OK]
amd64/centos	The official build of CentOS.	2	
mcnaughton/centos-base	centos base image	1	[OK]
blacklabelops/centos	CentOS Base Image! Built and Updates Daily!	1	[OK]
starlabio/centos-native-build	Our CentOS image for native builds	0	[OK]
smartentry/centos	centos with smartentry	0	[OK]

3.　下载一个镜像

使用 docker pull 拉取镜像到本地，使用 docker images 可以查看本地镜像列表。

```
[root@localhost ~]# docker pull centos
Using default tag: latest
latest: Pulling from library/centos
a1d0c7532777: Pull complete
Digest: sha256:a27fd8080b517143cbbbab9dfb7c8571c40d67d534bbdee55bd-
```

```
6c473f432b177
    Status: Downloaded newer image for centos:latest
    [root@localhost ~]# docker.io/library/centos:latest
    [root@localhost ~]#  docker pull centos:centos7
    centos7: Pulling from library/centos
    2d473b07cdd5: Pull complete
    Digest:  sha256:9d4bcbbb213dfd745b58be38b13b996ebb5ac315fe75711b-
d618426a630e0987
    Status: Downloaded newer image for centos:centos7
    docker.io/library/centos:centos7
    [root@localhost ~]# docker images
    REPOSITORY       TAG          IMAGE ID        CREATED          SIZE
    hello-world      latest       feb5d9fea6a5    3 months ago     13.3KB
    centos           centos7      eeb6ee3f44bd    3 months ago     204MB
    centos           latest       5d0da3dc9764    3 months ago     231MB
    [root@localhost  ~]# docker tag eeb6ee3f44bd   docker.io/test123/
centos:centos7
    [root@localhost ~]# docker images
    REPOSITORY       TAG          IMAGE ID        CREATED          SIZE
    hello-world      latest       feb5d9fea6a5    3 months ago     13.3KB
    test123/centos   centos7      eeb6ee3f44bd    3 months ago     204MB
    centos           centos7      eeb6ee3f44bd    3 months ago     204MB
    centos           latest       5d0da3dc9764    3 months ago     231MB
```

4. 推送镜像

使用 docker push 推送镜像到仓库中。镜像的名称遵循以下结构 "${registry_name}/${repository_name}/${image_name}:${tag_name}"。其中，${registry_name}/${repository_name} 是仓库名称，${image_name} 是镜像名称，${tag_name} 是版本号。例如，一个镜像的名称为 docker.io/library/centos:latest，指的就是 docker.io/library 中的 centos 镜像，版本名称为 latest；docker.io/test123/ubuntu:v1.1.0，指的就是 docker.io/test123 中 ubuntu 镜像，版本名称为 v1.1.0。

```
    [root@localhost ~]# docker push docker.io/test123/centos:centos7
    The push refers to repository [docker.io/test123/centos]
    174f56854903: Layer already exists
    centos7:  digest:  sha256:dead07b4d8ed7e29e98de0f4504d87e8880d-
4347859d839686a31da35a3b532f size: 529
```

推送完成后，登录 Docker Hub 查看，可以看到本地镜像已经推送到了仓库中，如图 4-11、图 4-12 所示。

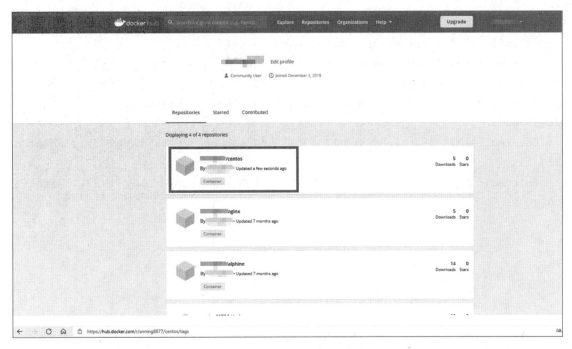

图 4-11　仓库中的镜像列表

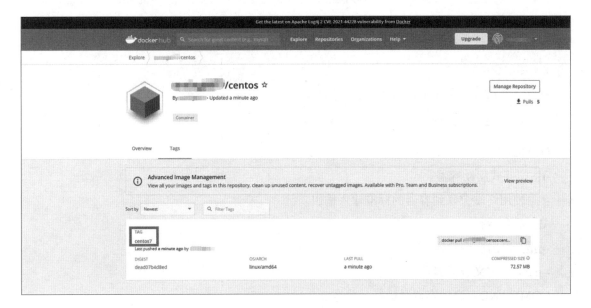

图 4-12　镜像详细信息

5. 删除镜像

使用 docker rmi 命令删除本地镜像。

```
[root@localhost ~]# docker rmi docker.io/test123/centos:centos7
Untagged: test123/centos:centos7
Untagged: test123/centos@sha256:dead07b4d8ed7e29e98de0f4504d87e8880d-
```

```
4347859d839686a31da35a3b532f
    [root@localhost ~]# docker rmi -f eeb6ee3f44bd
    Untagged: test123/centos:centos7
    Untagged: centos:centos7
    Untagged: centos@sha256:9d4bcbbb213dfd745b58be38b13b996ebb5ac315fe75711bd618
426a630e0987
    Deleted: sha256:eeb6ee3f44bd0b5103bb561b4c16bcb82328cfe5809ab675bb17ab3a16c5
17c9
    Deleted: sha256:174f5685490326fc0a1c0f5570b8663732189b327007e47ff13d2ca59673
db024.
```

任务 4　容器管理

1. 查看容器进程

```
[root@localhost ~]# docker ps
CONTAINER ID   IMAGE   COMMAND   CREATED   STATUS   PORTS   NAMES
[root@localhost ~]# docker ps -a
CONTAINER ID   IMAGE        COMMAND   CREATED      STATUS          PORTS   NAMES
5778103fc2bc   hello-world  "/hello"  20 hours ago Exited (0) 20 hours
ago            eloquent_hellman
```

2. 启动容器

启动容器的命令为 docker run，命令用法如下：

```
docker run [OPTIONS] IMAGE [COMMAND] [ARG...]
```

OPTIONS：

-i：交互式启动容器。

-t：使用终端关联到容器的标准输入输出。

-d：后台运行。

--rm：退出后删除容器。

--name：定义容器名称。

COMMAND：启动容器时运行的命令。

下面给出一个使用交互式的方式启动 centos 容器示例，启动后进入系统的 bash 界面。使用交互式的方式可以进入容器内部进行操作。

```
[root@localhost ~]# docker run -it centos:latest /bin/bash
[root@7a1565402db7 /]# ls
bin  dev  etc  home  lib  lib64  lost+found  media  mnt  opt  proc  root
run  sbin  srv  sys  tmp  usr  var
```

```
[root@localhost ~]# docker ps -a
CONTAINER ID    IMAGE           COMMAND        CREATED        STATUS        PORTS        NAMES
ccbd2941c4f7    centos:latest  "/bin/bash"  6 seconds ago       Up 5 seconds
heuristic_gates
5778103fc2bc    hello-world "/hello" 20 hours ago      Exited (0) 20 hours ago
eloquent_hellman
```

使用 --rm 参数可以设定退出容器后删除容器。在很多场景中，容器里只是临时运行一个简单的进程，运行完成后容器不再使用，就可以使用这种方式进行自动删除。

```
[root@localhost ~]# docker run --rm centos /bin/echo hello
hello
[root@localhost ~]# docker ps -a
CONTAINER ID    IMAGE           COMMAND        CREATED        STATUS        PORTS        NAMES
ccbd2941c4f7     centos:latest "/bin/bash" About a minute ago  Exited (0) 34
seconds ago              heuristic_gates
5778103fc2bc      hello-world  "/hello"      20 hours ago      Exited (0) 20 hours
ago               eloquent_hellman
```

当需要容器在后台挂起持续运行时，可以使用 -d 参数在后台启动容器。

```
[root@localhost ~]# docker run -d --name test1 centos /bin/sleep 300
7aee3bb74a5dc395a7a46262cfd7fd60c83044bf6f42aaf86f436a7842309973
[root@localhost ~]# docker ps
CONTAINER ID IMAGE    COMMAND      CREATED       STATUS       PORTS       NAMES
7aee3bb74a5d centos  "/bin/sleep 300"  4 seconds ago  Up 3 seconds     test1
[root@localhost ~]# ps aux|grep sleep
root    65969 0.1 0.0 23032    936 ?        Ss  23:45 0:00 /usr/bin/coreutils
--coreutils-prog-shebang=sleep /bin/sleep 300
```

3. 进入容器

使用 docker exec 命令进入容器。

```
[root@localhost ~]# docker ps
CONTAINER ID  IMAGE  COMMAND  CREATED  STATUS  PORTS  NAMES
7aee3bb74a5d centos "/bin/sleep 300" 3 minutes ago Up 3 minutes     test1
[root@localhost ~]# docker exec -it 7aee3bb74a5d /bin/bash
[root@7aee3bb74a5d /]# ps aux
USER    PID %CPU %MEM    VSZ    RSS TTY       STAT START    TIME COMMAND
root      1  0.0 0.0  23032    936 ?        Ss 15:45 0:00/usr/bin/coreutils
--coreutils-prog-shebang=sleep /bin/sleep 300
root      8  0.5 0.1  12036   2096 pts/0     Ss  15:49   0:00 /bin/bash
root     22 0.0  0.0  44652   1776 pts/0     R+  15:49   0:00 ps aux
```

4. 停止容器

使用 docker stop 命令停止容器。

```
[root@localhost ~]# docker ps
CONTAINER ID   IMAGE    COMMAND        CREATED       STATUS        PORTS      NAMES
7aee3bb74a5d   centos   "/bin/sleep 300"  4 minutes ago  Up 4 minutes             test1
[root@localhost ~]# docker stop 7aee3bb74a5d
7aee3bb74a5d
[root@localhost ~]# docker ps
CONTAINER ID   IMAGE    COMMAND        CREATED       STATUS        PORTS      NAMES
```

5. 重启容器

使用 docker restart 命令重启容器。

```
[root@localhost ~]# docker ps
CONTAINER ID   IMAGE    COMMAND        CREATED       STATUS        PORTS      NAMES
[root@localhost ~]# docker ps -a
CONTAINER ID   IMAGE       COMMAND        CREATED       STATUS          PORTS       NAMES
7aee3bb74a5d   centos      "/bin/sleep 300" 6 minutes ago  Exited (0) About a
minute ago            test1
5778103fc2bc   hello-world  "/hello"       21 hours ago   Exited (0) 21 hours
ago                 eloquent_hellman
[root@localhost ~]# docker restart 7aee3bb74a5d
7aee3bb74a5d
[root@localhost ~]# docker ps
CONTAINER ID   IMAGE    COMMAND        CREATED       STATUS        PORTS      NAMES
7aee3bb74a5d   centos   "/bin/sleep 300"  6 minutes ago  Up 2 seconds            test1
```

6. 删除容器

使用 docker rm 命令删除容器，如果删除正在运行的容器，加 -f 进行强制删除。

```
[root@localhost ~]# docker rm test1
Error response from daemon: You cannot remove a running container 7aee3bb
74a5dc395a7a46262cfd7fd60c83044bf6f42aaf86f436a7842309973. Stop the container
before attempting removal or force remove
[root@localhost ~]# docker rm -f test1
test1
```

有时需要批量删除已退出的容器时，可以借助 shell 中文本处理工具 awk 来获取处于退出
exited 状态的容器 ID，进一步使用循环语句进行批量删除。

```
[root@localhost ~]# docker ps -a
CONTAINER ID   IMAGE    COMMAND        CREATED       STATUS        PORTS      NAMES
```

```
    dd05fc14a462    centos      "/bin/bash"   10 minutes ago    Exited (0) 10 minutes
ago      test
    ccbd2941c4f7    centos:latest  "/bin/bash"   13 minutes ago      Exited (0) 12
minutes ago              heuristic_gates
    70cc518f6fc1    centos:latest  "pwd"      14 minutes ago    Exited (0) 14 minutes
ago              busy_feynman
    7a1565402db7    centos:latest  "/bin/bash"  15 minutes ago    Exited (0) 14
minutes ago              beautiful_noether
    80bea167a41c    centos:latest  "pwd"      15 minutes ago    Exited (0) 15 minutes
ago              reverent_hermann
    92db7da2f9cf    centos:latest  "/bin/bash"   15 minutes ago      Exited (0) 15
minutes ago              objective_hugle
    306b4ab346cd    centos:latest  "/bin/bash"   15 minutes ago      Exited (0) 15
minutes ago              blissful_lehmann
    5778103fc2bc    hello-world   "/hello"     21 hours ago    Exited (0) 21 hours
ago              eloquent_hellman
    [root@localhost ~]# for i in `docker ps -a|grep -i exit|awk '{print $1}'`;do
docker rm -f $i;done
    dd05fc14a462
    ccbd2941c4f7
    70cc518f6fc1
    7a1565402db7
    80bea167a41c
    92db7da2f9cf
    306b4ab346cd
    5778103fc2bc
    [root@localhost ~]# docker ps -a
    CONTAINER ID   IMAGE     COMMAND    CREATED    STATUS    PORTS    NAMES
```

7. 修改容器

有时候希望把对基础镜像的修改固化到只读层，构建新的镜像。这时可以通过 docker commit 命令重新构建镜像。在本例当中，在 centos 镜像的基础上，将编写的 txt 文档固化到镜像中，用新的镜像启动容器时，可以看到容器启动后，添加的 txt 文档已经被固化到了只读层。

```
    [root@localhost ~]# docker run -it centos /bin/bash
    [root@4c8eed52f1c5 /]# vi 1.txt
    hello world!
    [root@4c8eed52f1c5 /]# ls -l
    total 4
    -rw-r--r--   1 root root  13 Jan  8 17:38 1.txt
    lrwxrwxrwx   1 root root   7 Nov  3 2020 bin -> usr/bin
```

```
drwxr-xr-x    5 root root 360 Jan  8 17:38 dev
drwxr-xr-x    1 root root  66 Jan  8 17:38 etc
drwxr-xr-x    2 root root   6 Nov  3  2020 home
lrwxrwxrwx    1 root root   7 Nov  3  2020 lib -> usr/lib
lrwxrwxrwx    1 root root   9 Nov  3  2020 lib64 -> usr/lib64
drwx------    2 root root   6 Sep 15 14:17 lost+found
drwxr-xr-x    2 root root   6 Nov  3  2020 media
drwxr-xr-x    2 root root   6 Nov  3  2020 mnt
drwxr-xr-x    2 root root   6 Nov  3  2020 opt
dr-xr-xr-x  127 root root   0 Jan  8 17:38 proc
dr-xr-x---    2 root root 162 Sep 15 14:17 root
drwxr-xr-x   11 root root 163 Sep 15 14:17 run
lrwxrwxrwx    1 root root   8 Nov  3  2020 sbin -> usr/sbin
drwxr-xr-x    2 root root   6 Nov  3  2020 srv
dr-xr-xr-x   13 root root   0 Jan  8 17:38 sys
drwxrwxrwt    7 root root 171 Sep 15 14:17 tmp
drwxr-xr-x   12 root root 144 Sep 15 14:17 usr
drwxr-xr-x   20 root root 262 Sep 15 14:17 var
[root@4c8eed52f1c5 /]# exit
exit
[root@localhost ~]# docker commit -p 4c8eed52f1c5 test123/centos_with_1.txt
sha256:5268290bcab95fabfbaaeb7a602395e885a3803c595db362418adfae485d3ad3
[root@localhost ~]# docker images
REPOSITORY                 TAG      IMAGE ID       CREATED          SIZE
test123/centos_with_1.txt  latest   5268290bcab9   18 seconds ago   231MB
hello-world                latest   feb5d9fea6a5   3 months ago     13.3KB
centos                     latest   5d0da3dc9764   3 months ago     231MB
[root@localhost ~]# docker run -it test123/centos_with_1.txt /bin/bash
[root@3bf91a5992b2 /]# ls
1.txt  bin  dev  etc  home  lib  lib64  lost+found  media  mnt  opt  proc
root  run  sbin  srv  sys  tmp  usr  var
[root@3bf91a5992b2 /]# cat 1.txt
hello world!
```

8. 导出镜像

可以借助打包的方式导出镜像。使用 docker save 保存镜像，并打包为 tar 文件。

```
[root@localhost ~]# docker images
REPOSITORY                 TAG      IMAGE ID       CREATED          SIZE
test123/centos_with_1.txt  latest   5268290bcab9   6 minutes ago    231MB
hello-world                latest   feb5d9fea6a5   3 months ago     13.3KB
```

```
centos                        latest   5d0da3dc9764   3 months ago    231MB
[root@localhost ~]# docker save 5268290bcab9 centos_with_1.txt.tar
cowardly refusing to save to a terminal. Use the -o flag or redirect
[root@localhost ~]# docker save 5268290bcab9 > centos_with_1.txt.tar
[root@localhost ~]# ll
total 233016
-rw-r--r-- 1 root root       11 Sep 14 00:20 1
-rw-r--r-- 1 root root        7 Sep 14 00:21 2
drwxr-xr-x 2 root root        6 Sep 13 18:15 abc
-rw-r--r-- 1 root root        8 Sep 14 00:24 a.exe
-rw-------. 1 root root     1381 Apr  8  2021 anaconda-ks.cfg
-rw-r--r-- 1 root root        0 Sep 14 00:18 a.txt
-rw-r--r-- 1 root root        0 Sep 13 19:36 b.txt
-rw-r--r-- 1 root root 238586880 Jan  9 01:47 centos_with_1.txt.tar
-rw-r--r-- 1 root root        6 Sep 13 19:37 t5
```

删除镜像后，可以从 tar 包恢复镜像。

```
[root@localhost ~]# docker rmi -f 5268290bcab9
Untagged: test123/centos_with_1.txt:latest
Deleted: sha256:5268290bcab95fabfbaaeb7a602395e885a3803c595db362418adfae485d3ad3
[root@localhost ~]# docker images
REPOSITORY      TAG        IMAGE ID        CREATED         SIZE
hello-world     latest     feb5d9fea6a5    3 months ago    13.3KB
centos          latest     5d0da3dc9764    3 months ago    231MB
[root@localhost ~]# docker load < centos_with_1.txt.tar
Loaded image ID: sha256:5268290bcab95fabfbaaeb7a602395e885a3803c595db362418a
dfae485d3ad3
[root@localhost ~]# docker images
REPOSITORY      TAG        IMAGE ID        CREATED         SIZE
<none>          <none>     5268290bcab9    8 minutes ago   231MB
hello-world     latest     feb5d9fea6a5    3 months ago    13.3KB
centos          latest     5d0da3dc9764    3 months ago    231MB
[root@localhost ~]# docker tag 5268290bcab9 test123/centos_with_1.txt
[root@localhost ~]# docker images
REPOSITORY                  TAG      IMAGE ID        CREATED         SIZE
test123/centos_with_1.txt   latest   5268290bcab9    8 minutes ago   231MB
hello-world                 latest   feb5d9fea6a5    3 months ago    13.3KB
centos                      latest   5d0da3dc9764    3 months ago    231MB
[root@localhost ~]# docker run -it 5268290bcab9 /bin/bash
[root@89190dbb29b9 /]# ls
1.txt  dev  home  lib64     media  opt   root  sbin  sys  usr
```

```
bin    etc  lib    lost+found  mnt    proc  run  srv  tmp  var
[root@89190dbb29b9 /]# cat 1.txt
hello world!
```

9. 查看容器的操作日志

可以通过 docker logs + 容器 ID 号的方式查看容器的操作日志。

```
[root@localhost ~]# docker ps -a
CONTAINER ID    IMAGE         COMMAND        CREATED       STATUS        PORTS       NAMES
89190dbb29b9    5268290bcab9     "/bin/bash" 40 hours ago  Exited (0) 40 hours
ago           funny_engelbart
3bf91a5992b2    test123/centos_with_1.txt  "/bin/bash"  41 hours ago  Exited
(130) 41 hours ago           goofy_gates
4c8eed52f1c5    centos        "/bin/bash"  41 hours ago  Exited (0) 41 hours ago
romantic_shockley
[root@localhost ~]# docker logs 89190dbb29b9
[root@89190dbb29b9 /]# ls
1.txt  dev  home  lib64      media  opt    root  sbin  sys  usr
bin    etc  lib   lost+found  mnt    proc  run  srv  tmp  var
[root@89190dbb29b9 /]# cat 1.txt
hello world!
[root@89190dbb29b9 /]# exit
exit
```

10. 本地系统与容器交互

容器往往承担了提供服务的作用，这就要求容器暴露自身端口来对外提供服务。例如，此时运行一个 nginx 的容器，将容器的 80 端口映射到宿主机的 81 端口上，就可以通过"宿主机 IP:81"来访问 nginx 服务了。

```
[root@localhost ~]# docker pull nginx
Using default tag: latest
latest: Pulling from library/nginx
a2abf6c4d29d: Pull complete
a9edb18cadd1: Pull complete
589b7251471a: Pull complete
186b1aaa4aa6: Pull complete
b4df32aa5a72: Pull complete
a0bcbecc962e: Pull complete
Digest: sha256:0d17b565c37bcbd895e9d92315a05c1c3c9a29f762b011a10c54a66cd53c9b31
Status: Downloaded newer image for nginx:latest
docker.io/library/nginx:latest
[root@localhost ~]# docker images
```

```
REPOSITORY                    TAG        IMAGE ID       CREATED          SIZE
test123/centos_with_1.txt     latest     5268290bcab9   41 hours ago     231MB
nginx                         latest     605c77e624dd   11 days ago      141MB
hello-world                   latest     feb5d9fea6a5   3 months ago     13.3KB
centos                        latest     5d0da3dc9764   3 months ago     231MB
[root@localhost ~]# docker tag 605c77e624dd test123/nginx
[root@localhost ~]# docker images
REPOSITORY                    TAG        IMAGE ID       CREATED          SIZE
test123/centos_with_1.txt     latest     5268290bcab9   41 hours ago     231MB
test123/nginx                 latest     605c77e624dd   11 days ago      141MB
nginx                         latest     605c77e624dd   11 days ago      141MB
hello-world                   latest     feb5d9fea6a5   3 months ago     13.3KB
centos                        latest     5d0da3dc9764   3 months ago     231MB
[root@localhost ~]# docker run --rm --name mynginx -d -p81:80 test123/nginx
7bba3aa0b54aba442e639fd27f4218a9c0229a1767990fc64938c96008f832cb
[root@localhost ~]# docker ps -a
CONTAINER ID   IMAGE          COMMAND          CREATED       STATUS       PORTS        NAMES
7bba3aa0b54a   test123/nginx     "/docker-entrypoint.…"   4 seconds ago   Up 3
seconds              0.0.0.0:81->80/tcp, :::81->80/tcp  mynginx
    89190dbb29b9   5268290bcab9      "/bin/bash"   41 hours ago     Exited (0) 41
hours ago      funny_engelbart
    3bf91a5992b2   test123/centos_with_1.txt   "/bin/bash"   41 hours ago    Exited
(130) 41 hours ago      goofy_gates
    4c8eed52f1c5   centos      "/bin/bash"   41 hours ago  Exited (0) 41 hours ago
romantic_shockley
```

访问宿主机的 81 端口，可以看到 nginx 访问界面如图 4-13 所示。

图 4-13 nginx 访问界面

在操作容器时，可以将本地数据卷映射到容器内部。例如，现有一个容器正在提供nginx服务，假设网站代码有更新，可以在本地创建一个目录作为站点代码存储目录，然后将其通过挂载数据卷的方式映射到容器内部。这样一来，如果站点代码有更新，只需要更新本地目录中的文件，容器就会自动进行同步映射，无须停掉服务就可以实现站点内容的更新了。

```
[root@localhost ~]# mkdir html
[root@localhost ~]# cd html/
[root@localhost html]# wget www.taobao.com -O index.html
--2022-01-10 18:38:43--  http://www.taobao.com/
Resolving www.taobao.com (www.taobao.com)... 219.144.108.233,
219.144.108.232, 240e:bf:c800:2912:3::3fa, ...
Connecting to www.taobao.com (www.taobao.com)|219.144.108.233|:80...
connected.
HTTP request sent, awaiting response... 301 Moved Permanently
Location: https://www.taobao.com/ [following]
--2022-01-10 18:38:43--  https://www.taobao.com/
Connecting to www.taobao.com (www.taobao.com)|219.144.108.233|:443...
connected.
HTTP request sent, awaiting response... 200 OK
Length: unspecified [text/html]
Saving to: 'index.html'

    [ <=>                          ] 113,640      --.-K/s    in 0.03s

2022-01-10 18:38:43 (3.95 MB/s) - 'index.html' saved [113640]


[root@localhost html]# docker run -d --rm --name mytaobao -p82:80 -v/root/
html:/usr/share/nginx/html test123/nginx
660265c5e6adb442ef71bd4d3c34729cd98d2d11878da7f6aafc99c8feda5e16
[root@localhost html]# docker run -d --rm --name mytaobao -p82:80 -v/root/
html:/usr/share/nginx/html test123/nginx
660265c5e6adb442ef71bd4d3c34729cd98d2d11878da7f6aafc99c8feda5e16
[root@localhost html]# docker exec -it mytaobao /bin/bash
[root@660265c5e6ad:/# ls /usr/share/nginx/html/
index.html
```

访问宿主机的 82 端口，可以看到容器中运行的 web 页面，如图 4-14 所示。

图 4-14　容器中运行的 web 页面

4.4　项目实战——基于 Docker Compose 定义和运行多容器

项目背景

在一个使用 Docker 运行服务的业务场景中，通常都需要不止一个容器来完成整个业务，如果一个一个单独对容器进行定义和运行，工作量较大，Docker Compose 提供了一种定义和运行多容器的方式。使用 Docker Compose 首先要在 Dockerfile 中定义应用程序所需要的环境；然后编写 yml 文件来配置整个业务场景当中所需要的服务；最后，使用 docker-compose up 命令来一次性创建并启动所有服务。

现公司需要搭建一个网站用来记录网站的访问量，要求用容器来实现业务场景。

项目实施

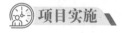

任务 1　安装 docker-compose

安装 docker-compose 所需要的依赖包。

```
[root@localhost ~]# yum -y install epel-release
[root@localhost ~]# yum -y install python-pip python3-devel libffi-devel
openssl-dev gcc glibc-devel rust cargo make
```

　　下载 docker-compose 应用程序，并赋予可执行权限。将可执行文件链接到 /usr/bin/ 下。安装成功后，使用 docker-compose --version 查看 docker-compose 相关信息。

```
[root@localhost ~]# curl -L "https://github.com/docker/compose/releases/
download/v1.29.2/docker-compose-$(uname -s)-$(uname -m)" -o /usr/local/bin/
docker-compose
[root@localhost ~]# chmod +x /usr/local/bin/docker-compose
[root@localhost ~]# ln -s /usr/local/bin/docker-compose /usr/bin/docker-
compose
[root@localhost ~]# docker-compose --version
docker-compose version 1.29.2, build 5becea4c
```

<h2 style="text-align:center">任务 2　搭建网站记录网站访问量</h2>

1. 基础环境设置

在管理员目录下新建一个单元目录 composetest。

```
[root@localhost ~]# mkdir composetest
[root@localhost ~]# cd composetest
```

编写单元应用程序 app.py。使用经典的 web 框架 flask，数据库使用 redis 数据库。

```
[root@localhost composetest]# vi app.py
import time

import redis
from flask import Flask

app = Flask(__name__)
cache = redis.Redis(host='redis', port=6379)

def get_hit_count():
    retries = 5
    while True:
        try:
            return cache.incr('hits')
        except redis.exceptions.ConnectionError as exc:
            if retries == 0:
                raise exc
            retries -= 1
            time.sleep(0.5)
```

```
@app.route('/')
def hello():
    count = get_hit_count()
    return 'Hello World! I have been seen {} times.\n'.format(count)
```

设置所需安装的依赖软件包 flask 和 redis。

```
[root@localhost composetest]# vi requirements.txt
flask
redis
```

2. 创建 Dockerfile

编写 Dockerfile，从 python3.7 开始构建，将工作目录设置为 /code，并设置运行 flask 的环境变量。

```
[root@localhost composetest]# vi Dockerfile
FROM python:3.7-alpine
WORKDIR /code
ENV FLASK_APP=app.py
ENV FLASK_RUN_HOST=0.0.0.0
RUN echo https://mirrors.ustc.edu.cn/alpine/latest-stable/main > /etc/apk/
repositories; \
    echo https://mirrors.ustc.edu.cn/alpine/latest-stable/community >> /etc/
apk/repositories
RUN apk add --update --no-cache gcc musl-dev linux-headers
COPY requirements.txt requirements.txt
RUN pip install -r requirements.txt
EXPOSE 5000
COPY . .
CMD ["flask", "run"]
```

3. 编写 yml 文件定义服务

定义两个服务 web 和 redis 并定义它们的初始镜像及端口等相关配置。

```
[root@localhost composetest]# vi docker-compose.yml

version: "3.9"
services:
  web:
    build: .
    ports:
      - "5000:5000"
  redis:
    image: "redis:alpine"
```

4. 使用 docker-compose up 运行应用

```
[root@localhost composetest]# docker-compose up
Building web
Sending build context to Docker daemon  18.43kB
Step 1/11 : FROM python:3.7-alpine
 ---> a1034fd13493
Step 2/11 : WORKDIR /code
 ---> Using cache
 ---> 3d40488c8b7d
Step 3/11 : ENV FLASK_APP=app.py
 ---> Using cache
 ---> d34305afff76
Step 4/11 : ENV FLASK_RUN_HOST=0.0.0.0
 ---> Using cache
 ---> 2ad191618735
Step 5/11 : RUN echo https://mirrors.ustc.edu.cn/alpine/latest-stable/main
> /etc/apk/repositories;  echo https://mirrors.ustc.edu.cn/alpine/latest-stable/
community >> /etc/apk/repositories
 ---> Running in afac6d11bafd
Removing intermediate container afac6d11bafd
 ---> 8fe22683fb43
Step 6/11 : RUN apk add --update --no-cache gcc musl-dev linux-headers
 ---> Running in 100502fca72e
fetch https://mirrors.ustc.edu.cn/alpine/latest-stable/main/x86_64/APKINDEX.
tar.gz
fetch https://mirrors.ustc.edu.cn/alpine/latest-stable/community/x86_64/
APKINDEX.tar.gz
(1/13) Installing libgcc (10.3.1_git20211027-r0)
(2/13) Installing libstdc++ (10.3.1_git20211027-r0)
(3/13) Installing binutils (2.37-r3)
(4/13) Installing libgomp (10.3.1_git20211027-r0)
(5/13) Installing libatomic (10.3.1_git20211027-r0)
(6/13) Installing libgphobos (10.3.1_git20211027-r0)
(7/13) Installing gmp (6.2.1-r0)
(8/13) Installing isl22 (0.22-r0)
(9/13) Installing mpfr4 (4.1.0-r0)
(10/13) Installing mpc1 (1.2.1-r0)
(11/13) Installing gcc (10.3.1_git20211027-r0)
(12/13) Installing linux-headers (5.10.41-r0)
(13/13) Installing musl-dev (1.2.2-r7)
```

```
Executing busybox-1.34.1-r3.trigger
OK: 139 MiB in 48 packages
Removing intermediate container 100502fca72e
 ---> d0eb07ac0f8c
Step 7/11 : COPY requirements.txt requirements.txt
 ---> 13243085a68f
Step 8/11 : RUN pip install -r requirements.txt
 ---> Running in a7ab2f639ab5
Collecting flask
  Downloading Flask-2.0.2-py3-none-any.whl (95 KB)
Collecting redis
  Downloading redis-4.1.1-py3-none-any.whl (173 KB)
Collecting click>=7.1.2
  Downloading click-8.0.3-py3-none-any.whl (97 KB)
Collecting Jinja2>=3.0
  Downloading Jinja2-3.0.3-py3-none-any.whl (133 KB)
Collecting Werkzeug>=2.0
  Downloading Werkzeug-2.0.2-py3-none-any.whl (288 KB)
Collecting itsdangerous>=2.0
  Downloading itsdangerous-2.0.1-py3-none-any.whl (18 KB)
Collecting deprecated>=1.2.3
  Downloading Deprecated-1.2.13-py2.py3-none-any.whl (9.6 KB)
Collecting importlib-metadata>=1.0
  Downloading importlib_metadata-4.10.1-py3-none-any.whl (17 KB)
Collecting packaging>=20.4
  Downloading packaging-21.3-py3-none-any.whl (40 KB)
Collecting wrapt<2,>=1.10
  Downloading wrapt-1.13.3-cp37-cp37m-musllinux_1_1_x86_64.whl (78 KB)
Collecting typing-extensions>=3.6.4
  Downloading typing_extensions-4.0.1-py3-none-any.whl (22 KB)
Collecting zipp>=0.5
  Downloading zipp-3.7.0-py3-none-any.whl (5.3 KB)
Collecting MarkupSafe>=2.0
  Downloading MarkupSafe-2.0.1-cp37-cp37m-musllinux_1_1_x86_64.whl (30 KB)
Collecting pyparsing!=3.0.5,>=2.0.2
  Downloading pyparsing-3.0.6-py3-none-any.whl (97 KB)
Installing collected packages: zipp, typing-extensions, wrapt, pyparsing,
MarkupSafe, importlib-metadata, Werkzeug, packaging, Jinja2, itsdangerous,
deprecated, click, redis, flask
Successfully installed Jinja2-3.0.3 MarkupSafe-2.0.1 Werkzeug-2.0.2 click-8.0.3
```

deprecated-1.2.13 flask-2.0.2 importlib-metadata-4.10.1 itsdangerous-2.0.1 packaging-21.3 pyparsing-3.0.6 redis-4.1.1 typing-extensions-4.0.1 wrapt-1.13.3 zipp-3.7.0

WARNING: Running pip as the 'root' user can result in broken permissions and conflicting behaviour with the system package manager. It is recommended to use a virtual environment instead: https://pip.pypa.io/warnings/venv

WARNING: You are using pip version 21.2.4; however, version 21.3.1 is available.

You should consider upgrading via the '/usr/local/bin/python -m pip install --upgrade pip' command.

Removing intermediate container a7ab2f639ab5

 ---> 9bc52916561f

Step 9/11 : EXPOSE 5000

 ---> Running in e74001bcf9e7

Removing intermediate container e74001bcf9e7

 ---> 0bd1e7c23eb2

Step 10/11 : COPY . .

 ---> b02f061ec235

Step 11/11 : CMD ["flask", "run"]

 ---> Running in 1bf7da5bd459

Removing intermediate container 1bf7da5bd459

 ---> 278691cb207f

Successfully built 278691cb207f

Successfully tagged composetest_web:latest

WARNING: Image for service web was built because it did not already exist. To rebuild this image you must use 'docker-compose build' or 'docker-compose up --build'.

Pulling redis (redis:alpine)...

alpine: Pulling from library/redis

59bf1c3509f3: Already exists

719adce26c52: Pull complete

b8f35e378c31: Pull complete

d034517f789c: Pull complete

3772d4d76753: Pull complete

211a7f52febb: Pull complete

Digest: sha256:4bed291aa5efb9f0d77b76ff7d4ab71eee410962965d052552db1fb80576431d

Status: Downloaded newer image for redis:alpine

Creating composetest_web_1   ... done

Creating composetest_redis_1 ... done

Attaching to composetest_redis_1, composetest_web_1

redis_1 | 1:C 18 Jan 2022 10:08:48.953 # oO0OoO00oO00o Redis is starting oO0OoO00oO00o

```
    redis_1  | 1:C 18 Jan 2022 10:08:48.954 # Redis version=6.2.6, bits=64,
commit=00000000, modified=0, pid=1, just started
    redis_1  | 1:C 18 Jan 2022 10:08:48.954 # Warning: no config file specified,
using the default config. In order to specify a config file use redis-server /path/
to/redis.conf
    redis_1  | 1:M 18 Jan 2022 10:08:48.955 * monotonic clock: POSIX clock_gettime
    redis_1  | 1:M 18 Jan 2022 10:08:48.956 * Running mode=standalone, port=6379.
    redis_1  | 1:M 18 Jan 2022 10:08:48.956 # WARNING: The TCP backlog setting
of 511 cannot be enforced because /proc/sys/net/core/somaxconn is set to the
lower value of 128.
    redis_1  | 1:M 18 Jan 2022 10:08:48.956 # Server initialized
    redis_1  | 1:M 18 Jan 2022 10:08:48.956 # WARNING overcommit_memory is set
to 0! Background save may fail under low memory condition. To fix this issue
add 'vm.overcommit_memory = 1' to /etc/sysctl.conf and then reboot or run the
command 'sysctl vm.overcommit_memory=1' for this to take effect.
    redis_1  | 1:M 18 Jan 2022 10:08:48.956 * Ready to accept connections
    web_1    | * Serving Flask app 'app.py' (lazy loading)
    web_1    | * Environment: production
    web_1    |   WARNING: This is a development server. Do not use it in a
production deployment.
    web_1    |   Use a production WSGI server instead.
    web_1    | * Debug mode: off
    web_1    | * Running on all addresses.
    web_1    |   WARNING: This is a development server. Do not use it in a
production deployment.
    web_1    | * Running on http://172.17.0.3:5000/ (Press CTRL+C to quit)
    web_1    | 192.168.100.1 - - [18/Jan/2022 10:09:20] "GET / HTTP/1.1" 200 -
    web_1    | 192.168.100.1 - - [18/Jan/2022 10:09:20] "GET /favicon.ico
HTTP/1.1" 404 -
    web_1    | 192.168.100.1 - - [18/Jan/2022 10:09:31] "GET / HTTP/1.1" 200 -
    web_1    | 192.168.100.1 - - [18/Jan/2022 10:09:32] "GET / HTTP/1.1" 200 -
    web_1    | 192.168.100.1 - - [18/Jan/2022 10:09:32] "GET / HTTP/1.1" 200 -
    web_1    | 192.168.100.1 - - [18/Jan/2022 10:09:33] "GET / HTTP/1.1" 200 -
    web_1    | 192.168.100.1 - - [18/Jan/2022 10:14:03] "GET / HTTP/1.1" 200 -
    redis_1  | 1:M 18 Jan 2022 11:08:49.052 * 1 changes in 3600 seconds.
Saving...
    redis_1  | 1:M 18 Jan 2022 11:08:49.072 * Background saving started by pid 13
    redis_1  | 13:C 18 Jan 2022 11:08:49.089 * DB saved on disk
    redis_1  | 13:C 18 Jan 2022 11:08:49.089 * RDB: 4 MB of memory used by copy-on-write
    redis_1  | 1:M 18 Jan 2022 11:08:49.183 * Background saving terminated with success
```

5. 测试

在浏览器中输入 http://IP:5000/，就可以看到网站页面了，如图 4-15 所示。

图 4-15　容器中运行 web 服务页面

每刷新一次页面，计数器加 1，如图 4-16 所示。

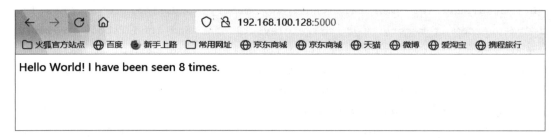

图 4-16　刷新页面计数器递增

6. 更新服务

当需要更新服务时，可以通过挂载数据卷的方式来实时更新应用程序。

首先停止服务。

```
[root@localhost composetest]# docker-compose down
Stopping composetest_redis_1 ... done
Stopping composetest_web_1   ... done
Removing composetest_redis_1 ... done
Removing composetest_web_1   ... done
Removing network composetest_default
```

修改 yml 文件，为服务绑定挂载，以便修改应用程序即时生效，无须重新构建镜像。

```
[root@localhost composetest]# vi docker-compose.yml
version: "3.9"
services:
  web:
    build: .
    ports:
      - "5000:5000"
    volumes:
```

```
      - .:/code
    environment:
      FLASK_ENV: development
  redis:
    image: "redis:alpine"
```

修改应用程序内容，查看即时生效效果。

```
[root@localhost composetest]# vi app.py
import time

import redis
from flask import Flask

app = Flask(__name__)
cache = redis.Redis(host='redis', port=6379)

def get_hit_count():
    retries = 5
    while True:
        try:
            return cache.incr('hits')
        except redis.exceptions.ConnectionError as exc:
            if retries == 0:
                raise exc
            retries -= 1
            time.sleep(0.5)

@app.route('/')
def hello():
    count = get_hit_count()
    return 'Hello World@@@@@! I have been seen {} times.\n'.format(count)
```

访问宿主机的 5000 端口，可以看到对页面内容的修改实时生效，如图 4-17 所示。

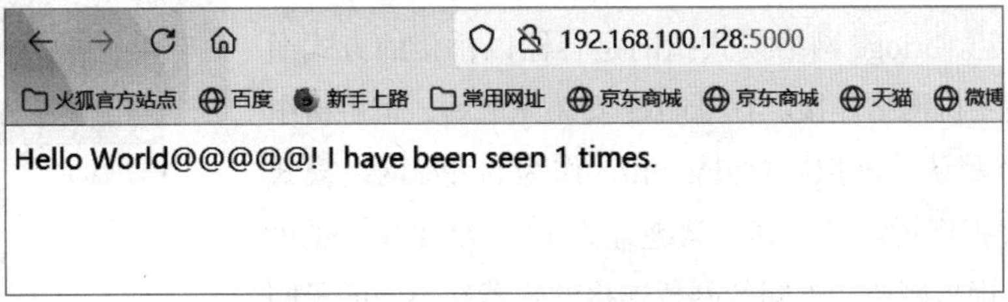

图 4-17 更新后应用程序页面

4.5 Docker 网络

当使用容器搭建业务时，有时需要保证容器间网络的连通性来实现容器间的相互访问，或者需要将容器与非容器环境中的工作负载连通，这时就需要为容器部署网络环境来实现对容器的访问。

Docker 的网络模式分为以下三类：

1. none 网络模式

none 网络模式顾名思义就是指容器内没有任何网络配置，使用 --net=none 来指定。容器在 none 网络模式下有独立的 network namespace，但是没有网络配置。这种网络模式一般应用于对安全性要求较高，并且没有通信要求的容器中。none 网络模式中容器与宿主机的网络关系如图 4-18 所示。

2. host 网络模式

host 网络模式指的是容器和宿主机共享同一个网络栈，共享 network namespace，使用 --net=host 来指定。共享网络栈的优势是容器网络传输效率高，但相应灵活性较差，并且同一个网络栈可能会导致映射端口冲突的问题。host 网络模式中容器与宿主机的网络关系如图 4-19 所示。

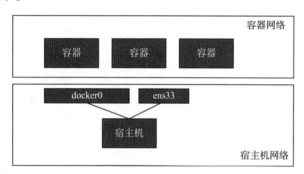

图 4-18　none 网络模式中容器与
宿主机的网络关系

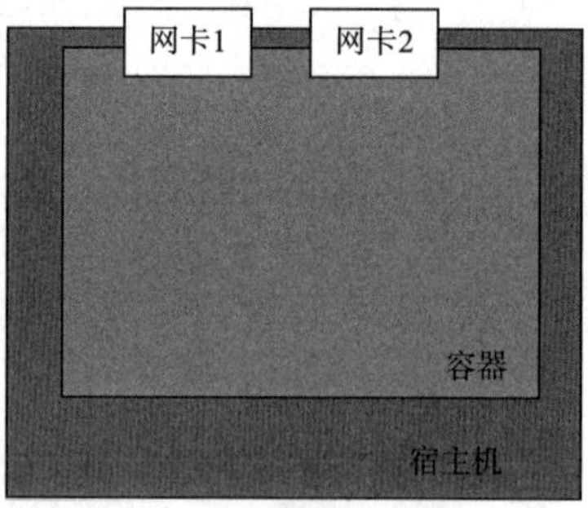

图 4-19　host 网络模式中容器与
宿主机的网络关系

3. bridge 网络模式

bridge 网络模式是 Docker 网络的默认使用模式，可以通过 --net=bridge 来指定。bridge 网络模式指的是容器以桥接的方式通过虚拟网桥与宿主机网络连接。在 docker 安装完成后，宿主机中会自动创建一个默认的虚拟网桥 docker0，作为 bridge 网络模式下容器连接的虚拟网桥设备，其默认地址为 172.17.0.1/16。也可以根据需求创建虚拟网桥，不同的网桥连接的容器在不同的子网中。bridge 网络模式中容器与宿主机的网络关系如图 4-20 所示。

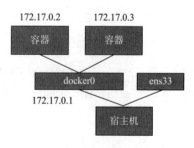

图 4-20　bridge 网络模式中容器与
宿主机的网络关系

4.6　项目实战——Docker 网络运维管理

项目背景

为容器设置合适的网络模式是容器对外提供业务、容器内数据流量转发的重要基础。某公司为保证容器中业务的功能，需为不同类型的容器配置不同的网络模式并相应进行运维管理。

项目实施

任务 1　构建 none 网络模式的容器

使用 busybox 镜像，构建一个 none 网络模式的容器，查看其网络配置，发现没有容器、没有任何网络配置。

```
root@localhost html]# docker run -it --network=none  busybox
Unable to find image 'busybox:latest' locally
latest: Pulling from library/busybox
5cc84ad355aa: Pull complete
Digest: sha256:5acba83a746c7608ed544dc1533b87c737a0b0fb730301639a0179f9344b1678
Status: Downloaded newer image for busybox:latest
/ # ip a
1: lo: <LOOPBACK,UP,LOWER_UP> mtu 65536 qdisc noqueue qlen 1000
    link/loopback 00:00:00:00:00:00 brd 00:00:00:00:00:00
    inet 127.0.0.1/8 scope host lo
       valid_lft forever preferred_lft forever
/ #
```

任务 2　构建 host 网络模式的容器

使用 busybox 镜像，构建一个 host 网络模式的容器，容器启动成功后，查看其网络配置，发现容器中的网络配置与宿主机完全一致。

```
[root@localhost html]# ip a
1: lo: <LOOPBACK,UP,LOWER_UP> mtu 65536 qdisc noqueue state UNKNOWN group
default qlen 1000
    link/loopback 00:00:00:00:00:00 brd 00:00:00:00:00:00
    inet 127.0.0.1/8 scope host lo
       valid_lft forever preferred_lft forever
    inet6 ::1/128 scope host
       valid_lft forever preferred_lft forever
```

```
    2: ens33: <BROADCAST,MULTICAST,UP,LOWER_UP> mtu 1500 qdisc pfifo_fast state
UP group default qlen 1000
        link/ether 00:0c:29:6f:bf:6d brd ff:ff:ff:ff:ff:ff
         inet 10.0.0.137/24 brd 10.0.0.255 scope global noprefixroute dynamic
ens33
           valid_lft 1356sec preferred_lft 1356sec
        inet6 fe80::20c:29ff:fe6f:bf6d/64 scope link
           valid_lft forever preferred_lft forever
    3: docker0: <BROADCAST,MULTICAST,UP,LOWER_UP> mtu 1500 qdisc noqueue state
UP group default
        link/ether 02:42:8b:e2:1b:29 brd ff:ff:ff:ff:ff:ff
        inet 172.17.0.1/16 brd 172.17.255.255 scope global docker0
           valid_lft forever preferred_lft forever
        inet6 fe80::42:8bff:fee2:1b29/64 scope link
           valid_lft forever preferred_lft forever
    27: vethd942024@if26: <BROADCAST,MULTICAST,UP,LOWER_UP> mtu 1500 qdisc
noqueue master docker0 state UP group default
        link/ether da:b0:db:95:26:bf brd ff:ff:ff:ff:ff:ff link-netnsid 0
        inet6 fe80::d8b0:dbff:fe95:26bf/64 scope link
           valid_lft forever preferred_lft forever
    29: veth2b63158@if28: <BROADCAST,MULTICAST,UP,LOWER_UP> mtu 1500 qdisc
noqueue master docker0 state UP group default
        link/ether be:27:26:fe:d9:f9 brd ff:ff:ff:ff:ff:ff link-netnsid 1
        inet6 fe80::bc27:26ff:fefe:d9f9/64 scope link
           valid_lft forever preferred_lft forever
[root@localhost html]# docker run -it --network=host busybox
/ # hostname
localhost.localdomain
/ # ip a
1: lo: <LOOPBACK,UP,LOWER_UP> mtu 65536 qdisc noqueue qlen 1000
        link/loopback 00:00:00:00:00:00 brd 00:00:00:00:00:00
        inet 127.0.0.1/8 scope host lo
           valid_lft forever preferred_lft forever
        inet6 ::1/128 scope host
           valid_lft forever preferred_lft forever
    2: ens33: <BROADCAST,MULTICAST,UP,LOWER_UP> mtu 1500 qdisc pfifo_fast qlen
1000
        link/ether 00:0c:29:6f:bf:6d brd ff:ff:ff:ff:ff:ff
         inet 10.0.0.137/24 brd 10.0.0.255 scope global dynamic noprefixroute
ens33
```

```
        valid_lft 1336sec preferred_lft 1336sec
    inet6 fe80::20c:29ff:fe6f:bf6d/64 scope link
        valid_lft forever preferred_lft forever
  3: docker0: <BROADCAST,MULTICAST,UP,LOWER_UP> mtu 1500 qdisc noqueue
    link/ether 02:42:8b:e2:1b:29 brd ff:ff:ff:ff:ff:ff
    inet 172.17.0.1/16 brd 172.17.255.255 scope global docker0
        valid_lft forever preferred_lft forever
    inet6 fe80::42:8bff:fee2:1b29/64 scope link
        valid_lft forever preferred_lft forever
  27: vethd942024@if26: <BROADCAST,MULTICAST,UP,LOWER_UP,M-DOWN> mtu 1500
qdisc noqueue master docker0
    link/ether da:b0:db:95:26:bf brd ff:ff:ff:ff:ff:ff
    inet6 fe80::d8b0:dbff:fe95:26bf/64 scope link
        valid_lft forever preferred_lft forever
  29: veth2b63158@if28: <BROADCAST,MULTICAST,UP,LOWER_UP,M-DOWN> mtu 1500
qdisc noqueue master docker0
    link/ether be:27:26:fe:d9:f9 brd ff:ff:ff:ff:ff:ff
    inet6 fe80::bc27:26ff:fefe:d9f9/64 scope link
        valid_lft forever preferred_lft forever
  / #
```

任务 3　构建 bridge 网络模式的容器

　　使用 busybox 镜像，构建一个 bridge 网络模式的容器，容器启动成功后，查看其网络配置，发现容器的 IP 地址与 docker0 在同一子网当中。这里容器是以桥接的方式连接到 docker0 这个虚拟网桥上的。

```
  3: docker0: <BROADCAST,MULTICAST,UP,LOWER_UP> mtu 1500 qdisc noqueue
    link/ether 02:42:8b:e2:1b:29 brd ff:ff:ff:ff:ff:ff
    inet 172.17.0.1/16 brd 172.17.255.255 scope global docker0
        valid_lft forever preferred_lft forever
    inet6 fe80::42:8bff:fee2:1b29/64 scope link
        valid_lft forever preferred_lft forever
[root@localhost html]# docker run -it busybox
/ # ip a
1: lo: <LOOPBACK,UP,LOWER_UP> mtu 65536 qdisc noqueue qlen 1000
    link/loopback 00:00:00:00:00:00 brd 00:00:00:00:00:00
    inet 127.0.0.1/8 scope host lo
        valid_lft forever preferred_lft forever
  30: eth0@if31: <BROADCAST,MULTICAST,UP,LOWER_UP,M-DOWN> mtu 1500 qdisc
noqueue
```

```
        link/ether 02:42:ac:11:00:04 brd ff:ff:ff:ff:ff:ff
        inet 172.17.0.4/16 brd 172.17.255.255 scope global eth0
            valid_lft forever preferred_lft forever
/ #
[root@localhost html]#  yum install bridge-utils
[root@localhost html]# brctl show
bridge name        bridge id              STP enabled      interfaces
docker0            8000.02428be21b29      no               veth2b63158
                                                           vethd942024
```

任务 4　构建 user-defined 用户自定义网络

　　除了默认的 docker0 以外，用户还可以根据具体需求自行添加虚拟网桥。添加成功后，使用 ip a 命令查看，可以看到主机中多了一个网络设备 br-9291c369eef2，这是一个虚拟网桥，在没有指定其 IP 地址的情况下，默认地址为 172.18.0.1/16。

```
[root@localhost html]# docker network create --driver bridge mynet
9291c369eef252bb602a32f20cea7d443c7f092725322bec4afffbd24e48d345
[root@localhost html]# ip a
1: lo: <LOOPBACK,UP,LOWER_UP> mtu 65536 qdisc noqueue state UNKNOWN group
default qlen 1000
        link/loopback 00:00:00:00:00:00 brd 00:00:00:00:00:00
        inet 127.0.0.1/8 scope host lo
            valid_lft forever preferred_lft forever
        inet6 ::1/128 scope host
            valid_lft forever preferred_lft forever
2: ens33: <BROADCAST,MULTICAST,UP,LOWER_UP> mtu 1500 qdisc pfifo_fast state
UP group default qlen 1000
        link/ether 00:0c:29:6f:bf:6d brd ff:ff:ff:ff:ff:ff
         inet 10.0.0.137/24 brd 10.0.0.255 scope global noprefixroute dynamic
ens33
            valid_lft 1592sec preferred_lft 1592sec
        inet6 fe80::20c:29ff:fe6f:bf6d/64 scope link
            valid_lft forever preferred_lft forever
3: docker0: <BROADCAST,MULTICAST,UP,LOWER_UP> mtu 1500 qdisc noqueue state
UP group default
        link/ether 02:42:8b:e2:1b:29 brd ff:ff:ff:ff:ff:ff
        inet 172.17.0.1/16 brd 172.17.255.255 scope global docker0
            valid_lft forever preferred_lft forever
        inet6 fe80::42:8bff:fee2:1b29/64 scope link
            valid_lft forever preferred_lft forever
```

```
    27: vethd942024@if26: <BROADCAST,MULTICAST,UP,LOWER_UP> mtu 1500 qdisc
noqueue master docker0 state UP group default
        link/ether da:b0:db:95:26:bf brd ff:ff:ff:ff:ff:ff link-netnsid 0
        inet6 fe80::d8b0:dbff:fe95:26bf/64 scope link
            valid_lft forever preferred_lft forever
    29: veth2b63158@if28: <BROADCAST,MULTICAST,UP,LOWER_UP> mtu 1500 qdisc
noqueue master docker0 state UP group default
        link/ether be:27:26:fe:d9:f9 brd ff:ff:ff:ff:ff:ff link-netnsid 1
        inet6 fe80::bc27:26ff:fefe:d9f9/64 scope link
            valid_lft forever preferred_lft forever
    32: br-9291c369eef2: <NO-CARRIER,BROADCAST,MULTICAST,UP> mtu 1500 qdisc
noqueue state DOWN group default
        link/ether 02:42:73:b2:9b:1e brd ff:ff:ff:ff:ff:ff
        inet 172.18.0.1/16 brd 172.18.255.255 scope global br-9291c369eef2
            valid_lft forever preferred_lft forever
```

如果需要指定 IP 地址，可以使用 --gateway 参数来指定。例如，指定新建网桥的 IP 地址为 172.100.100.1，子网为 172.100.100.0/24，这时可以看到主机中新增了一个虚拟网桥 br-c1bf0d60c7d0，其地址就是指定的 172.100.100.1。

```
    [root@localhost html]# docker network create --driver bridge --subnet
172.100.100.0/24 --gateway 172.100.100.1 mynet2
    c1bf0d60c7d02bb74ee4eea60f6b3500707e06068cd77be1bf5411477de92079
    [root@localhost html]# ip a
    1: lo: <LOOPBACK,UP,LOWER_UP> mtu 65536 qdisc noqueue state UNKNOWN group
default qlen 1000
        link/loopback 00:00:00:00:00:00 brd 00:00:00:00:00:00
        inet 127.0.0.1/8 scope host lo
            valid_lft forever preferred_lft forever
        inet6 ::1/128 scope host
            valid_lft forever preferred_lft forever
    2: ens33: <BROADCAST,MULTICAST,UP,LOWER_UP> mtu 1500 qdisc pfifo_fast state
UP group default qlen 1000
        link/ether 00:0c:29:6f:bf:6d brd ff:ff:ff:ff:ff:ff
        inet 10.0.0.137/24 brd 10.0.0.255 scope global noprefixroute dynamic ens33
            valid_lft 1475sec preferred_lft 1475sec
        inet6 fe80::20c:29ff:fe6f:bf6d/64 scope link
            valid_lft forever preferred_lft forever
    3: docker0: <BROADCAST,MULTICAST,UP,LOWER_UP> mtu 1500 qdisc noqueue state
UP group default
        link/ether 02:42:8b:e2:1b:29 brd ff:ff:ff:ff:ff:ff
```

```
     inet 172.17.0.1/16 brd 172.17.255.255 scope global docker0
        valid_lft forever preferred_lft forever
     inet6 fe80::42:8bff:fee2:1b29/64 scope link
        valid_lft forever preferred_lft forever
  27: vethd942024@if26: <BROADCAST,MULTICAST,UP,LOWER_UP> mtu 1500 qdisc
noqueue master docker0 state UP group default
     link/ether da:b0:db:95:26:bf brd ff:ff:ff:ff:ff:ff link-netnsid 0
     inet6 fe80::d8b0:dbff:fe95:26bf/64 scope link
        valid_lft forever preferred_lft forever
  29: veth2b63158@if28: <BROADCAST,MULTICAST,UP,LOWER_UP> mtu 1500 qdisc
noqueue master docker0 state UP group default
     link/ether be:27:26:fe:d9:f9 brd ff:ff:ff:ff:ff:ff link-netnsid 1
     inet6 fe80::bc27:26ff:fefe:d9f9/64 scope link
        valid_lft forever preferred_lft forever
  32: br-9291c369eef2: <NO-CARRIER,BROADCAST,MULTICAST,UP> mtu 1500 qdisc
noqueue state DOWN group default
     link/ether 02:42:73:b2:9b:1e brd ff:ff:ff:ff:ff:ff
     inet 172.18.0.1/16 brd 172.18.255.255 scope global br-9291c369eef2
        valid_lft forever preferred_lft forever
  33: br-c1bf0d60c7d0: <NO-CARRIER,BROADCAST,MULTICAST,UP> mtu 1500 qdisc
noqueue state DOWN group default
     link/ether 02:42:50:8b:47:3b brd ff:ff:ff:ff:ff:ff
     inet 172.100.100.1/24 brd 172.100.100.255 scope global br-c1bf0d60c7d0
        valid_lft forever preferred_lft forever
```

自定义网络创建成功后，使用 busybox 镜像启动一个容器，指定其网络使用自定义网络 mynet2。启动成功后，查看其 IP 地址，可以看到容器获取到的 IP 地址在 mynet2 网络当中。

```
[root@localhost html]# docker run -it --network=mynet2 busybox
/ # ip a
1: lo: <LOOPBACK,UP,LOWER_UP> mtu 65536 qdisc noqueue qlen 1000
   link/loopback 00:00:00:00:00:00 brd 00:00:00:00:00:00
   inet 127.0.0.1/8 scope host lo
      valid_lft forever preferred_lft forever
36: eth0@if37: <BROADCAST,MULTICAST,UP,LOWER_UP,M-DOWN> mtu 1500 qdisc
noqueue
   link/ether 02:42:ac:64:64:02 brd ff:ff:ff:ff:ff:ff
   inet 172.100.100.2/24 brd 172.100.100.255 scope global eth0
      valid_lft forever preferred_lft forever
```

任务 5　实现不同容器之间的通信

在容器中不同的 network 之间是相互隔离的，因此若想实现不同容器之间网络的连通性，必须确保不同容器具有同一网络的网卡才能相互通信。因此，以 docker0 为网关的容器若想与 mynet2 网络中的容器通信，必须在容器中添加一张 mynet2 网络的网卡，可以使用 docker network connect mynet2 实现。

```
# docker network connect mynet2 3086a330c2e9
[root@localhost html]# docker exec -it 3086a330c2e9 /bin/sh
/ # ip a
1: lo: <LOOPBACK,UP,LOWER_UP> mtu 65536 qdisc noqueue qlen 1000
    link/loopback 00:00:00:00:00:00 brd 00:00:00:00:00:00
    inet 127.0.0.1/8 scope host lo
       valid_lft forever preferred_lft forever
 38: eth0@if39: <BROADCAST,MULTICAST,UP,LOWER_UP,M-DOWN> mtu 1500 qdisc
noqueue
    link/ether 02:42:ac:11:00:04 brd ff:ff:ff:ff:ff:ff
    inet 172.17.0.4/16 brd 172.17.255.255 scope global eth0
       valid_lft forever preferred_lft forever
 40: eth1@if41: <BROADCAST,MULTICAST,UP,LOWER_UP,M-DOWN> mtu 1500 qdisc
noqueue
    link/ether 02:42:ac:64:64:02 brd ff:ff:ff:ff:ff:ff
    inet 172.100.100.2/24 brd 172.100.100.255 scope global eth1
       valid_lft forever preferred_lft forever
```

4.7　项目实战——搭建容器私有仓库

项目背景

在生产环境中，为了提高安全性，方便单元组共享镜像，现决定搭建私有仓库。常用的私有仓库有两种：一种是 Docker 官方提供的 Registry，另一种是目前较为流行的 Docker Harbor。

项目实施

任务 1　搭建 Docker Registry

1. 使用官方 Registry 镜像运行容器

将容器的 5000 端口暴露出来。

```
[root@localhost ~]# docker run -d -p 5000:5000 --restart=always --name
registry registry
```

默认情况下，仓库会被创建在容器的 /var/lib/registry 目录下。可以通过 -v 参数将镜像文件存放在本地的指定路径。例如下面的例子，将上传的镜像文件存放到本地的 /opt/data/registry 目录。

```
[root@localhost ~]# docker run -d \
    -p 5000:5000 \
    -v /opt/data/registry:/var/lib/registry \
    registry
```

2. 在私有仓库上传、搜索、下载镜像

创建好私有仓库之后，就可以使用 docker tag 来标记一个镜像，然后推送它到仓库。例如，推送到地址为 127.0.0.1:5000 的 registry 仓库。

先在本机查看已有的镜像。

```
[root@localhost ~]# docker images
REPOSITORY      TAG        IMAGE ID        CREATED        SIZE
<none>          <none>     4565a1830fd7    12 days ago    231MB
busybox         latest     beae173ccac6    3 months ago   1.24MB
nginx           latest     605c77e624dd    3 months ago   141MB
httpd           latest     dabbfbe0c57b    3 months ago   144MB
registry        latest     b8604a3fe854    4 months ago   26.2MB
mariadb         latest     e2278f24ac88    5 months ago   410MB
hello-world     latest     feb5d9fea6a5    6 months ago   13.3KB
centos          latest     5d0da3dc9764    6 months ago   231MB
[root@localhost ~]# docker tag busybox:latest 127.0.0.1:5000/busybox:latest
[root@localhost ~]# docker images
REPOSITORY              TAG        IMAGE ID        CREATED        SIZE
<none>                  <none>     4565a1830fd7    12 days ago    231MB
127.0.0.1:5000/busybox  latest     beae173ccac6    3 months ago   1.24MB
busybox                 latest     beae173ccac6    3 months ago   1.24MB
nginx                   latest     605c77e624dd    3 months ago   141MB
httpd                   latest     dabbfbe0c57b    3 months ago   144MB
registry                latest     b8604a3fe854    4 months ago   26.2MB
mariadb                 latest     e2278f24ac88    5 months ago   410MB
hello-world             latest     feb5d9fea6a5    6 months ago   13.3KB
centos                  latest     5d0da3dc9764    6 months ago   231MB
```

使用 docker push 上传标记的镜像。

```
[root@localhost ~]# docker push 127.0.0.1:5000/busybox:latest
The push refers to repository [127.0.0.1:5000/busybox]
01fd6df81c8e: Pushed
latest: digest: sha256:62ffc2ed7554e4c6d360bce40bbcf196573dd27c4ce080641a2c5
9867e732dee size: 527
```

用 curl 查看仓库中的镜像。

```
[root@localhost ~]# curl 127.0.0.1:5000/v2/_catalog
{"repositories":["busybox"]}
```

这里可以看到 {"repositories":["busybox"]}，表明镜像已经被成功上传了。

接下来验证是否可以从私有仓库拉去并使用我们自己上传的镜像。先删除本地镜像，再尝试从私有仓库中下载这个镜像。

```
[root@localhost ~]# docker image rm 127.0.0.1:5000/busybox
Untagged: 127.0.0.1:5000/busybox:latest
Untagged: 127.0.0.1:5000/busybox@sha256:62ffc2ed7554e4c6d360bce40bbcf196573d-
d27c4ce080641a2c59867e732dee
[root@localhost ~]# docker pull 127.0.0.1:5000/busybox:latest
latest: Pulling from busybox
Digest: sha256:62ffc2ed7554e4c6d360bce40bbcf196573dd27c4ce080641a2c59867e732dee
Status: Downloaded newer image for 127.0.0.1:5000/busybox:latest
127.0.0.1:5000/busybox:latest
[root@localhost ~]# docker images
REPOSITORY              TAG         IMAGE ID        CREATED         SIZE
<none>                  <none>      4565a1830fd7    12 days ago     231MB
busybox                 latest      beae173ccac6    3 months ago    1.24MB
127.0.0.1:5000/busybox  latest      beae173ccac6    3 months ago    1.24MB
nginx                   latest      605c77e624dd    3 months ago    141MB
httpd                   latest      dabbfbe0c57b    3 months ago    144MB
registry                latest      b8604a3fe854    4 months ago    26.2MB
mariadb                 latest      e2278f24ac88    5 months ago    410MB
hello-world             latest      feb5d9fea6a5    6 months ago    13.3KB
centos                  latest      5d0da3dc9764    6 months ago    231MB
```

3.　配置非 https 仓库地址

将私有仓库提供给局域网内集群使用，需将例如 192.168.199.100:5000 这样的内网地址作为私有仓库地址，这时会发现无法成功推送镜像。

```
[root@localhost ~]# docker tag busybox:latest 10.0.0.137:5000/busybox:latest
[root@localhost ~]# docker push 10.0.0.137:5000/busybox:latest
The push refers to repository [10.0.0.137:5000/busybox]
Get "https://10.0.0.137:5000/v2/": http: server gave HTTP response to HTTPS client
```

这是因为 Docker 默认不允许非 https 方式推送镜像。可以通过 Docker 的配置选项来取消这个限制。

```
[root@localhost ~]# cat /etc/docker/daemon.json
{
  "registry-mirrors": ["https://jctfhowk.mirror.aliyuncs.com"],
  "insecure-registries": [
    "10.0.0.137:5000"
  ]
}
[root@localhost ~]# systemctl daemon-reload
[root@localhost ~]# systemctl restart docker
[root@localhost ~]# docker push 10.0.0.137:5000/busybox:latest
The push refers to repository [10.0.0.137:5000/busybox]
01fd6df81c8e: Layer already exists
latest: digest: sha256:62ffc2ed7554e4c6d360bce40bbcf196573dd27c4ce080641a2c5
9867e732dee size: 527
```

任务 2　搭建 Harbor

1. 为服务器签发证书

Harbor 需要通过 https 访问，因此先要为服务器签发证书。

```
[root@localhost data]# mkdir -pv /data/ssl
mkdir: created directory '/data/ssl'
[root@localhost data]# cd ssl/
[root@localhost ssl]# openssl genrsa -out server.key 1024
Generating RSA private key, 1024 bit long modulus
..................................++++++
...............++++++
e is 65537 (0x10001)
[root@localhost ssl]# chmod 600 server.key
[root@localhost ssl]# openssl req -new -key server.key -out server.csr
You are about to be asked to enter information that will be incorporated
into your certificate request.ser
What you are about to enter is what is called a Distinguished Name or a DN.
There are quite a few fields but you can leave some blank
For some fields there will be a default value,
If you enter '.', the field will be left blank.
-----
Country Name (2 letter code) [XX]:CN
```

```
State or Province Name (full name) []:Sichuan
Locality Name (eg, city) [Default City]:Chengdu
Organization Name (eg, company) [Default Company Ltd]:cdp
Organizational Unit Name (eg, section) []:cloud
Common Name (eg, your name or your server's hostname) []:harbor.cdp.com
Email Address []:

Please enter the following 'extra' attributes
to be sent with your certificate request
A challenge password []:
An optional company name []:
[root@localhost ssl]# cd /etc/pki/CA/
[root@localhost CA]# openssl genrsa -out private/cakey.pem 2048
Generating RSA private key, 2048 bit long modulus
..................................................................+++
...............+++
e is 65537 (0x10001)
[root@localhost CA]# chmod 600 private/cakey.pem
[root@localhost CA]# openssl req -new -x509 -key private/cakey.pem -out
cacert.pem -days 365
You are about to be asked to enter information that will be incorporated
into your certificate request.
What you are about to enter is what is called a Distinguished Name or a DN.
There are quite a few fields but you can leave some blank
For some fields there will be a default value,
If you enter '.', the field will be left blank.
-----
Country Name (2 letter code) [XX]:CN
State or Province Name (full name) []:Sichuan
Locality Name (eg, city) [Default City]:Chengdu
Organization Name (eg, company) [Default Company Ltd]:cdp
Organizational Unit Name (eg, section) []:cloud
Common Name (eg, your name or your server's hostname) []:
Email Address []:
[root@localhost ~]# cd /etc/pki/CA/
[root@localhost CA]# touch index.txt
[root@localhost CA]# echo "01" > serial
[root@localhost CA]# cd /data/ssl/
[root@localhost ssl]# openssl ca -in server.csr -out server.crt -days 365
Using configuration from /etc/pki/tls/openssl.cnf
Check that the request matches the signature
```

```
Signature ok
Certificate Details:
        Serial Number: 1 (0x1)
        Validity
            Not Before: Jan 10 16:47:20 2022 GMT
            Not After : Jan 10 16:47:20 2023 GMT
        Subject:
            countryName = CN
            stateOrProvinceName = Sichuan
            organizationName = cdp
            organizationalUnitName = cloud
            commonName = harbor.cdp.com
        X509v3 extensions:
            X509v3 Basic Constraints:
                CA:FALSE
            Netscape Comment:
                OpenSSL Generated Certificate
            X509v3 Subject Key Identifier:
                6D:15:28:38:4B:E1:6A:24:FB:C3:ED:C1:BF:F2:69:72:89:89:5E:15
            X509v3 Authority Key Identifier:
                keyid:FA:F2:CE:1A:C5:CA:D7:47:1A:AE:07:89:8B:52:5F:15:8F:CF:43:6
Certificate is to be certified until Jan 10 16:47:20 2023 GMT (365 days)
Sign the certificate? [y/n]:y


1 out of 1 certificate requests certified, commit? [y/n]y
Write out database with 1 new entries
Data Base Updated
```

2. 安装 docker-compose

```
[root@localhost ssl]# yum -y install docker-compose
```

3. 下载 Harbor 离线安装包

编辑 harbor.cfg 配置服务器地址，使用协议、证书所在路径等，通过 docker-compose 部署 Harbor。

```
[root@localhost ssl]# cd /opt/
[root@localhost opt]# wget https://storage.googleapis.com/harbor-releases/releas
e-1.7.0/harbor-offline-installer-v1.7.1.tgz
    --2022-01-11 00:50:22--  https://storage.googleapis.com/harbor-releases/release-
1.7.0/harbor-offline-installer-v1.7.1.tgz
```

```
     Resolving storage.googleapis.com (storage.googleapis.com)... 172.217.160.112, 14
2.251.43.16, 172.217.160.80, ...
     Connecting to storage.googleapis.com (storage.googleapis.com)|172.217.160.112|:4
43... connected.
     HTTP request sent, awaiting response... 200 OK
     Length: 597857483 (570M) [application/x-tar]
     Saving to: 'harbor-offline-installer-v1.7.1.tgz'

     100%[====================================================================
==========================================>] 597,857,483 10.5MB/s    in 53s

     2022-01-11 00:51:15 (10.8 MB/s) - 'harbor-offline-installer-v1.7.1.tgz' saved
[597857483/597857483]
     [root@localhost opt]# tar -xvf harbor-offline-installer-v1.7.1.tgz -C /opt/
     [root@localhost opt]# ll
     total 583848
     drwx--x--x 4 root root         28 Jan  8 00:20 containerd
     drwxr-xr-x 3 root root        270 Jan 11 00:54 harbor
     -rw-r--r-- 1 root  root 597857483 Jan  7  2019 harbor-offline-installer-
v1.7.1.tgz
     [root@localhost harbor]# cd harbor
     [root@localhost harbor]# vi harbor.cfg
     hostname = 10.0.0.100
     ui_url_protocol = https
     ssl_cert = /data/ssl/server.crt
     ssl_cert_key = /data/ssl/server.key
     [root@localhost harbor]# ./prepare
     Generated and saved secret to file: /data/secretkey
     Generated configuration file: ./common/config/nginx/nginx.conf
     Generated configuration file: ./common/config/adminserver/env
     Generated configuration file: ./common/config/core/env
     Generated configuration file: ./common/config/registry/config.yml
     Generated configuration file: ./common/config/db/env
     Generated configuration file: ./common/config/jobservice/env
     Generated configuration file: ./common/config/jobservice/config.yml
     Generated configuration file: ./common/config/log/logrotate.conf
     Generated configuration file: ./common/config/registryctl/env
     Generated configuration file: ./common/config/core/app.conf
     Generated certificate, key file: ./common/config/core/private_key.pem, cert
file: ./common/config/registry/root.crt
     The configuration files are ready, please use docker-compose to start the service.
```

```
[root@localhost harbor]# ./install.sh

[Step 0]: checking installation environment ...

Note: docker version: 20.10.12

Note: docker-compose version: 1.18.0

[Step 1]: loading Harbor images ...
Loaded image: goharbor/registry-photon:v2.6.2-v1.7.1
Loaded image: goharbor/harbor-migrator:v1.7.1
Loaded image: goharbor/harbor-adminserver:v1.7.1
Loaded image: goharbor/harbor-core:v1.7.1
Loaded image: goharbor/harbor-log:v1.7.1
Loaded image: goharbor/harbor-jobservice:v1.7.1
Loaded image: goharbor/notary-server-photon:v0.6.1-v1.7.1
Loaded image: goharbor/clair-photon:v2.0.7-v1.7.1
Loaded image: goharbor/harbor-portal:v1.7.1
Loaded image: goharbor/harbor-db:v1.7.1
Loaded image: goharbor/redis-photon:v1.7.1
Loaded image: goharbor/nginx-photon:v1.7.1
Loaded image: goharbor/harbor-registryctl:v1.7.1
Loaded image: goharbor/notary-signer-photon:v0.6.1-v1.7.1
Loaded image: goharbor/chartmuseum-photon:v0.7.1-v1.7.1


[Step 2]: preparing environment ...
Clearing the configuration file: ./common/config/adminserver/env
Clearing the configuration file: ./common/config/core/env
Clearing the configuration file: ./common/config/core/app.conf
Clearing the configuration file: ./common/config/core/private_key.pem
Clearing the configuration file: ./common/config/db/env
Clearing the configuration file: ./common/config/jobservice/env
Clearing the configuration file: ./common/config/jobservice/config.yml
Clearing the configuration file: ./common/config/registry/config.yml
Clearing the configuration file: ./common/config/registry/root.crt
Clearing the configuration file: ./common/config/registryctl/env
Clearing the configuration file: ./common/config/registryctl/config.yml
Clearing the configuration file: ./common/config/nginx/cert/server.crt
Clearing the configuration file: ./common/config/nginx/cert/server.key
Clearing the configuration file: ./common/config/nginx/nginx.conf
```

```
Clearing the configuration file: ./common/config/log/logrotate.conf
loaded secret from file: /data/secretkey
Generated configuration file: ./common/config/nginx/nginx.conf
Generated configuration file: ./common/config/adminserver/env
Generated configuration file: ./common/config/core/env
Generated configuration file: ./common/config/registry/config.yml
Generated configuration file: ./common/config/db/env
Generated configuration file: ./common/config/jobservice/env
Generated configuration file: ./common/config/jobservice/config.yml
Generated configuration file: ./common/config/log/logrotate.conf
Generated configuration file: ./common/config/registryctl/env
Generated configuration file: ./common/config/core/app.conf
Generated certificate, key file: ./common/config/core/private_key.pem, cert
file: ./common/config/registry/root.crt
The configuration files are ready, please use docker-compose to start the service.


[Step 3]: checking existing instance of Harbor ...

Note: stopping existing Harbor instance ...
Stopping harbor-jobservice  ... done
Stopping harbor-portal       ... done
Stopping harbor-core         ... done
Stopping redis               ... done
Stopping harbor-db           ... done
Stopping harbor-adminserver  ... done
Stopping registryctl         ... done
Stopping registry            ... done
Stopping harbor-log          ... done
Removing nginx               ... done
Removing harbor-jobservice   ... done
Removing harbor-portal       ... done
Removing harbor-core         ... done
Removing redis               ... done
Removing harbor-db           ... done
Creating harbor-log ... done
Removing registryctl         ... done
Removing registry            ... done
Removing harbor-log          ... done
Removing network harbor_harbor
```

```
Creating harbor-adminserver ... done
Creating harbor-core ... done
[Step 4]: starting Harbor ...
Creating harbor-portal ... done
Creating nginx ... done
Creating registry ...
Creating registryctl ...
Creating redis ...
Creating harbor-adminserver ...
Creating harbor-db ...
Creating harbor-core ...
Creating harbor-portal ...
Creating harbor-jobservice ...
Creating nginx ...

√ ----Harbor has been installed and started successfully.----

Now you should be able to visit the admin portal at https://10.0.0.100.
For more details, please visit https://github.com/goharbor/harbor .
```

部署完毕，在浏览器中使用 https://IP/harbor 访问 Harbor 的 web 界面，默认用户为 admin，密码为 Harbor12345，如图 4-21 所示。

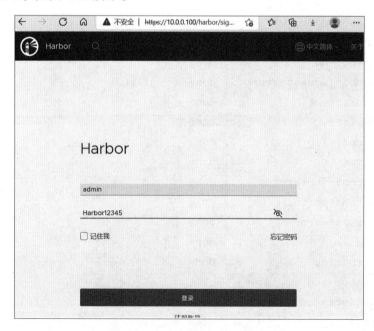

图 4-21　Harbor 登录界面

在 Harbor 管理平台中可以管理单元、镜像等，并监控仓库状态信息，如图 4-22 所示。

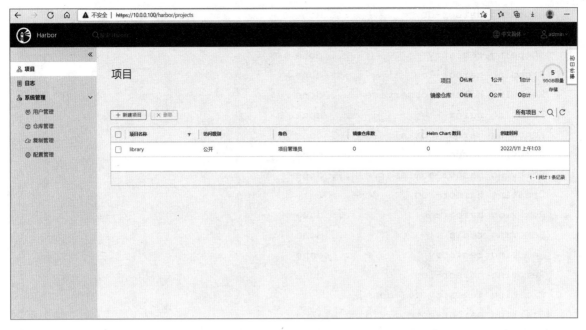

图 4-22　Harbor 管理界面

4. 使用 Harbor，将 Harbor 设置为本地私有仓库

编辑 daemon.json，添加私有镜像仓库。

```
[root@localhost harbor]# cat /etc/docker/daemon.json
{
    "graph": "/data/docker",
    "storage-driver": "overlay",
     "insecure-registries": ["registry.access.redhat.com","quay.io",
"10.0.0.100","harbor.od.com"],
    "registry-mirrors": ["https://jctfhowk.mirror.aliyuncs.com"],
    "bip": "172.7.21.1/24",
    "exec-opts": ["native.cgroupdriver=systemd"],
    "live-restore": true
}
```

重启 Harbor。

```
[root@localhost harbor]# docker-compose down
Stopping harbor-jobservice   ... done
Stopping nginx               ... done
Stopping harbor-portal       ... done
Stopping harbor-core         ... done
Stopping harbor-adminserver  ... done
Stopping registryctl         ... done
```

```
Stopping harbor-db           ... done
Stopping redis               ... done
Stopping registry            ... done
Stopping harbor-log          ... done
Removing harbor-jobservice   ... done
Removing nginx               ... done
Removing harbor-portal       ... done
Removing harbor-core         ... done
Removing harbor-adminserver ... done
Creating harbor-log ... done
Removing harbor-db           ... done
Removing redis               ... done
Removing registry            ... done
Removing harbor-log          ... done
Creating registry ... done
Creating harbor-core ... done
[root@localhost harbor]# systemctl restart docker
[root@localhost harbor]# docker-compose up -d
Creating harbor-portal ... done
Creating nginx ... done
Creating registry ...
Creating harbor-db ...
Creating harbor-adminserver ...
Creating redis ...
Creating registryctl ...
Creating harbor-core ...
Creating harbor-portal ...
Creating harbor-jobservice ...
Creating nginx ...
```

登录 Harbor。

```
[root@localhost harbor]# docker login https://10.0.0.100
Username: admin
Password:
WARNING! Your password will be stored unencrypted in /root/.docker/config.json.
Configure a credential helper to remove this warning. See
https://docs.docker.com/engine/reference/commandline/login/#credentials-store

Login Succeeded
```

将镜像推送到 Harbor 仓库中。

```
[root@localhost harbor]# docker tag 605c77e624dd 10.0.0.100/library/nginx
[root@localhost harbor]# docker push 10.0.0.100/library/nginx
Using default tag: latest
The push refers to repository [10.0.0.100/library/nginx]
d874fd2bc83b: Pushed
32ce5f6a5106: Pushed
f1db227348d0: Pushed
b8d6e692a25e: Pushed
e379e8aedd4d: Pushed
2edcec3590a4: Pushed
latest: digest: sha256:ee89b00528ff4f02f2405e4ee221743ebc3f8e8dd0bfd5c4c20a-
2fa2aaa7ede3 size: 1570
```

在 Harbor 的 web 管理界面就可以看到推送进去的镜像了，如图 4-23 所示。

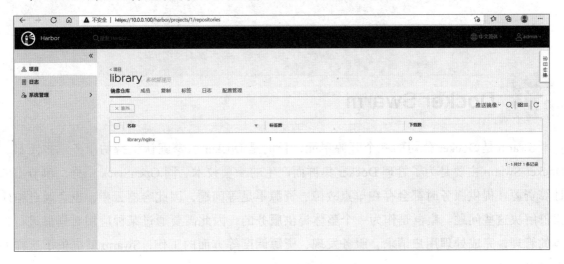

图 4-23　Harbor 镜像仓库列表

单元 5

容器集群管理

5.1 Docker Swarm

Swarm 是 Docker 公司的一个开源单元，目前是 Docker 引擎默认支持的服务编排工具。Docker Swarm 主要是为了管理 Docker 集群而产生的一项技术。同 OpenStack 一样，单节点的计算资源在提供服务时都会存在单点故障、资源不足等问题，因此容器云中，也会通过构建集群解决这些问题。集群是作为一个整体提供服务的，因此需要通过某种应用对集群进行整体的管理，完成处理用户请求、服务发现、资源调度等方面的工作，Swarm 就提供了这样的功能。

在一个 Docker Swarm 管理的集群当中，节点分为两种角色，分别是 manager 和 node。集群通过选举策略选出 manager，manager 节点负责整个集群的管理工作；其余的 node 节点是容器运行的实际地点，node 节点接受 manager 的统一管理。图 5-1 为 Swarm 集群的架构图。

在集群中，由于容器会频繁重启和销毁，其 IP 地址也在不停地变化，因此真正对用户提供访问的是服务 service，服务通过对外暴露端口提供给用户访问接口。Docker Swarm 还提供了默认的负载均衡功能，可以自动实现负载均衡，其示意图如图 5-2 所示。

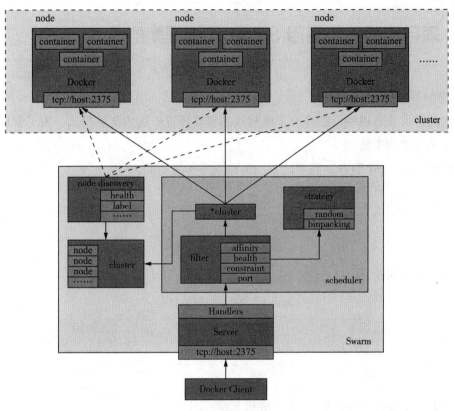

图 5-1　Swarm 集群的架构图

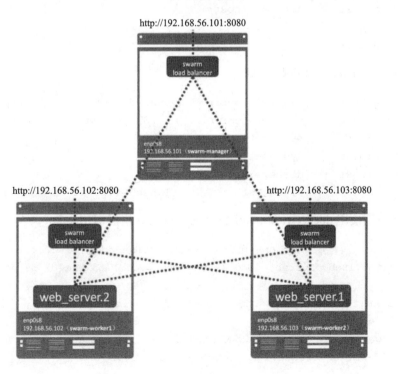

图 5-2　Swarm 负载均衡功能示意图

5.2　项目实战——使用 Swarm 管理集群

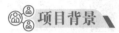 **项目背景**

现公司因扩大业务规模，需将原来单节点的容器环境扩展为集群模式。为了管理方便，需要安装 Web 界面管理工具。

在集群中启动 nginx 服务，并测试 swarm 的服务伸缩功能。

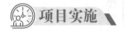 **项目实施**

任务 1　基础环境搭建

集群 IP 地址规划见表 5-1。

表 5-1　集群 IP 地址规划

节点名称	IP 地址
master	10.0.0.37
node	10.0.0.8

修改集群当中每台主机的 hosts 主机解析文件。

```
[root@localhost ~]# cat /etc/hosts
10.0.0.37 master
10.0.0.8 node
```

配置 master 和 node 节点 docker API。

```
[root@localhost ~]# cat /lib/systemd/system/docker.service|grep ExecStart
ExecStart=/usr/bin/dockerd -H tcp://0.0.0.0:2375 -H unix:///var/run/docker.sock
[root@localhost ~]# systemctl daemon-reload
[root@localhost ~]# systemctl restart docker
[root@localhost ~]# docker images|grep swarm
swarm              latest       1a5eb59a410f   16 months ago   12.7MB
```

在 master 节点使用 docker swarm inti 命令进行 swarm 集群初始化。

```
[root@master ~]# docker swarm init --advertise-addr 10.0.0.37
Swarm initialized: current node (x4rzm3l6wkwyod717t3xmhc4u) is now a manager.

To add a worker to this swarm, run the following command:

docker swarm join --token SWMTKN-1-3abvt79csrclmad83q384otm7p3udsupvmur-
j4r9vpfqbssoj9-0bfn4wv5zeqsc1zjq7m5m8o8i 10.0.0.37:2377
```

初始化完成后根据上述提示，使用集群的 token 将 node 节点加入集群。

```
[root@node ~]# docker swarm join --token SWMTKN-1-3abvt79csrclmad83q384otm-
7p3udsupvmurj4r9vpfqbssoj9-0bfn4wv5zeqsc1zjq7m5m8o8i 10.0.0.37:2377
  This node joined a swarm as a worker.
```

集群创建完成后，使用 docker node ls 命令验证 swarm 集群，可以看到当前集群中有两个节点，分别是 master 和 node，它们的状态都为就绪状态。

```
[root@master ~]# docker node ls
ID                          HOSTNAME    STATUS   AVAILABILITY   MANAGER STATUS   ENGINE VERSION
x4rzm3l6wkwyod717t3xmhc4u * master      Ready    Active         Leader           20.10.12
vlh80zt9f3vijilv5d06sm8hv   node        Ready    Active                          20.10.12
```

任务 2　安装 Web 界面管理工具 portainer

在容器中启动 portainer 服务。首先创建一个提供给 portainer 使用的数据卷，接下来创建名称为 portainer 的服务，将服务的工作端口映射到宿主机的 9000 端口，定义副本数为 1，容器工作在 manager 节点上。

```
[root@master ~]# docker volume create portainer_data
portainer_data
[root@master ~]# docker service create --name portainer --publish 9000:9000
--replicas=1 --constraint 'node.role==manager' --mount type=bind,src=/var/
run/docker.sock,dst=/var/run/docker.sock --mount type=volume,src=portainer_
data,dst=/data portainer/portainer -H unix:///var/run/docker.sock
b2xwx0yrmh0hazf9o2048eiw3
overall progress: 1 out of 1 tasks
1/1: running   [==================================================>]
verify: Service converged
```

容器启动成功后，用浏览器访问 http:// 宿主机 IP:9000，就可以看到 portainer 的主页界面了。在 portainer 主页界面中，可以通过界面操作的方式管理整个集群，如图 5-3、图 5-4 所示。

图 5-3　portainer 主页界面

图 5-4　portainer 管理页面

任务 3　管理 Docker Service

1. service 伸缩

启动 nginx 服务，默认副本数量为 2。

```
[root@master ~]# docker service create --name nginx --replicas 2 nginx
z7myxrt4qfxckuc5fmfd66b8f
overall progress: 2 out of 2 tasks
1/2: running
2/2: running
verify: Service converged
[root@master ~]# docker service ls
ID               NAME        MODE         REPLICAS    IMAGE              PORTS
z7myxrt4qfxc     nginx       replicated   2/2         nginx:latest
xmbiumsjtyxl     protainer   replicated   0/1                            portainer/portainer:
latest    *:9000->9000/tcp
[root@master ~]# docker service ps nginx
ID               NAME        IMAGE         NODE    DESIRED STATE CURRENT STATE ERROR PORTS
1frin78pprg6     nginx.1     nginx:latest  node    Running       Running       52 seconds ago
3081v2i8bxro     nginx.2     nginx:latest  node    Running       Running       52 seconds ago
```

将服务规模扩大到五个副本数量，查看服务状态，可以看到容器数量由原来的 2 扩展至 5。

```
[root@master ~]# docker service scale nginx=5
nginx scaled to 5
```

```
overall progress: 5 out of 5 tasks
1/5: running
2/5: running
3/5: running
4/5: running
5/5: running
verify: Service converged
[root@master ~]# docker service ls
ID                    NAME         MODE          REPLICAS      IMAGE
PORTS
z7myxrt4qfxc    nginx       replicated    5/5           nginx:latest
xmbiumsjtyxl    protainer    replicated    0/1           portainer/
portainer:latest    *:9000->9000/tcp
[root@master ~]# docker service ps nginx
ID               NAME        IMAGE         NODE      DESIRED STATE    CURRENT
STATE           ERROR      PORTS
1frin78pprg6    nginx.1     nginx:latest    node      Running          Running 2
minutes ago
3081v2i8bxro    nginx.2     nginx:latest    node      Running          Running 2
minutes ago
me8aig85vx0v    nginx.3     nginx:latest    master    Running          Running
33 seconds ago
fw9tf8jdxtg5    nginx.4     nginx:latest    node      Running          Running 33
seconds ago
63w2yl6aooxs    nginx.5     nginx:latest    node      Running          Running 33
seconds ago
```

如果希望容器只被调度到非 manager 的节点上，可以设置禁止在 manager 节点上运行容器。

```
[root@master ~]# docker node update --availability active master
master
[root@master ~]# docker node ls
ID                              HOSTNAME    STATUS    AVAILABILITY    MANAGER
STATUS    ENGINE VERSION
gordckzydp01ig1lzhxbabh92 *    master      Ready     Drain           Leader
20.10.12
tffdxg32b6399oep360wdwc6t            node        Ready     Active
20.10.12
[root@master ~]# docker service ps nginx
ID               NAME        IMAGE         NODE      DESIRED STATE    CURRENT
STATE           ERROR      PORTS
```

```
    1frin78pprg6    nginx.1    nginx:latest    node    Running    Running 3
minutes ago
    3081v2i8bxro    nginx.2    nginx:latest    node    Running    Running 3
minutes ago
    me8aig85vx0v    nginx.3    nginx:latest    node    Running         Running
about a minute ago
    fw9tf8jdxtg5    nginx.4    nginx:latest    node    Running         Running
about a minute ago
    63w2yl6aooxs    nginx.5    nginx:latest    node    Running         Running
about a minute ago
```

　　暴露 service 端口到外部，以便通过浏览器访问 web 站点页面，使用 http://manager 的 IP 地址 :8080 访问站点，此时 Dockers Swarm 会自动提供负载均衡功能，如图 5-5 所示。

```
[root@master ~]# docker service update --publish-add 8080:80 nginx
nginx
overall progress: 5 out of 5 tasks
1/5: running
2/5: running
3/5: running
4/5: running
5/5: running
verify: Service converged
```

图 5-5　容器集群 web 服务页面

Cloud Computing Platform
Operation, Maintenance and
Application (Bilingual)

An Ning◎Chief Editor

Peng Tianwei, Zhang Chunrong, Song Mu◎Associate Editor

中国铁道出版社有限公司
CHINA RAILWAY PUBLISHING HOUSE CO., LTD.

Introduction

Developed on the basis of real project cases and close to the production environment, this book, as a Chinese-English textbook for higher vocational education, boasts strong significance in practical guidance, including the contents of such five units as virtualization technology, cloud computing technology, private cloud technology, docker cloud technology and docker cluster management, integrating the vocational skill level certificate, "cloud computing platform operation and development" 1+X certificate, "cloud computing" competition items of the national vocational college skills competition, "cloud computing" competition items of the BRICS National Vocational Skills Competition and other related contents, and closely focusing on the development of knowledge and skills in all aspects of the "post course certificate" in the direction of cloud computing.

This book can be used as bilingual teaching materials for higher vocational education courses related to cloud computing, as well as a course for guiding cloud computing operation and maintenance technicians.

Preface

In recent years, the Internet, big data, cloud computing, artificial intelligence, blockchain and other technologies have accelerated innovation and increasingly integrated into the whole process of economic and social development in all fields. The digital economy is developing and expanding at an unprecedented speed with a far-reaching impact, becoming a key force in restructuring global factor resources, reshaping the global economic structure and changing the global competition pattern. The "Twelfth Five-Year Plan" has made it clear that strategic emerging industries are the key objects of national support in the future. Inclusively, information technology has been identified as one of the seven strategic emerging industries, and cloud computing is one of the important contents of the new generation of information technology. The report of 20th National Congress of the Communist Party of China clearly proposed to build a new generation of information technology, artificial intelligence and other new growth engine. Cloud computing is a new technology and a new standard in the IT field. At present, there are relatively few matching textbooks on the market, especially those for vocational education. Most of the existing textbooks are still focusing on explaining the basic knowledge of cloud computing with the technical contents relatively lagging behind, poor in actual contents or not deep enough in actual contents, which is far from production practice and market demand.

This book is a Chinese-English textbook for higher vocational education. Based on the

actual operation of projects and close to the production environment, this textbook does not just explain the profound and obscure theoretical knowledge, nor does it seek to explain all aspects of cloud computing technology. Instead, it focuses more on how to let a learner interested in cloud computing quickly use cloud computing to build his own business system. This book includes five units: virtualization technology, cloud computing technology, private cloud technology, docker cloud technology and docker cluster management. Each unit combines the corresponding contents to design actual projects against the background of the project requirements in the real production. This book, presented in bilingual and loose-leaf format, is not only applicable to the teaching of professional and bilingual courses related to cloud computing, but also serves as a guide course for cloud computing operation and maintenance technicians.

An Ning, a teacher of computer network technology at Chengdu Vocational and Technical College (Chengdu Polytechnic), is editor-in-chief of this book, followed by Peng Tianwei, Zhang Chunrong and Song Mu as associate editors. The members of the editor team are experienced in corporate work and actual projects, among them, Units 3 and 4 were written by An Ning, Unit 1 was written by Peng Tianwei, Unit 2 was written by Zhang Chunrong, and Unit 5 was written by Song Mu.

In the process of compiling this book, the editors consulted and mirrored many documents and here present sincere thanks to the relevant authors.

In the process of writing, the editors have invested a lot of time and energy, but in view of the limited time and know-how, this book may unavoidably have something inadequate and readers' criticism and comment are welcome.

An Ning

May, 2023

Contents

Unit 4   Container cloud technology—Docker 121

Unit 1

Virtualization technology

In traditional IT systems, all businesses rely on physical servers or data centers subject to limited energy utilization as the biggest problem. Relevant statistics show that the energy utilization rate of servers and data centers is only 5%, far from being fully utilized. Simultaneously, the physical host is high in hardware cost and poor in flexibility; moreover, a single host can only run a single operating system at the same moment.

After the emergence of virtualization technology, the whole IT system has had its energy utilization and flexibility greatly improved. With virtualization technology, a single host can run multiple operating systems at the same time. Each operating system is isolated from the kernel level, high in security. Combined with cloud computing technology, the virtual hosts can also achieve flexible scheduling, dynamic allocation and elastic flexibility. In many application scenarios such as bank outsourcing, call centers, computer rooms and Internet cafes, the electric charge alone can be greatly reduced if the desktop virtualization technology is applied.

1.1 Virtualization types

The physical host relies on some software called Hypervisor to virtualize physical resources for virtual machines. According to the destination of the Hypervisor in the host, virtualization can be divided into two types: Type-I virtualization and Type-II virtualization.

1. Type-I virtualization

Hypervisor is directly installed on the physical hardware, on which the virtual machine

(VM) runs. At this time, Hypervisor plays a role similar to the operating system. The host usually becomes a host dedicated to virtualization. This kind of virtualization method is called as Type-I virtualization. Typical Type-I virtualization technologies include VMware's ESX, ESXi and Xen. See Fig. 1-1 for the architecture of the Type-I virtualization system.

2. Type-II virtualization

Install an operating system on the physical hardware first, such as Redhat, Ubuntu and Windows, etc. and then install Hypervisor, which runs as an application in the host and provides virtualization and VM management functions. This kind of virtualization is called as Type-II virtualization. VMware Workstation and VirtualBox, commonly used in Windows operating system, belong to Type-II virtualization. In addition, it also includes the KVM virtualization technology commonly used in Linux operating system. See Fig. 1-2 for the architecture of Type-II virtualization system.

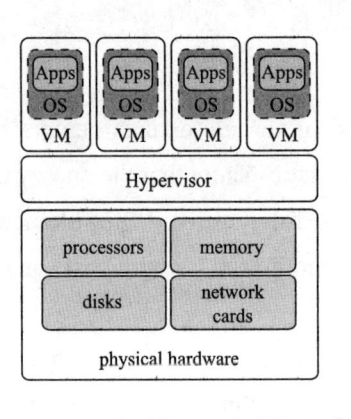

Fig. 1-1　Architecture of Type-I virtualization system

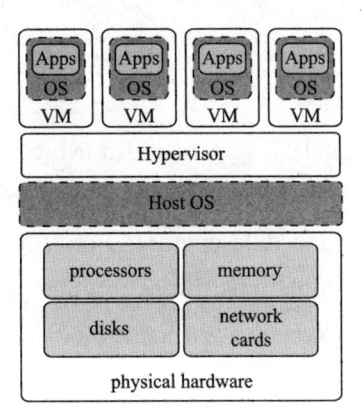

Fig. 1-2　Architecture of Type-II virtualization system

 ## Common virtualization software

At present, there are a variety of virtualization software products. In addition to the desktop-based VMware Workstation provided by VMware, there are also VMware vSphere (VMware ESXI) for servers and VMware Fusion (Mac) installed on Mac computers, as well as Oracle VM VirtualBox, XenServer, Microsoft Hyper-V, KVM and Huawei Fusion Sphere, etc.

With the continuous development of new technologies, virtualization applications and virtualization technologies are closely combined with cloud computing in today's era to provide a more flexible and self-service IT infrastructure. Fig. 1-3-Fig. 1-7 shows some common virtualization software.

Mac platform VM

Fig. 1-3　VMware Fusion

Fig. 1-4　VM VirtualBox

Fig. 1-5　XenServer

Fig. 1-6　Microsoft Hyper-V

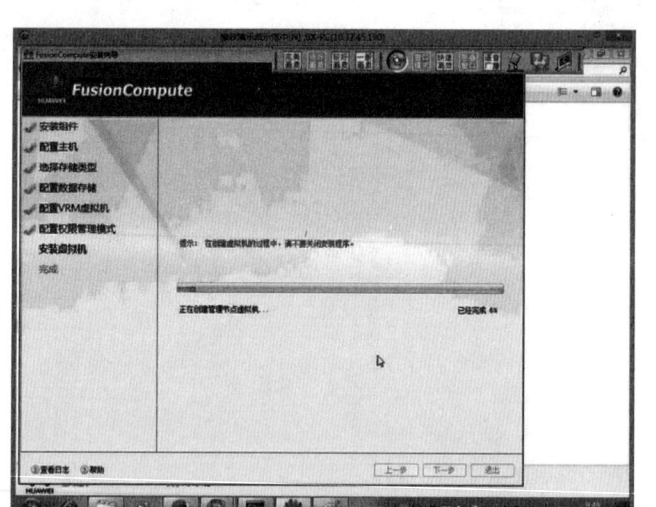

Fig. 1-7　Huawei virtualization software products

1.3　Project practice—host management based on KVM virtualization technology

 **Project background** ➤

A company needs to carry out development and testing in the Linux system for business

expansion. In order to reduce costs and make full use of existing hardware resources, we decided to use KVM virtualization technology to build VMs and install the Linux operating system to meet the requirements of the development and test environment.

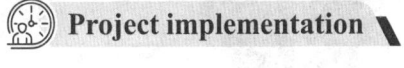 **Project implementation**

Task 1　Build a KVM virtual machine

1. Environmental preparation

(1) Preparation of virtualized environment

Create a VM in VMware Workstation, open CPU settings, and select the "Virtualize Intel VT-x/EPT or AMD-V/RVI" check box in the "Virtualization Engine" to support CPU virtualization, which is the prerequisite for using KVM, as shown in Fig. 1-8. Then install the CentOS7 operating system for the virtual machine.

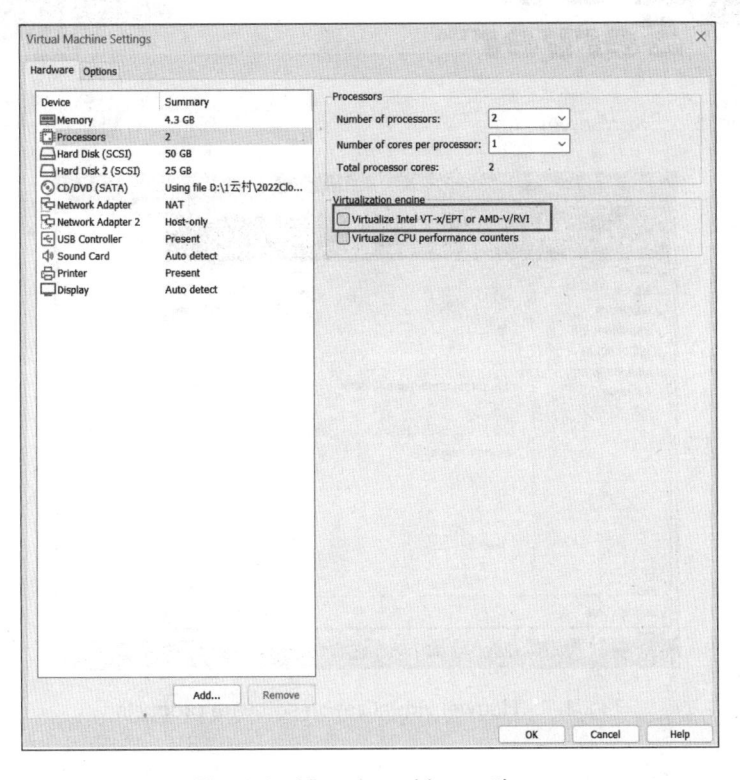

Fig. 1-8　Virtual machine settings

Verify whether the CPU supports virtualization in the system. Use the grep command to filter the contents of the cpuinfo file and check whether it contains the key fields of svm or vmx. If yes, it means that the current system supports CPU virtualization.

```
[root@localhost ~]# grep -o -E 'svm|vmx' /proc/cpuinfo
```

Verify whether the KVM module is loaded. Use the lsmod command to view the system modules loaded in the system. If the KVM module is contained, it means that the current system has loaded the KVM module.

```
[root@localhost ~]# lsmod|grep kvm
kvm_intel              174841  0
kvm                    578518  1 kvm_intel
irqbypass               13503  1 kvm
```

If the return is null, the KVM module needs to be loaded.

(2) Build yum source and close firewall and selinux

The yum source can use the warehouse that comes with the image, or aliyun and other warehouses. Here the aliyun image warehouse is taken as an example.

```
[root@localhost ~]# curl -o /etc/yum.repos.d/CentOS-Base.repo https://mir-
rors.aliyun.com/repo/Centos-7.repo
yum repolist
Loaded plugins: fastestmirror
Loading mirror speeds from cached hostfile
 * base: mirrors.aliyun.com
 * extras: mirrors.aliyun.com
 * updates: mirrors.aliyun.com
repo id                repo name                               status
base/7/x86_64          CentOS-7 - Base - mirrors.aliyun.com    10,072
extras/7/x86_64        CentOS-7 - Extras - mirrors.aliyun.com  509
local                  local                                   3,971
updates/7/x86_64       CentOS-7 - Updates - mirrors.aliyun.com 3,728
repolist: 18,280
[root@localhost ~]# systemctl stop firewalld
[root@localhost ~]# cat /etc/selinux/
config         final/          semanage.conf  targeted/      tmp/
[root@localhost ~]# cat /etc/selinux/config

# This file controls the state of SELinux on the system.
# SELINUX= can take one of these three values:
#       enforcing - SELinux security policy is enforced.
#       permissive - SELinux prints warnings instead of enforcing.
#       disabled - No SELinux policy is loaded.
SELINUX=disabled
# SELINUXTYPE= can take one of three two values:
#       targeted - Targeted processes are protected,
```

```
    #      minimum - Modification of targeted policy. Only selected processes are
protected.
    #      mls - Multi Level Security protection.
    SELINUXTYPE=targeted
```

(3) Install relevant applications

KVM is a virtualization technology based on the Linux kernel. It manages the virtual CPU and memory by loading the kernel module kvm.ko. The virtualization of I/O, storage and network needs to be implemented with the help of the qemu applications. Therefore, to use KVM virtualization technology, you need to install qemu-kvm, libvirt and virt-install applications. Inclusively, qemu-KVM is the core package of kvm and libvirt provides virtualization services and can be used to manage various Hypervisors such as KVM while virt-install is a utility for installing VMs. See Fig. 1-9 for the schematic diagram of relevant applications.

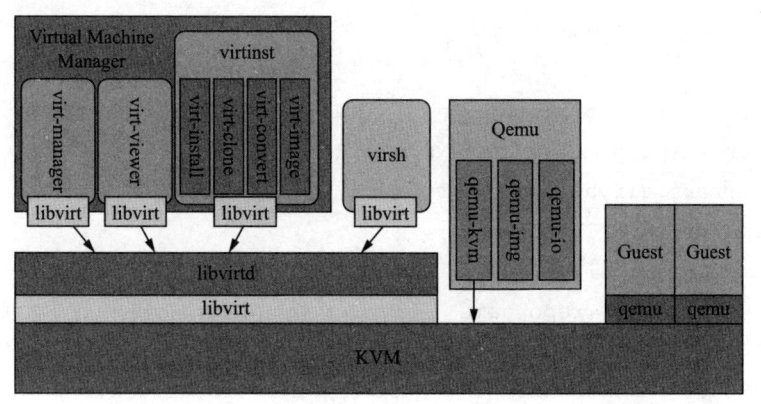

Fig. 1-9   Schematic diagram of relevant applications

Use the yum install command to install qemu-kvm, libvirt and virt-install.

```
[root@localhost ~]# yum -y install qemu-kvm libvirt virt-install
```

2. Start service

Start libvirtd and set the boot auto-start. Use the ip address to view the system network information and you can see the new virtual bridge virbr0 and virtual network card virbr0-nic.

```
[root@localhost ~]# systemctl start libvirtd
[root@localhost ~]# systemctl enable libvirtd
[root@localhost ~]# ip address
1: lo: <LOOPBACK,UP,LOWER_UP> mtu 65536 qdisc noqueue state UNKNOWN
    link/loopback 00:00:00:00:00:00 brd 00:00:00:00:00:00
    inet 127.0.0.1/8 scope host lo
      valid_lft forever preferred_lft forever
    inet6 ::1/128 scope host
```

```
            valid_lft forever preferred_lft forever
    2: eno16777736: <BROADCAST,MULTICAST,UP,LOWER_UP> mtu 1500 qdisc pfifo_fast
state UP qlen 1000
        link/ether 00:0c:29:2f:24:43 brd ff:ff:ff:ff:ff:ff
        inet 192.168.134.128/24 brd 192.168.134.255 scope global dynamic eno16777736
            valid_lft 1188sec preferred_lft 1188sec
        inet6 fe80::20c:29ff:fe2f:2443/64 scope link
            valid_lft forever preferred_lft forever
    3: virbr0: <NO-CARRIER,BROADCAST,MULTICAST,UP> mtu 1500 qdisc noqueue state
DOWN
        link/ether 52:54:00:6b:8e:4a brd ff:ff:ff:ff:ff:ff
        inet 192.168.122.1/24 brd 192.168.122.255 scope global virbr0
            valid_lft forever preferred_lft forever
    4: virbr0-nic: <BROADCAST,MULTICAST> mtu 1500 qdisc pfifo_fast master virbr0
state DOWN qlen 500
        link/ether 52:54:00:6b:8e:4a brd ff:ff:ff:ff:ff:ff
```

After the service is started successfully, check the network information and here appears a new network device, virtbr0, a virtual bridge device with the address of 192.168.122.1. The NAT network with virtbr0 as the gateway should be the default network used for creating the subsequent KVM virtual machines.

3. Use cirros disk images

In general, cirros can be downloaded on its official website. Place the downloaded image in the host/opt directory.

4. Manage virtual machines

(1) Create virtual machines

The system image uses cirros, and the virtual machine is named as cirros, 256 MB RAM, the default NAT network is used as the network mode.

```
    [root@localhost ~]#  virt-install --virt-type kvm --name cirros --ram 256
--boot hd --disk path=/opt/cirros-0.3.4-x86_64-disk.img --network network=default
--graphics vnc,listen=0.0.0.0 --noautoconsole
    [root@localhost ~]# virsh list --all
    Id    Name                          State
    ----------------------------------------------------
    3     cirros                        running
```

(2) Log in to the virtual machine console

Use the virsh console command to enter the virtual machine console. Then, check the IP address of the virtual machine and you can see that the IP address obtained by the virtual machine

is in the same network segment as virbr0 and that the cirros virtual machine uses virbr0 as the gateway.

```
[root@localhost opt]# virsh console cirros
=== network info ===
if-info: lo,up,127.0.0.1,8,::1
if-info: eth0,up,192.168.122.40,24,fe80::5054:ff:fec4:bc
ip-route:default via 192.168.122.1 dev eth0
ip-route:192.168.122.0/24 dev eth0   src 192.168.122.40
=== datasource: None None ===
=== cirros: current=0.3.4 uptime=277.42 ===

   ____               ____  ____
  / __/ __ ____  ____/ __ \/ __/
 / /__ / // __// __/ /_/ /\ \
 \___//_//_/  /_/   \____/___/
    http://cirros-cloud.net


login as 'cirros' user. default password: 'cubswin:)'. use 'sudo' for root.
cirros login: cirros
Password:
$ ip a
1: lo: <LOOPBACK,UP,LOWER_UP> mtu 16436 qdisc noqueue
    link/loopback 00:00:00:00:00:00 brd 00:00:00:00:00:00
    inet 127.0.0.1/8 scope host lo
    inet6 ::1/128 scope host
       valid_lft forever preferred_lft forever
2: eth0: <BROADCAST,MULTICAST,UP,LOWER_UP> mtu 1500 qdisc pfifo_fast qlen 1000
    link/ether 52:54:00:c4:00:bc brd ff:ff:ff:ff:ff:ff
    inet 192.168.122.40/24 brd 192.168.122.255 scope global eth0
    inet6 fe80::5054:ff:fec4:bc/64 scope link
       valid_lft forever preferred_lft forever
```

In the NAT mode, if the virtual machine needs to access the extranet, it needs to enable the routing and forwarding function of the host.

```
[root@localhost ~]# vi /etc/sysctl.conf
net.ipv4.ip_forward=1
[root@localhost ~]# sysctl -p
net.ipv4.ip_forward = 1
[root@localhost ~]# cat /proc/sys/net/ipv4/ip_forward
1
```

Verify the access to the extranet in the virtual environment.

```
$ ping www.baidu.com
PING www.a.shifen.com (14.215.177.39) 56(84) bytes of data.
64 bytes from 14.215.177.39 (14.215.177.39): icmp_seq=1 ttl=128 time=32.8 ms
64 bytes from 14.215.177.39 (14.215.177.39): icmp_seq=2 ttl=128 time=32.9 ms
64 bytes from 14.215.177.39 (14.215.177.39): icmp_seq=3 ttl=128 time=33.1 ms
64 bytes from 14.215.177.39 (14.215.177.39): icmp_seq=4 ttl=128 time=33.3 ms
^C
```

Task 2　Create a bridge network for KVM virtual machine

In Task 1 of this unit, the network mode used is the default NAT network. In some scenarios, the virtual and host machines may need to be in the same subnet. At this time, it is necessary to create a virtual bridge network for the virtual machine.

Create a centos7 virtual machine with the name of guest01 (1 GB RAM). The network mode uses the bridging pattern. After the creation is completed, use the VNC remote desktop protocol to log in to the virtual machine and access its desktop.

1. Create a virtual bridge network

Add a virtual bridge br0.

```
[root@localhost ~]# brctl addbr br0
```

Modify the network card configuration file.

```
[root@localhost ~]# cat /etc/sysconfig/network-scripts/ifcfg-ens33
TYPE=Ethernet
BOOTPROTO=static
NAME=ens33
DEVICE=ens33
ONBOOT=yes
BRIDGE=br0
[root@localhost ~]# cat /etc/sysconfig/network-scripts/ifcfg-br0
TYPE=Bridge
BOOTPROTO=none
NAME=br0
DEVICE=br0
ONBOOT=yes
IPADDR=192.168.134.128
NETMASK=255.255.255.0
GATEWAY=192.168.134.2
DNS1=192.168.134.2
```

Restart the network service.

```
[root@localhost ~]# systemctl restart network
[root@localhost ~]# ip a
1: lo: <LOOPBACK,UP,LOWER_UP> mtu 65536 qdisc noqueue state UNKNOWN group de-
fault qlen 1000
    link/loopback 00:00:00:00:00:00 brd 00:00:00:00:00:00
    inet 127.0.0.1/8 scope host lo
       valid_lft forever preferred_lft forever
    inet6 ::1/128 scope host
       valid_lft forever preferred_lft forever
2: ens33: <BROADCAST,MULTICAST,UP,LOWER_UP> mtu 1500 qdisc pfifo_fast master
br0 state UP group default qlen 1000
    link/ether 00:0c:29:e7:4d:35 brd ff:ff:ff:ff:ff:ff
3: virbr0: <BROADCAST,MULTICAST,UP,LOWER_UP> mtu 1500 qdisc noqueue state UP
group default qlen 1000
    link/ether 52:54:00:91:4f:b7 brd ff:ff:ff:ff:ff:ff
    inet 192.168.122.1/24 brd 192.168.122.255 scope global virbr0
       valid_lft forever preferred_lft forever
4: virbr0-nic: <BROADCAST,MULTICAST> mtu 1500 qdisc pfifo_fast master virbr0
state DOWN group default qlen 1000
    link/ether 52:54:00:91:4f:b7 brd ff:ff:ff:ff:ff:ff
12: vnet0: <BROADCAST,MULTICAST,UP,LOWER_UP> mtu 1500 qdisc pfifo_fast master
virbr0 state UNKNOWN group default qlen 1000
    link/ether fe:54:00:ed:80:55 brd ff:ff:ff:ff:ff:ff
    inet6 fe80::fc54:ff:feed:8055/64 scope link
       valid_lft forever preferred_lft forever
13: br0: <BROADCAST,MULTICAST,UP,LOWER_UP> mtu 1500 qdisc noqueue state UP
group default qlen 1000
    link/ether 00:0c:29:e7:4d:35 brd ff:ff:ff:ff:ff:ff
    inet 192.168.134.128/24 brd 192.168.134.255 scope global noprefixroute br0
       valid_lft forever preferred_lft forever
    inet6 fe80::20c:29ff:fee7:4d35/64 scope link
       valid_lft forever preferred_lft forever
[root@localhost ~]# brctl show
bridge name      bridge id           STP enabled      interfaces
br0              8000.000c29e74d35   no               ens33
virbr0           8000.525400914fb7   yes              virbr0-nic
                                                      vnet0
```

2. Create a virtual machine

Add the image file centos7-01. raw to simulate the virtual machine disk.

```
[root@localhost ~]# qemu-img create -f raw /opt/ guest01.raw 5G
```

Start the virtual machine.

```
[root@localhost ~]# virt-install --virt-type kvm --name guest01 --ram 1024
--cdrom=/opt/CentOS-7-x86_64-DVD-1908.iso --disk path=/opt/ guest01.raw --network
bridge=br0 --graphics vnc,listen=0.0.0.0  --noautoconsole
```

3. Build VNC remote desktop services

Install the VNC server program in CentOS and start the service.

```
[root@localhost opt]# yum -y install tigervnc-server
[root@localhost opt]# vncserver
Password:
Verify:
Would you like to enter a view-only password (y/n)? n
A view-only password is not used
```

Enter the host IP and VNC connector port number in the VNC client software. The port number is 5900 by default. If there are multiple VNC connections, the available ports will be enabled from 5900 successively. Here, you can use the ps aux and netstat-ntpl commands to confirm the port number corresponding to each VNC connection.

```
[root@localhost ~]# ps aux|grep vnc
root        4125   2.5   0.4 207884 37304 pts/0    Sl   10:58   0:00 /bin/Xvnc :1
-auth /root/.xauthmzSuvW -desktop localhost.localdomain:1 (root) -fp catalogue:/
etc/X11/fontpath.d -geometry 1024x768 -pn -rfbauth /root/.vnc/passwd -rfbport 5901
-rfbwait 30000
root        4132   0.0   0.0 113280  1204 pts/0    S    10:58   0:00 /bin/sh /root/.
vnc/xstartup
root        4313   2.3   0.2 193464 23520 pts/0    Sl   10:58   0:00 /bin/Xvnc :2
-auth /root/.xauthmzSuvW -desktop localhost.localdomain:2 (root) -fp catalogue:/
etc/X11/fontpath.d -geometry 1024x768 -pn -rfbauth /root/.vnc/passwd -rfbport 5902
-rfbwait 30000
root   4532   0.0   0.0 113280  1204 pts/0   S   10:58   0:00 /bin/sh /root/.vnc/xstartup
root   4943   0.0   0.0 112828   984 pts/0   R+  10:58   0:00 grep --color=auto vnc
[root@localhost ~]# netstat -ntpl
Active Internet connections (only servers)
Proto Recv-Q Send-Q Local Address   Foreign Address  State PID/Program name
tcp       0        0 127.0.0.1:631     0.0.0.0:*          LISTEN     1176/cupsd
tcp       0        0 127.0.0.1:25      0.0.0.0:*          LISTEN     1394/master
```

| tcp | 0 | 0 0.0.0.0:5901 | 0.0.0.0:* | LISTEN | 4125/Xvnc |
| tcp | 0 | 0 0.0.0.0:5902 | 0.0.0.0:* | LISTEN | 4313/Xvnc |

Access the centos 7 virtual machine desktop and view the IP address of the virtual machine. You can see that the IP address of the virtual machine is in the same network segment as that of the host ens33 network card, as shown in Fig. 1-10.

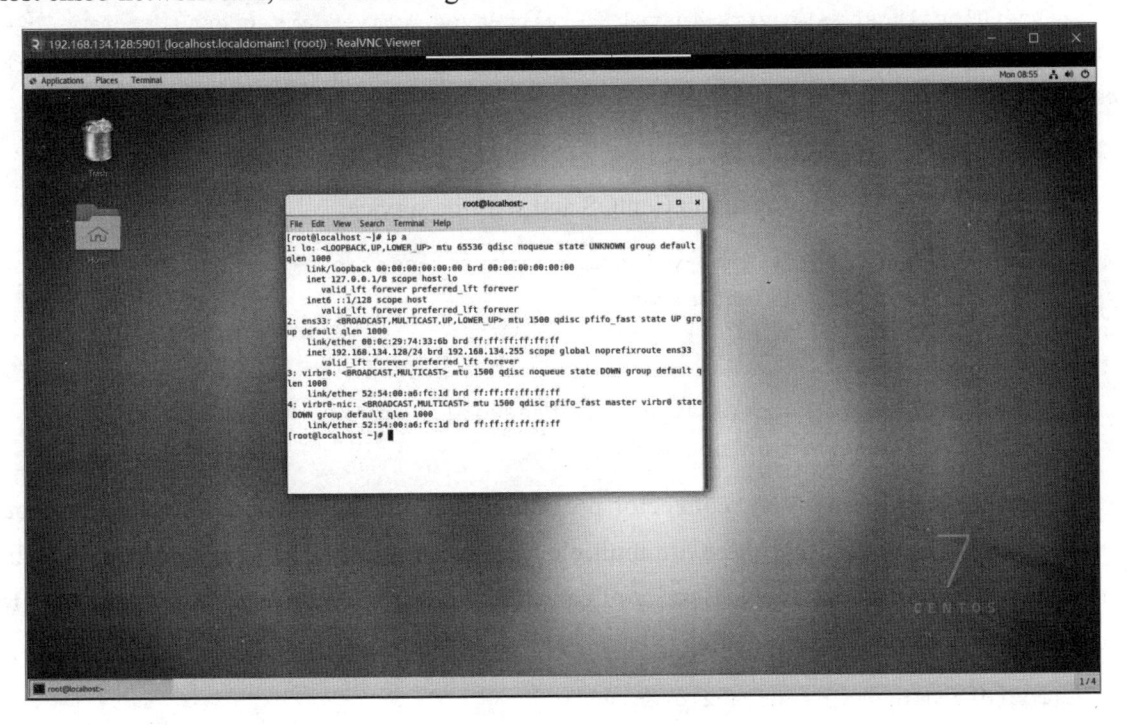

Fig. 1-10   View IP address

If you want to convert the network mode of the original virtual machine from NAT to bridge mode, you can modify the xml configuration file of the virtual machine. In the /etc/libvirt/qemu/ directory, the configuration file named after the virtual machine is the xml configuration file of the virtual machine, which defines the attribute parameters of all aspects of the virtual machine, and changes the network configuration part to bridge.

```
# virsh edit /etc/libvirt/qemu/ guest01
        <interface type='bridge'>
                <source bridge='br0'>
# virsh shutdown guest01
# virsh start guest01
```

1.4　Project practice—manage Linux server

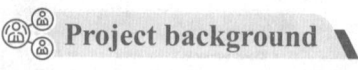
Project background

In order to improve the server efficiency, a company wanted to migrate the services in some original Windows servers to Linux servers. It decided to install the virtual software "VMware Workstation" on the Windows 10 operating system and install the Linux server of the CentOS 7 operating system.

Project implementation

Task 1　Get the Linux OS image

For technicians, they want to easily find the operating system images of each version for learning and research and fundamentally such images are most easily available through the official website. In addition, in many large communities, they also provide image stations containing various types and versions of operating systems, which are fairly friendly to learners or technicians. For example, in the image station of the aliyun developer community, you can easily obtain common operating system images. Here is an example of downloading the full version of CentOS7-1804 operating system image. Select the Mirror station, as shown in Fig. 1-11.

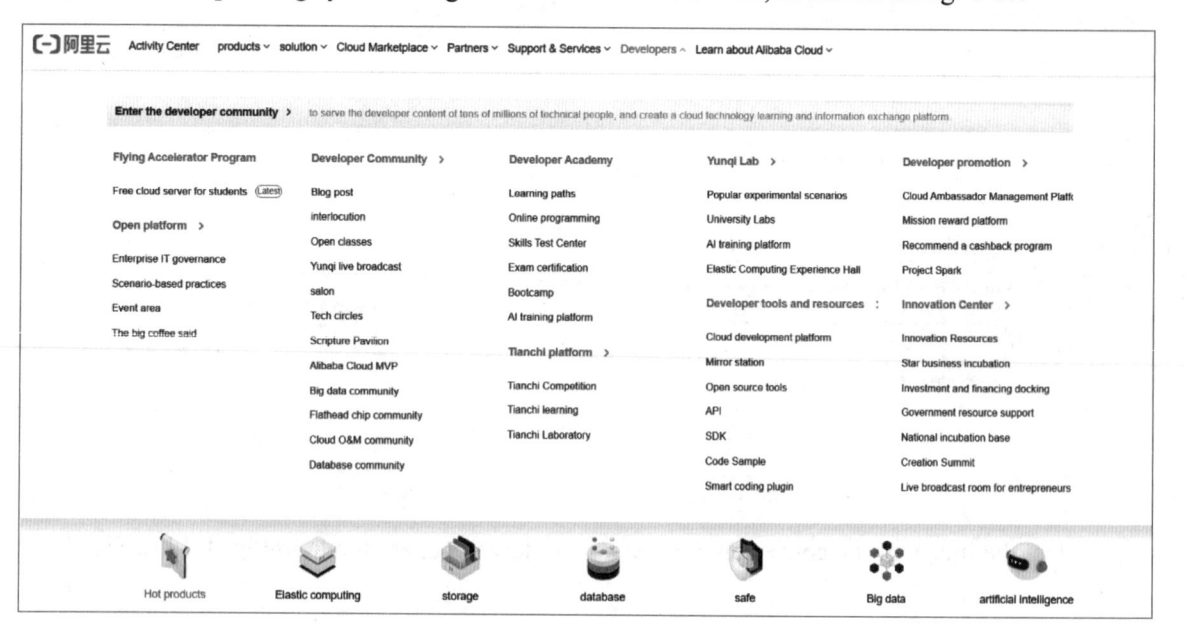

Fig. 1-11　Select the image station

Select the corresponding operating system, such as centos, as shown in Fig. 1-12.

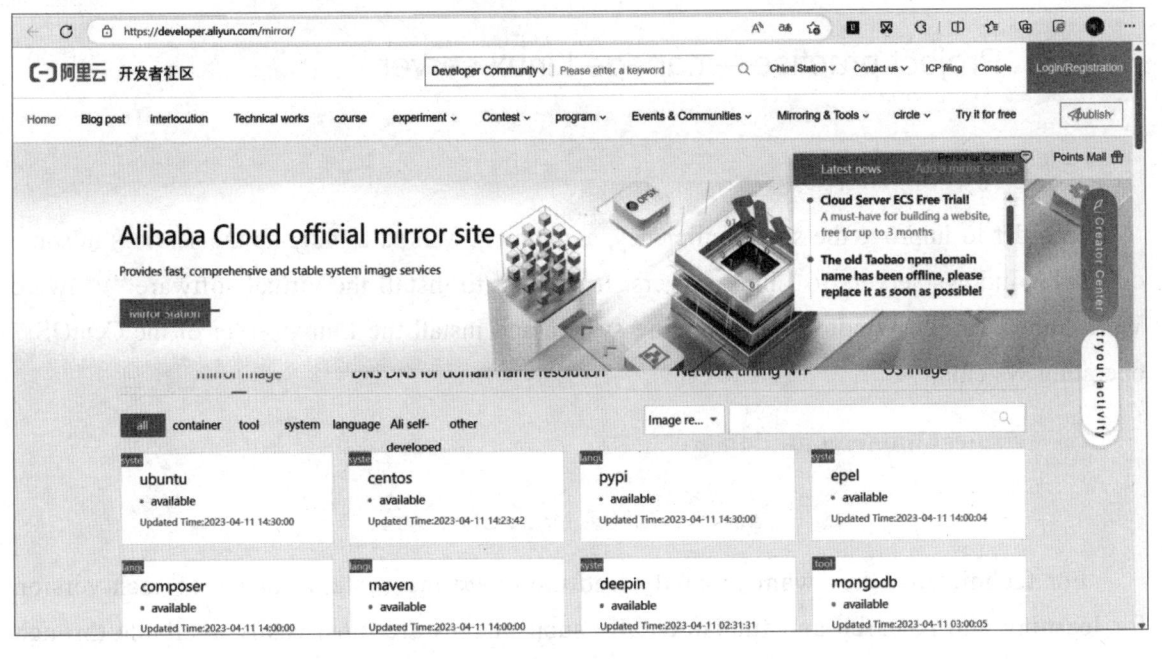

Fig. 1-12　Select centos

Enter the download directory, as shown in Fig. 1-13.

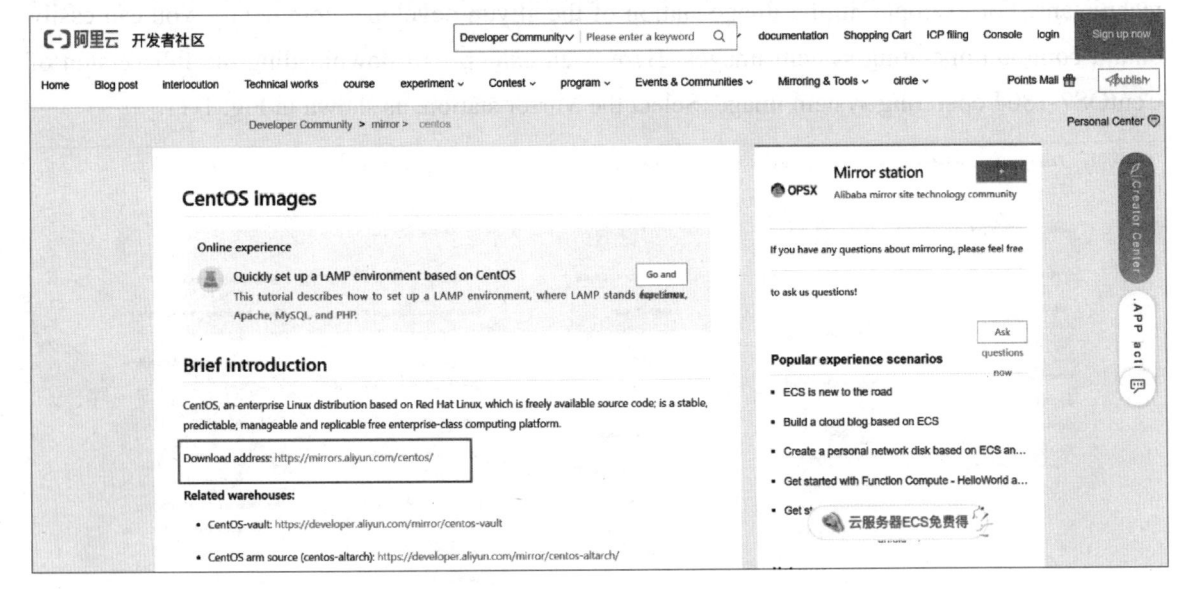

Fig. 1-13　Enter the download directory

Select the image of the corresponding version to download, as shown in Fig. 1-14 to Fig. 1-17.

```
../
2.1/                              19-Aug-2009 01:36           -
3.1/                              31-Jul-2005 16:05           -
3.3/                              17-Mar-2005 11:17           -
3.4/                              01-Mar-2005 01:38           -
3.5/                              28-Jul-2005 16:14           -
3.6/                              04-Apr-2006 16:59           -
3.7/                              06-May-2006 01:20           -
3.8/                              20-Apr-2012 10:55           -
3.9/                              20-Apr-2012 10:49           -
4.0/                              23-Jun-2005 12:09           -
4.0beta/                          15-Sep-2005 23:40           -
4.1/                              19-Oct-2005 12:09           -
4.2/                              04-Nov-2006 12:35           -
4.2beta/                          23-Oct-2005 11:17           -
4.3/                              27-Aug-2006 12:04           -
4.4/                              01-Nov-2006 19:32           -
4.5/                              08-Dec-2007 09:23           -
4.6/                              18-Sep-2008 14:35           -
4.7/                              14-Feb-2010 12:50           -
4.8/                              20-Apr-2012 21:13           -
4.9/                              20-Apr-2012 21:13           -
5.0/                              11-Apr-2007 17:53           -
5.1/                              27-Nov-2007 10:38           -
5.10/                             07-Oct-2014 13:31           -
5.11/                             27-Sep-2014 12:52           -
5.2/                              18-Jun-2008 23:34           -
5.3/                              14-Feb-2010 02:30           -
5.4/                              25-Jun-2010 14:38           -
5.5/                              14-May-2010 21:04           -
5.6/                              12-Aug-2011 19:51           -
5.7/                              13-Sep-2011 00:58           -
5.8/                              27-Feb-2012 19:57           -
5.9/                              12-Jan-2013 01:11           -
6.0/                              19-Dec-2011 18:49           -
6.1/                              08-Dec-2011 10:04           -
6.10/                             02-Jul-2018 15:32           -
6.2/                              19-Dec-2011 13:25           -
6.3/                              09-Jul-2012 20:10           -
6.4/                              20-Jun-2013 14:50           -
6.5/                              21-Dec-2013 14:05           -
6.6/                              31-Jul-2015 15:17           -
6.7/                              21-Jan-2016 13:22           -
6.8/                              24-May-2016 17:36           -
6.9/                              04-Apr-2017 19:38           -
7.0.1406/                         07-Apr-2015 14:36           -
7.1.1503/                         13-Nov-2015 13:01           -
7.2.1511/                         18-May-2016 16:48           -
7.3.1611/                         20-Feb-2017 22:21           -
7.4.1708/                         26-Feb-2018 14:32           -
7.5.1804/                         09-May-2018 20:39           -
7.6.1810/                         02-Dec-2018 14:34           -
7.7.1908/                         15-Sep-2019 01:00           -
7.8.2003/                         17-Jun-2020 17:55           -
7.9.2009/                         18-Jan-2021 14:14           -
8-stream/                         11-Aug-2021 18:51           -
8.0.1905/                         09-Sep-2020 07:43           -
8.1.1911/                         21-Oct-2020 07:53           -
8.2.2004/                         15-Jan-2021 09:07           -
8.3.2011/                         27-Apr-2021 16:00           -
8.4.2105/                         25-May-2021 21:18           -
altarch/                          12-Nov-2020 15:42           -
centos/                           09-Jul-2021 12:56           -
```

Fig. 1-14　Select the corresponding version

```
Index of /centos-vault/7.5.1804/

../
atomic/                           05-Jun-2015 11:33           -
centosplus/                       10-May-2018 12:48           -
cloud/                            03-Nov-2015 11:59           -
configmanagement/                 06-Oct-2017 15:49           -
cr/                               09-May-2018 20:43           -
dotnet/                           29-Sep-2017 12:33           -
extras/                           10-May-2018 04:12           -
fasttrack/                        01-Sep-2017 11:08           -
isos/                             04-May-2018 12:20           -
nfv/                              26-Feb-2018 14:32           -
opstools/                         13-Sep-2017 12:54           -
os/                               09-May-2018 14:54           -
paas/                             18-May-2016 15:36           -
rt/                               10-Feb-2017 21:18           -
sclo/                             04-Nov-2015 10:27           -
storage/                          13-Nov-2015 17:33           -
updates/                          13-Aug-2018 17:04           -
virt/                             12-Nov-2015 12:07           -
```

Fig. 1-15　Select the isos directory

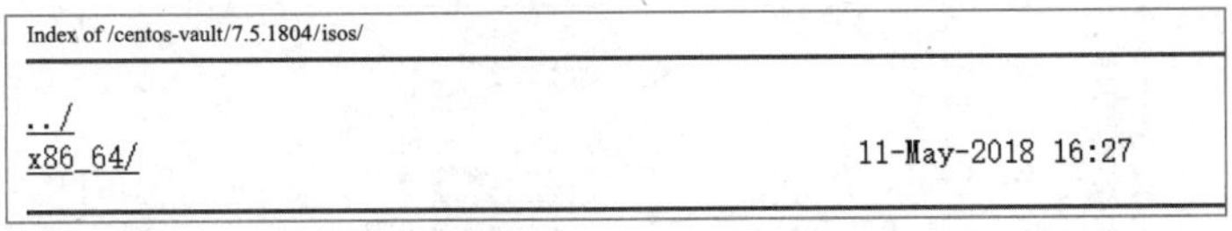

Index of /centos-vault/7.5.1804/isos/

../
x86_64/   11-May-2018 16:27

Fig. 1-16   Select the x86_ 64/ directory

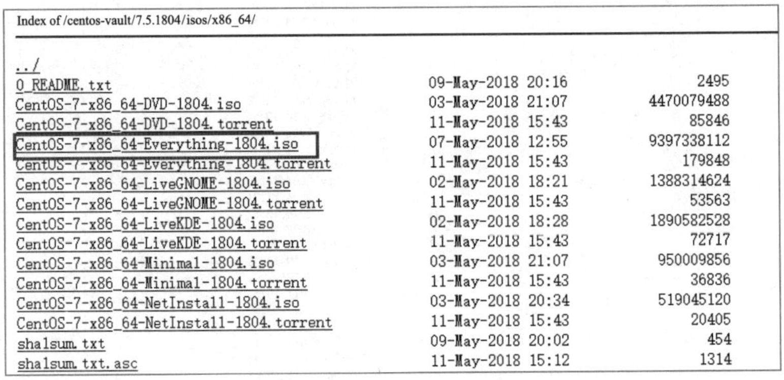

Index of /centos-vault/7.5.1804/isos/x86_64/

../
0_README.txt   09-May-2018 20:16   2495
CentOS-7-x86_64-DVD-1804.iso   03-May-2018 21:07   4470079488
CentOS-7-x86_64-DVD-1804.torrent   11-May-2018 15:43   85846
CentOS-7-x86_64-Everything-1804.iso   07-May-2018 12:55   9397338112
CentOS-7-x86_64-Everything-1804.torrent   11-May-2018 15:43   179848
CentOS-7-x86_64-LiveGNOME-1804.iso   02-May-2018 18:21   1388314624
CentOS-7-x86_64-LiveGNOME-1804.torrent   11-May-2018 15:43   53563
CentOS-7-x86_64-LiveKDE-1804.iso   02-May-2018 18:28   1890582528
CentOS-7-x86_64-LiveKDE-1804.torrent   11-May-2018 15:43   72717
CentOS-7-x86_64-Minimal-1804.iso   03-May-2018 21:07   950009856
CentOS-7-x86_64-Minimal-1804.torrent   11-May-2018 15:43   36836
CentOS-7-x86_64-NetInstall-1804.iso   03-May-2018 20:34   519045120
CentOS-7-x86_64-NetInstall-1804.torrent   11-May-2018 15:43   20405
sha1sum.txt   09-May-2018 20:02   454
sha1sum.txt.asc   11-May-2018 15:12   1314

Fig. 1-17   Select the CentOS-7 image

Task 2   Create a new virtual machine

Choose to use the software "VMware Workstation" to provide a virtualized environment. The software VMware Workstation is a powerful desktop virtual computer software, suitable for a single computer, enabling users to run different operating systems on a single desktop at the same time, and to develop, test, and deploy new applications. Moreover, the software VMware Workstation has become one of the mainstream desktop virtual computer software on the market with good flexibility and advanced technology.

Select the "New Virtual Machine" command, as shown in Fig. 1-18. Click the "Create @ New Virtual Machine" button.

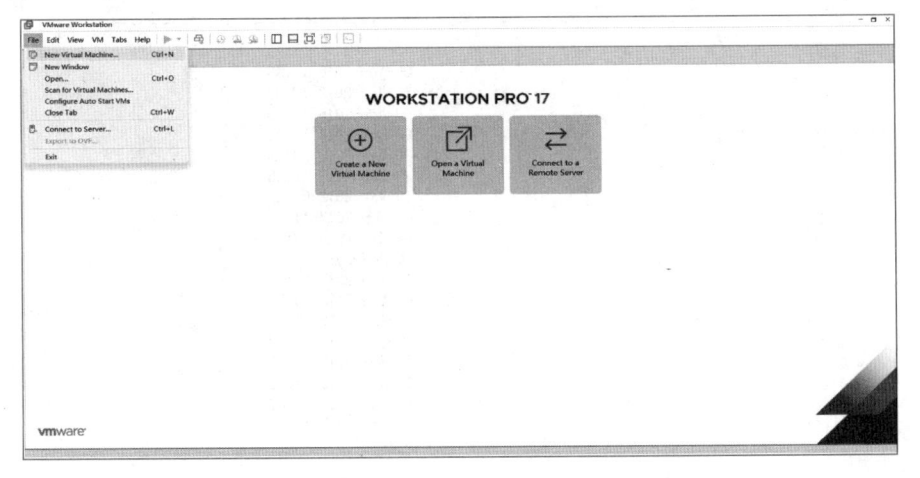

Fig. 1-18   Create a new virtual machine

Select the "Custom (advanced)" radio button, as shown in Fig. 1-19. Click the "Next" button.

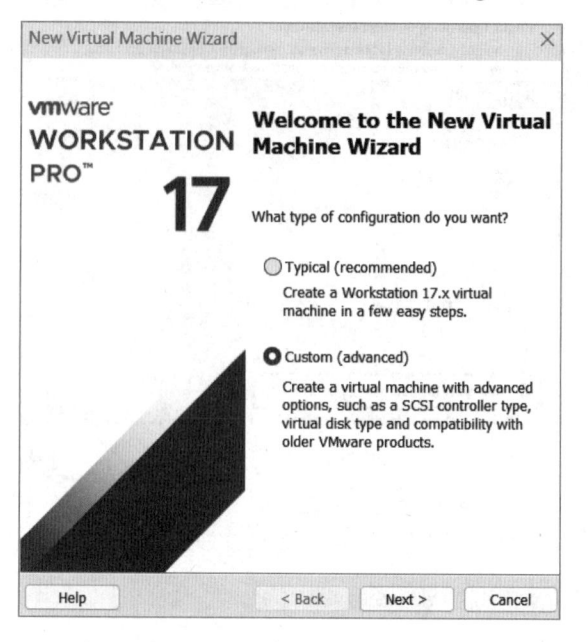

Fig. 1-19   Select the "Custom (advanced)" radio button

Select the virtual machine hardware compatibility and use the default options, as shown in Fig. 1-20. Click the "Next" button.

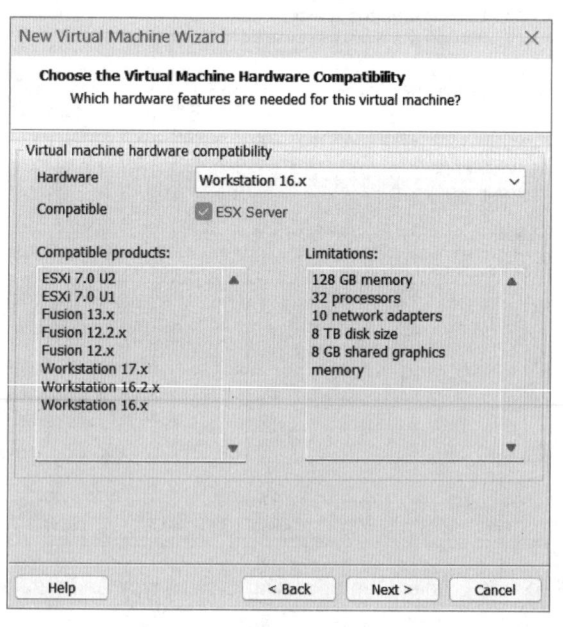

Fig. 1-20   Select virtual machine hardware compatibility

Select "I will install the operating system later." radio button. After the virtual machine is created, load the image file for system installation, as shown in Fig. 1-21. Click the "Next" button.

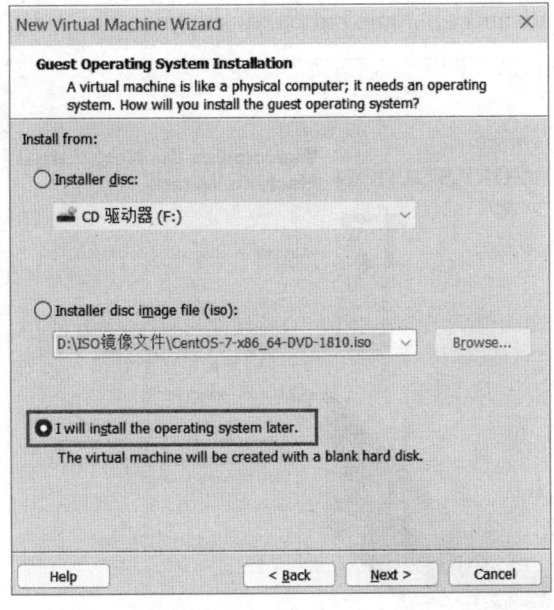

Fig. 1-21   Select "I will install the operating system later" radio button

Select the operating system type as Linux and the version as CentOS 7, as shown in Fig. 1-22. Click the "Next" button.

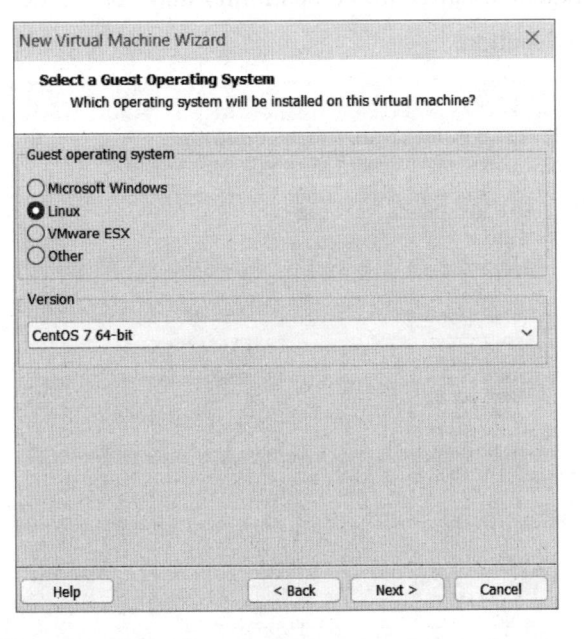

Fig. 1-22   Select the operating system type and version

Define the virtual machine name and installation destination, as shown in Fig. 1-23.

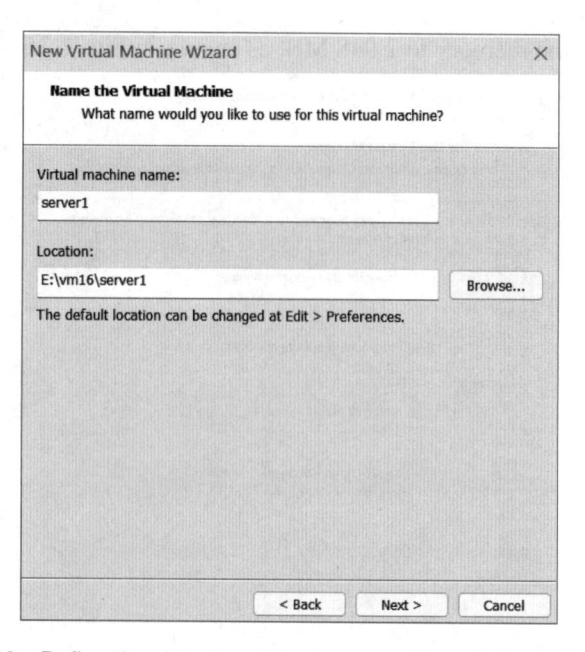

Fig. 1-23   Define the virtual machine name and installation destination

Configure the processor according to the principle that the virtual machine resource configuration does not exceed the upper limit of the physical host resource configuration. The relevant configuration of other hardware shall follow this principle. For example, if the physical host is a dual-core processor, the number of virtual machine processor cores cannot exceed 2, as shown in Fig. 1-24. Click the "Next" button.

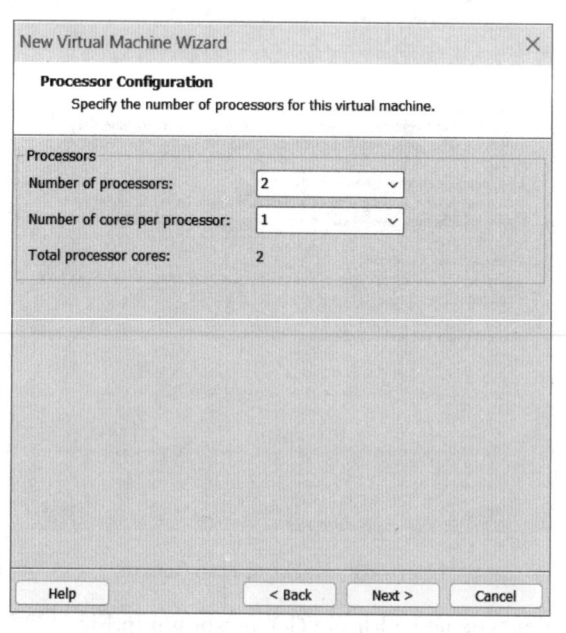

Fig. 1-24   Number of configured processors and cores

Set the virtual machine memory to 2 048 MB, as shown in Fig. 1-25. Click the "Next" button.

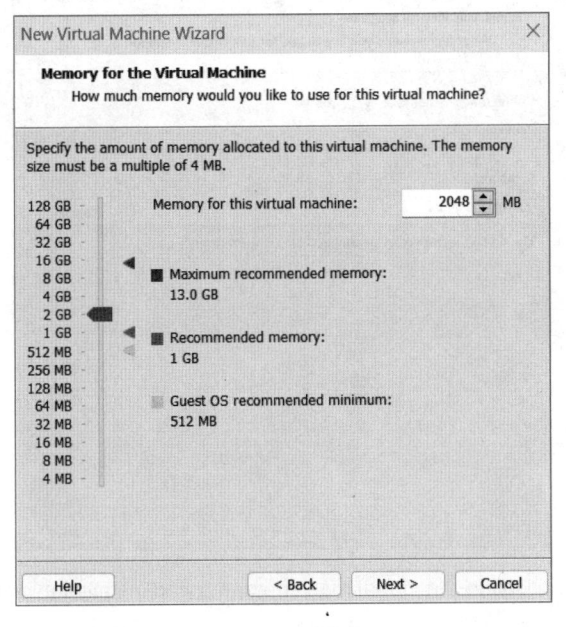

Fig. 1-25   Set virtual machine memory

Select the "Use network address translation (NAT)" radio button, as shown in Fig. 1-26. Click the "Next" button.

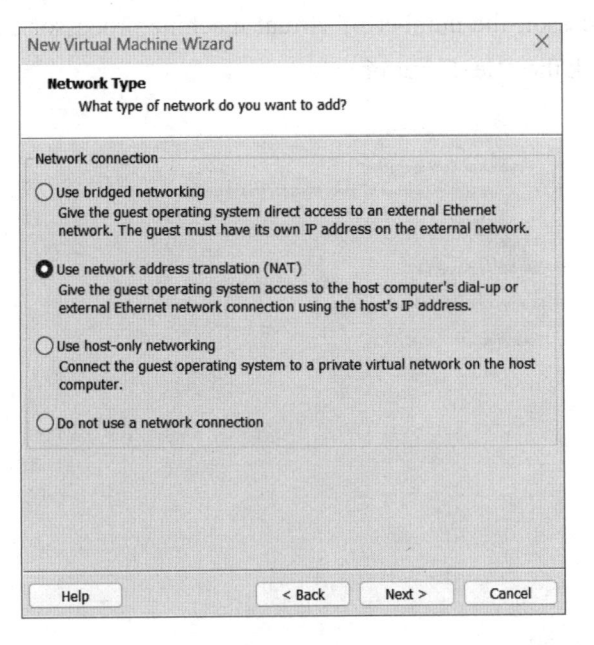

Fig. 1-26   Network type options

Select the I/O controller type as LSI logic (L), as shown in Fig. 1-27. Click the "Next" button.

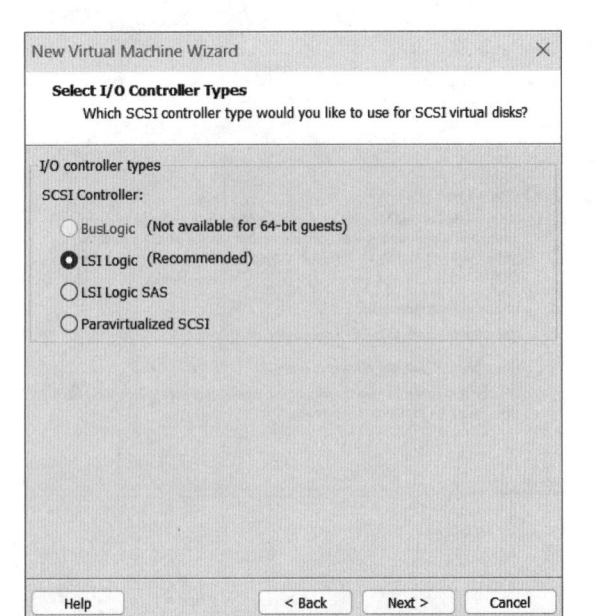

Fig. 1-27   Select I/O controller type

Select SCSI (S) as the disk type, as shown in Fig. 1-28. Click the "Next" button.

Fig. 1-28   Select disk type

Select the disk as "Create @ new virtual disk", as shown in Fig. 1-29. Click the "Next" button.

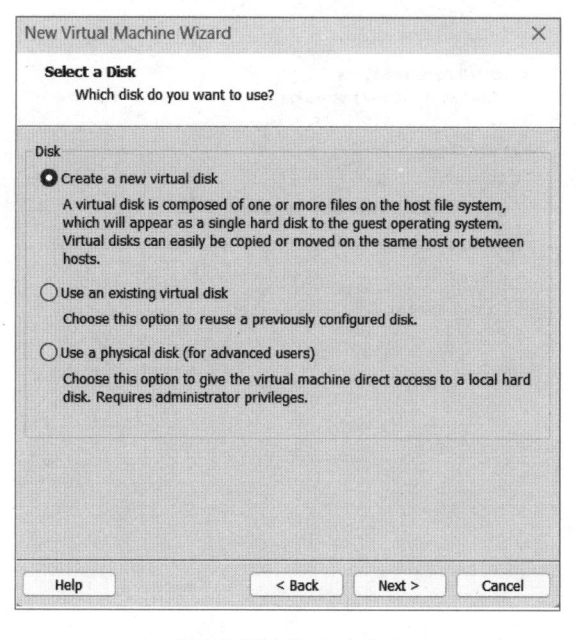

Fig. 1-29   Select disk

Do not check the "Allocate all disk space now" check box when configuring disks; otherwise, the disks of the physical host will be directly occupied with the corresponding space. On the contrary, the storage space of the physical host will increase dynamically with the use of virtual machines, which will save more resources. Set the disk size to 100 GB, as shown in Fig. 1-30. Click the "Next" button.

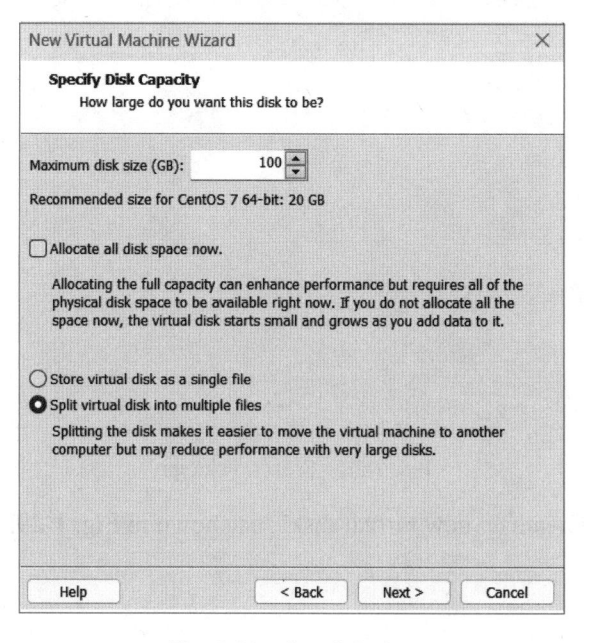

Fig. 1-30   Set disk size

Here, create a vmdk virtual disk image file on the physical host to simulate the disk device in the virtual machine, as shown in Fig. 1-31. Click the "Next" button.

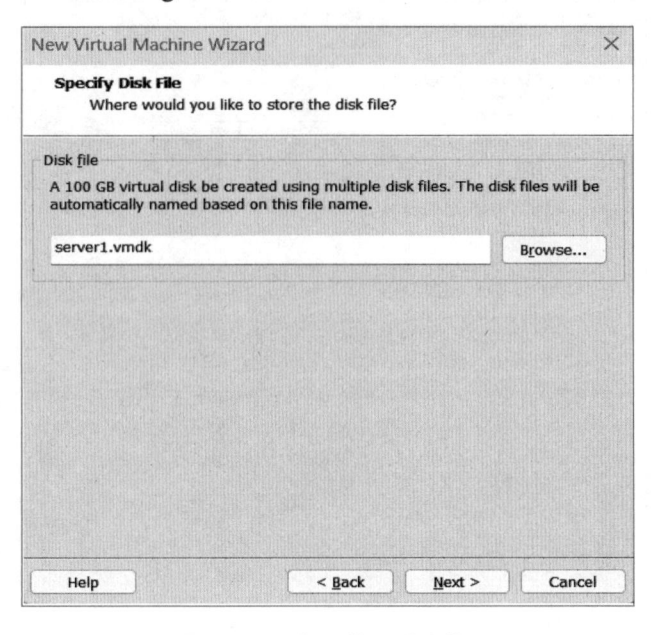

Fig. 1-31 Specify a disk file

While creating a virtual machine, you can adjust the hardware devices as needed and select custom hardware, as shown in Fig. 1-32.

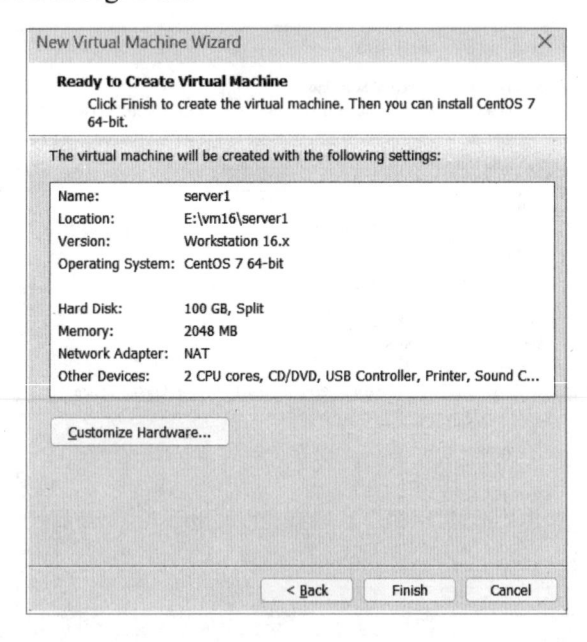

Fig. 1-32 Select customize hardware

Specify the image file used in the virtual optical drive as the CentOS 7 image file, as shown in Fig. 1-33.

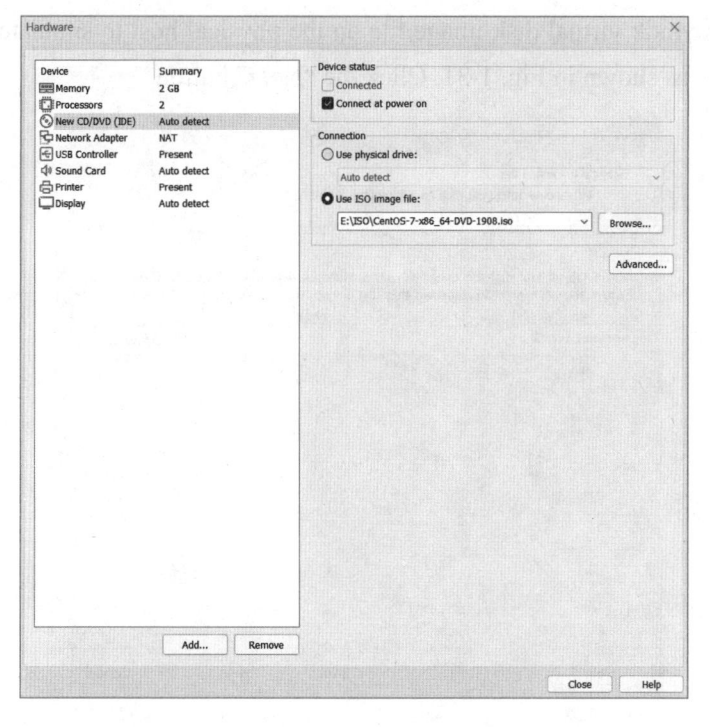

Fig. 1-33　Use ISO image file

Click the "Finish" button and you can create the virtual machine successfully, as shown in Fig. 1-34.

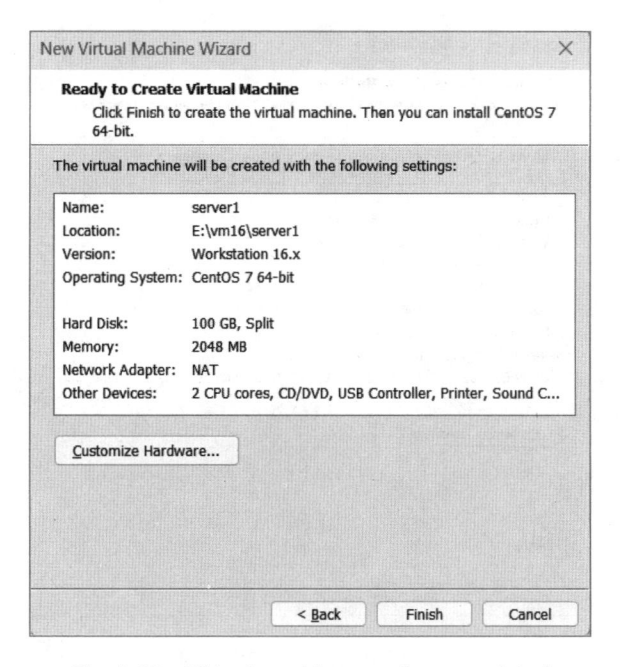

Fig. 1-34　Virtual machine creation completed

So far, the virtual machine has been created and we can complete OS installation and other operations just like using a real physical server.

Task 3   Install CentOS 7 operating system

Next, install the operating system for the created virtual machine and select "Install CentOS 7", as shown in Fig. 1-35.

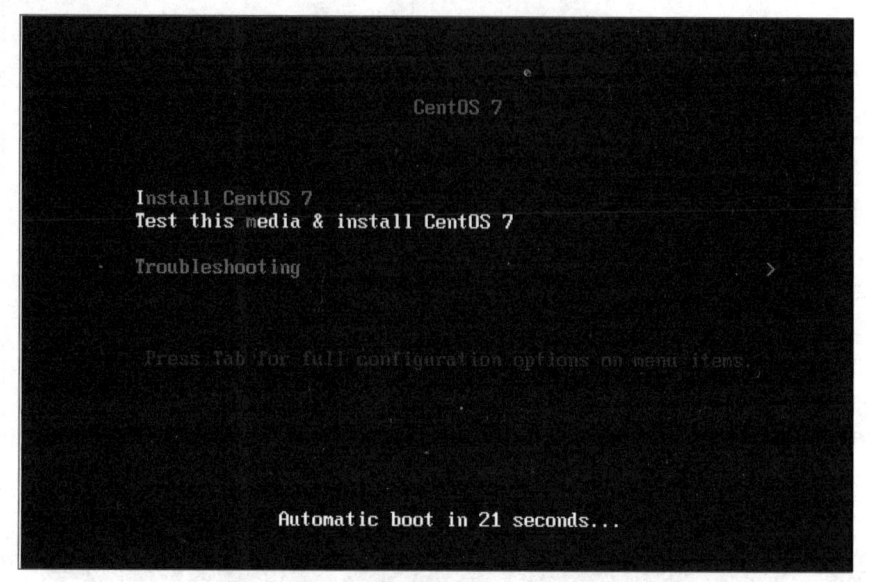

Fig. 1-35   Choose to install CentOS 7 directly

Select the language and time zone. While installing the system, partition the entire disk into the root partition (/), swap partition (swap) and boot partition (/boot) manually and set the partition size, as shown in Fig. 1-36 to Fig. 1-41.

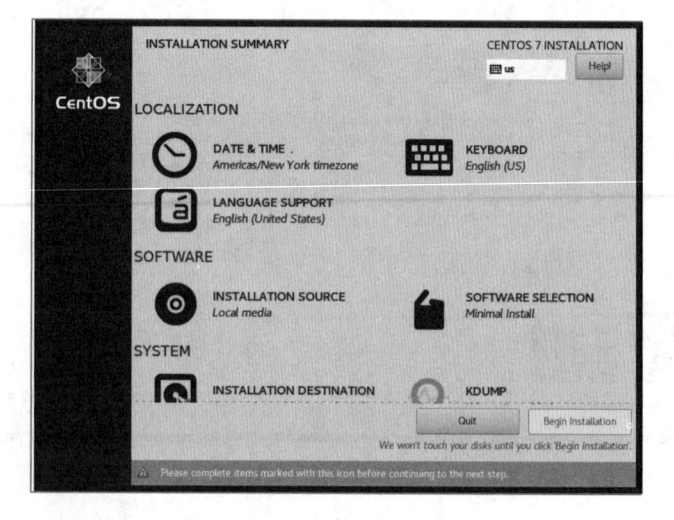

Fig. 1-36   Manual disk partition

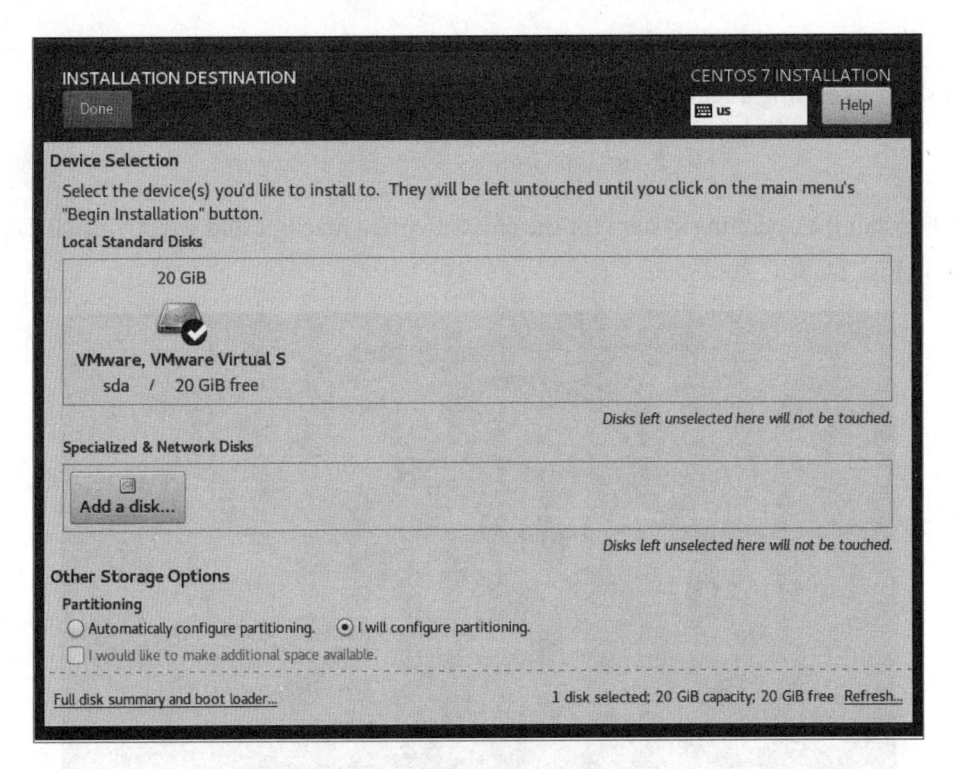

Fig. 1-37　Select manual partition

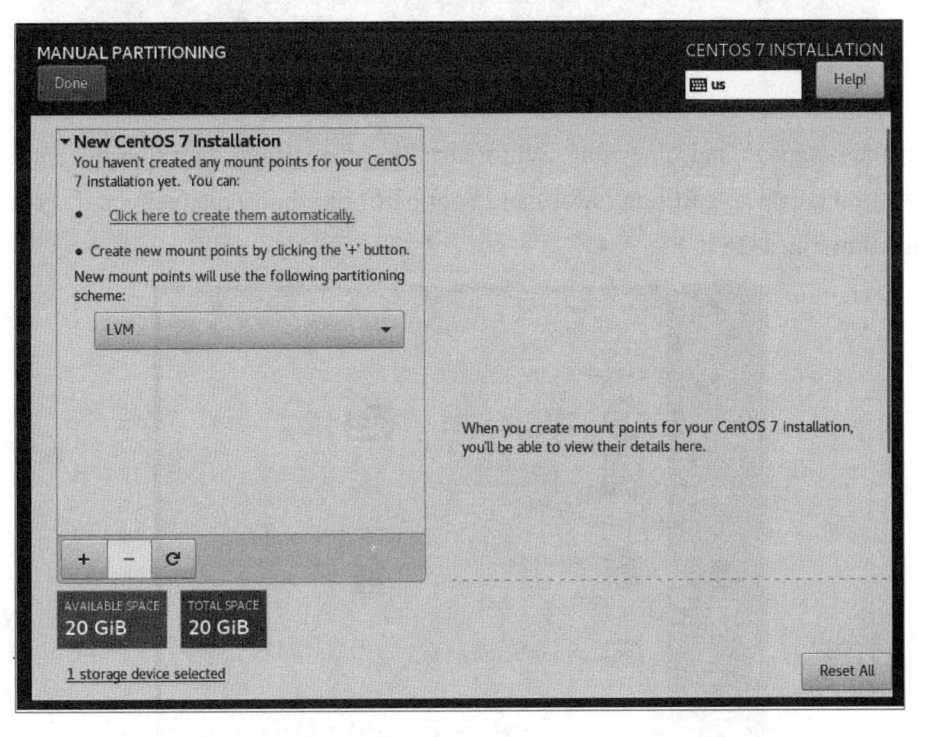

Fig. 1-38　New partition

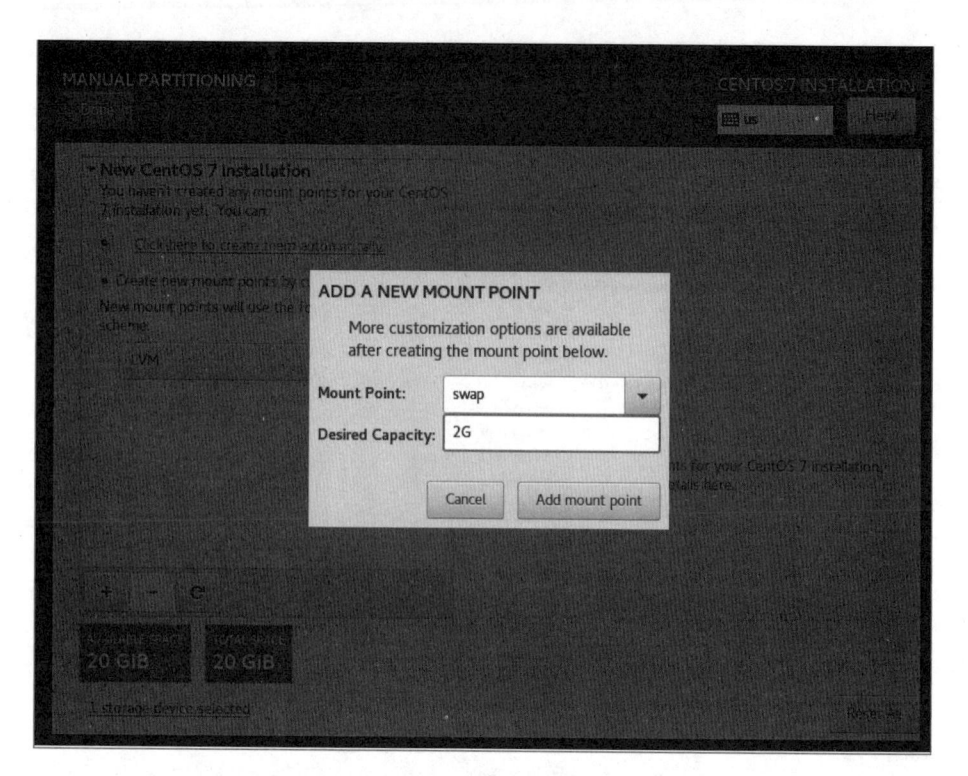

Fig. 1-39   Create a swap partition

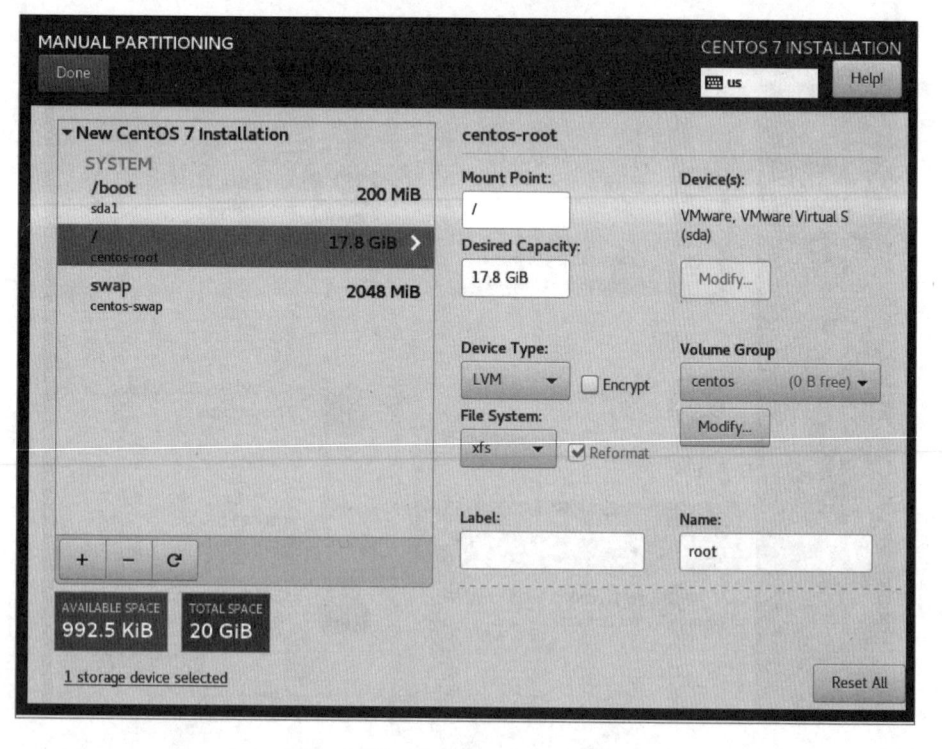

Fig. 1-40   Create a root partition

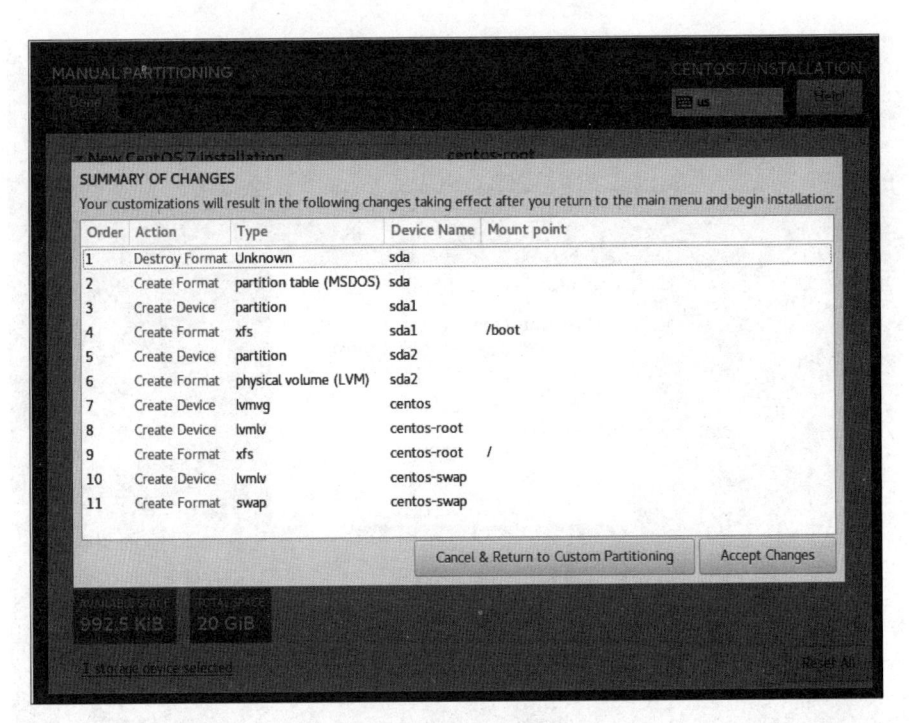

Fig. 1-41   Partition completed

You can disable the KDUMP function for better performance. Then click the "Begin Installation" button to start the installation, as shown in Fig. 1-42.

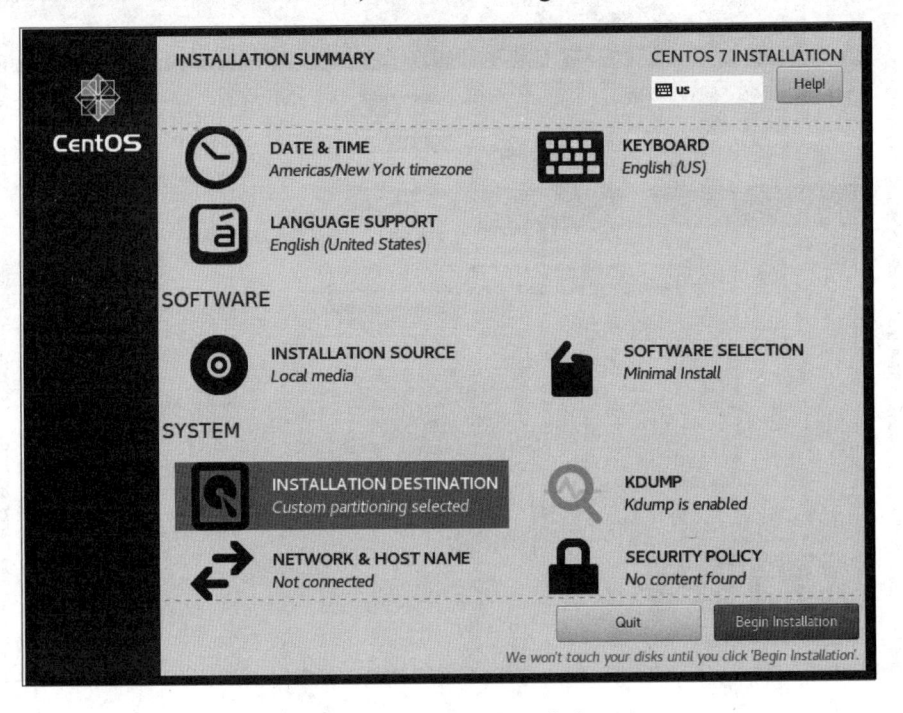

Fig. 1-42   Start the Installation

Set the root password, as shown in Fig. 1-43 and Fig. 1-44.

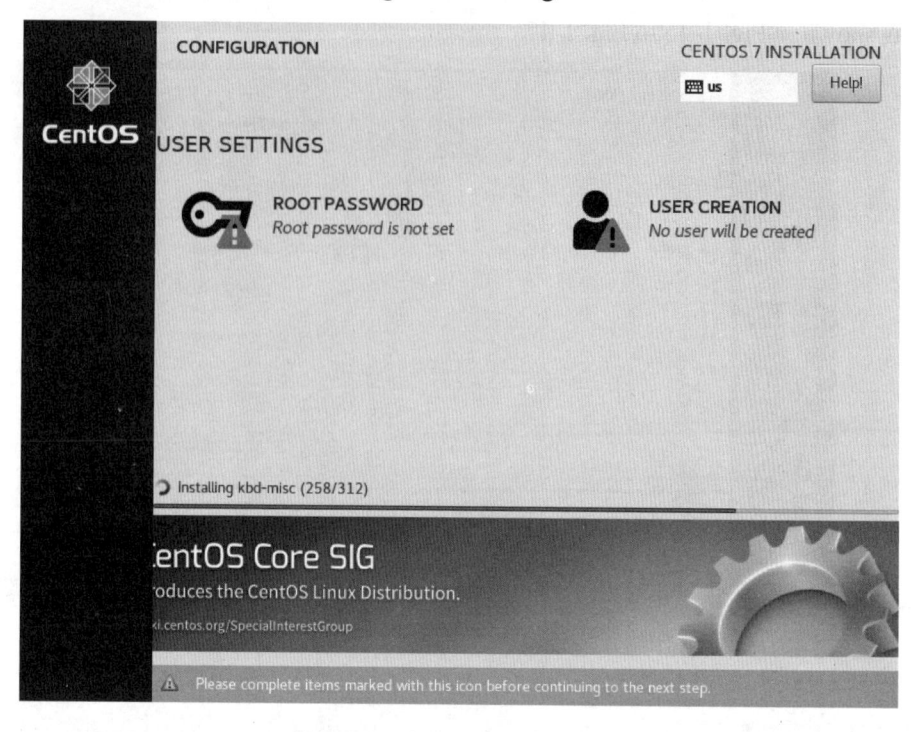

Fig. 1-43   Set the root password

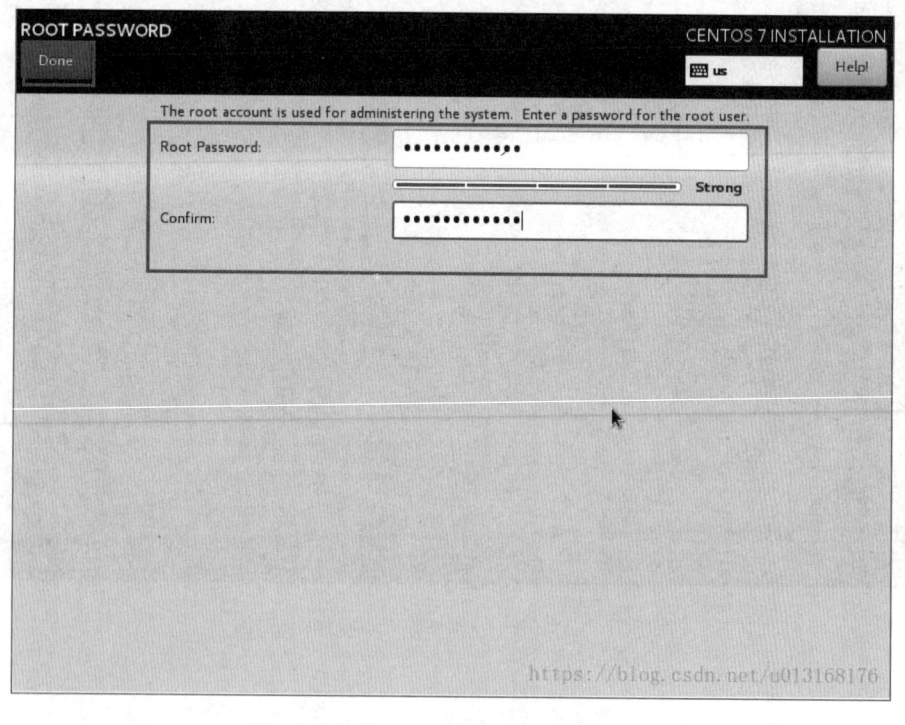

Fig. 1-44   Password setting completed

After installation, click Reboot to restart the system and log in with the root user and its password, as shown in Fig. 1-45 to Fig. 1-47.

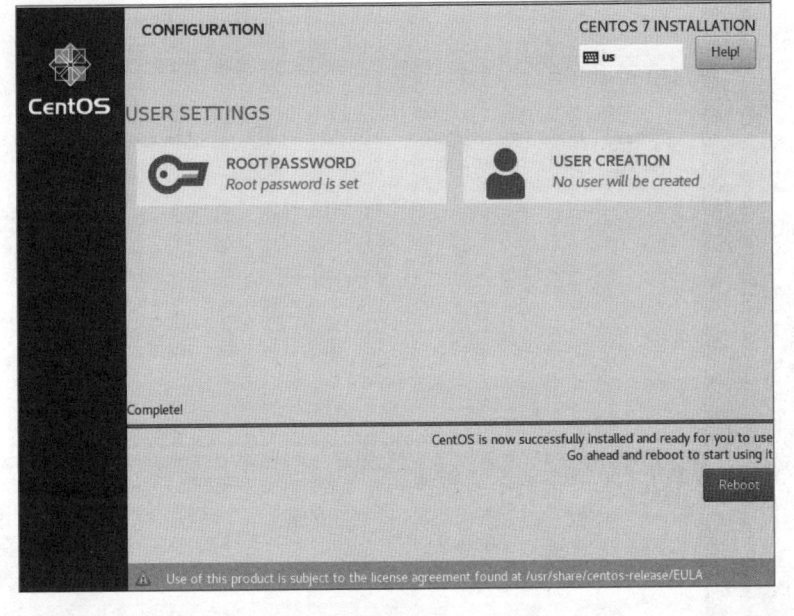

Fig. 1-45　Restart the system

Fig. 1-46　Enter user name and password

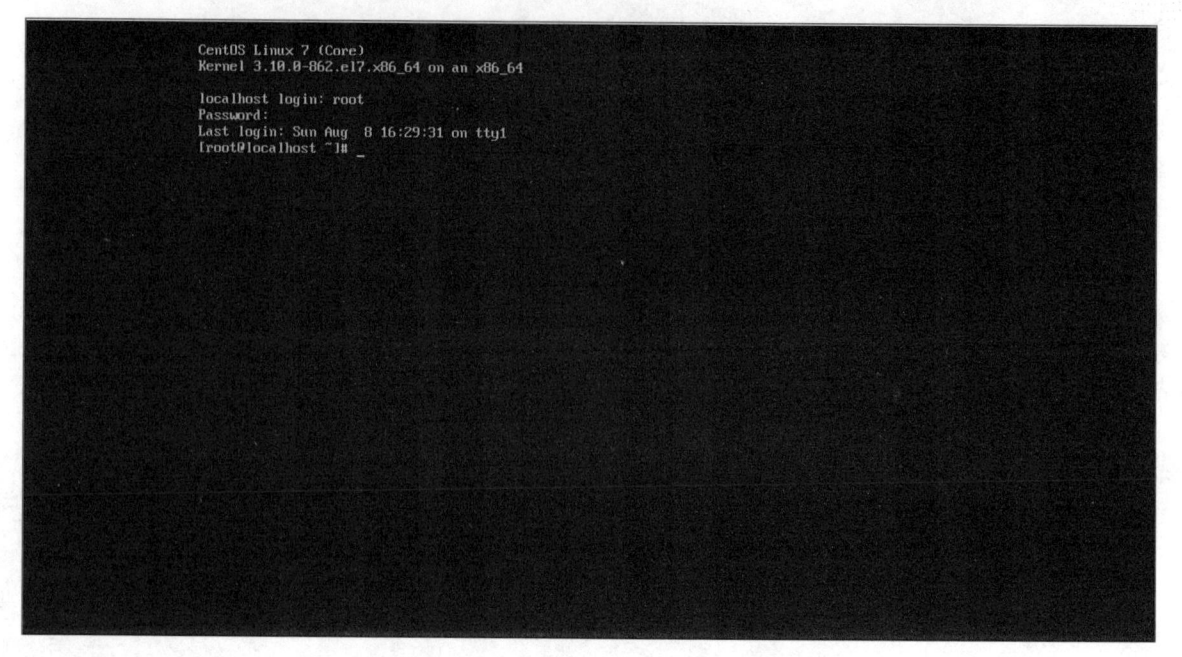

Fig. 1-47　Enter the character terminal interface

When you see the command prompt "[root@localhost ~] #"as shown in Fig. 1-47, the system is successfully logged in using root. So far, you have successfully installed a Linux server host with CentOS 7 operating system.

Unit 2

Cloud computing technology

On the basis of virtualization technology, the cloud computing technology has been developing rapidly, and more and more companies have begun to migrate their businesses to the cloud. Compared with traditional IT systems, cloud computing is to put computing resources on the cloud and provides them to users through network access. Users pay according to usage amount and do not need to care about operation or maintenance. Users can access cloud resources anytime and anywhere without worrying about insufficient capacity, hardware damage, and other relevant problems. Cloud computing is a model that provides and manages scalable, flexible and shared physical and virtual resource pools in an on-demand self-service manner, and provides network access.

 2.1   Five characteristics of cloud computing

1. Self-service

When using cloud resources, users can configure and obtain the required computing power according to their own demands. This feature can reduce users' time and operation costs, and users can have the ability to undertake the corresponding businesses according to their own needs when needed without additional human interaction.

2. Wide network access

Users can use various client devices, including mobile phones, tablet computers, notebook

computers and workstations to access resources from any places covered by the network.

3. Resource pooling

The computing resources can be physical servers or virtual computing resources, and they can be regarded as a collection, which is called as a resource pool. After resource pooling, the computing resources provided by cloud computing cannot be limited by the physical hardware and the resource pool can be integrated according to the user's needs for resource scheduling. In addition, when the underlying physical hardware on which the virtual resources depend fails, it is also transparent to users. The cloud service will automatically migrate the virtual resources to the underlying hardware available in the resource pool. Users do not need to carry out operation and maintenance and the service will not be interrupted.

4. Fast elasticity

For users'needs, cloud computing can provide fast, flexible, and scalable services. In the application scenarios of some large e-commerce websites, the advantage of cloud computing is particularly prominent. For example, when the shopping festival comes, the number of visits to Taobao and other websites to the sites will increase sharply. At this time, if based on the traditional IT system, we need to increase the number of nodes on the physical server to cope with it. After the shopping festival, these resources will be idle again, while the business deployed in the cloud can dynamically increase or release resources as the number of visits increases and decreases, so as to achieve the most efficient use of computing resources; physical or virtual resources can be supplied quickly, flexibly and sometimes automatically to achieve the purpose of rapid increase and decrease of resources. Moreover, in cloud computing, there is no limit to the computing resources available to users, so they can purchase any number of resources at any time. The rapidness and elasticity of cloud computing are very friendly to users because they do not need to plan the amount of resources and capacity in advance and they can make immediate adjustments when more or fewer resources are required.

5. On-demand billing

In cloud computing, users do not need to purchase hardware in advance, and do not need to pay additional fees without using computing resources. Instead, they only need to pay according to the use amount and time of resources.

2.2 Three service modes of cloud computing

From the bottom, the IT system is divided into base layer, network layer, storage layer, server layer, operating system layer, database layer, middleware layer, and application software layer,

on which the data information layer is located. Inclusively, the base layer, network layer, storage layer, and server layer belong to the infrastructure layer; the operating system layer, database layer, and middleware layer constitute the platform software layer, providing the operating system, library functions, databases, and other operating environments for the upper application software. According to the different IT system hierarchies provided by cloud server providers to users, the cloud computing can be divided into three service modes, namely IaaS (infrastructure as a service), PaaS (platform as a service), and SaaS (software as a service). See Fig. 2-1 for the IT system architecture.

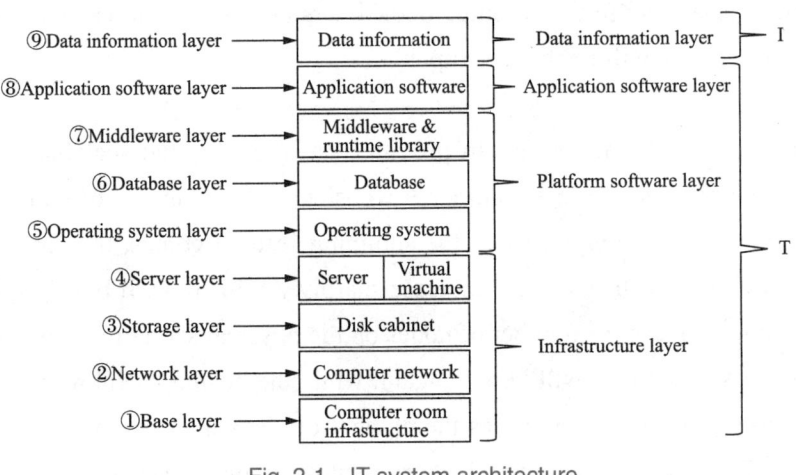

Fig. 2-1   IT system architecture

1. IaaS

The infrastructure architecture layer of the IT system is provided to users as a service. Cloud service providers carry out computer room infrastructure construction, deploy network, storage and physical servers, pool physical computing resources, and provide hardware servers or virtual servers, storage and networks (including public IP, DNS services, load balancing services, firewalls, etc.) to users. After obtaining the IaaS service, users need to install the operating system, deploy databases and middleware, and install application software on their own. Users who use IaaS are usually system technicians.

2. PaaS

After the infrastructure is built, the cloud service provider will install the operating system, database, and middleware on the server. At this time, the resources provided have had the environment to run the upper applications, so users who use PaaS are usually development technicians, and carry out development work on the basis of PaaS.

3. SaaS

All kinds of application system software are installed on the basis of the platform software layer, and users can access and use such software through the network access. For users, there is no need to do any maintenance or management but just use the software. At present, a lot of commonly used software, such as WeChat applet and Alipay, belong to SaaS products.

2.3 Four deployment models of cloud computing

1. Private cloud

Private cloud means that the users of cloud computing services are personnel from a unit organization, such as a school or an enterprise, or a public institution. The private cloud is usually deployed locally, but it can also be hosted on the public cloud and the unit organization uses VPN to access it. At present, the most widely used private cloud operating system is OpenStack, which is a source-open cloud operating system that uniformly manages the computing, storage, and network resources of the data center. Many companies have their private cloud platform products based on OpenStack.

2. Community cloud

Community cloud means that users of cloud computing services are personnel from multiple unit organizations.

3. Public cloud

Public cloud means that the users of cloud computing services are from the public. Generally, public cloud is provided to users by cloud computing companies through the Internet.

4. Hybrid cloud

Hybrid cloud means that cloud computing resources come from clouds with multiple deployment methods, for example, both private and public clouds are included.

2.4 Project practice—use cloud computing services to build a blog system

 **Project background**

A technical team is going to build its own blog system for technology sharing, but there is no suitable physical server available, so it has decided to use public cloud resources to construct the site. The products in each public cloud are different in types, features, and functions, but the basic

use method is the same and the operation and maintenance personnel of the project have selected Tencent Cloud as the public cloud resource.

 **Project implementation**

Task 1   Get an ECS

Get an ECS as the Web server of the website. Use a browser to access and log in to Tencent Cloud, as shown in Fig. 2-2.

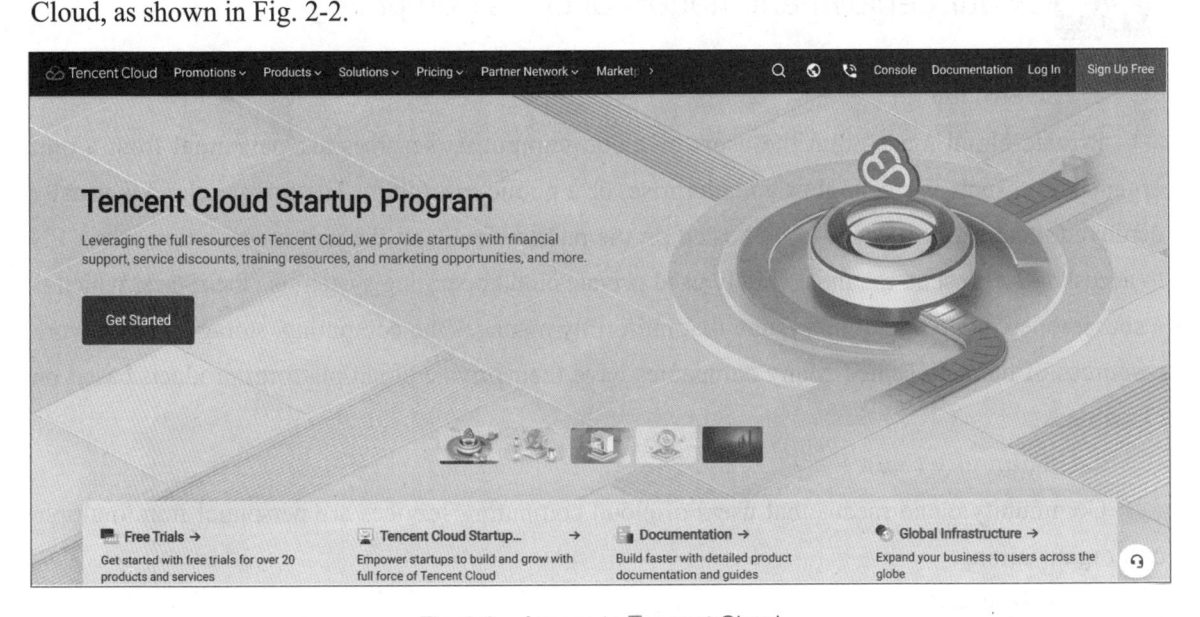

Fig. 2-2   Access to Tencent Cloud

Configure the ECS according to the business needs and confirm the purchase information.

Click the product, select "Cloud Virtual Machine" in the Compute and Container in the product list, and click the "Shop Now" button, as shown in Fig. 2-3 and Fig. 2-4.

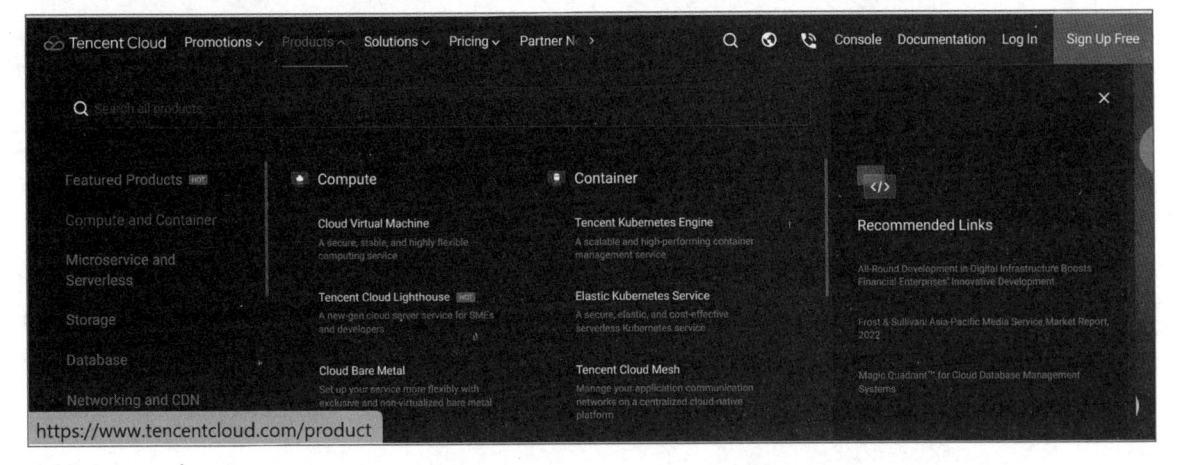

Fig. 2-3   Select "Cloud Virtual Machine" in Compute and Container

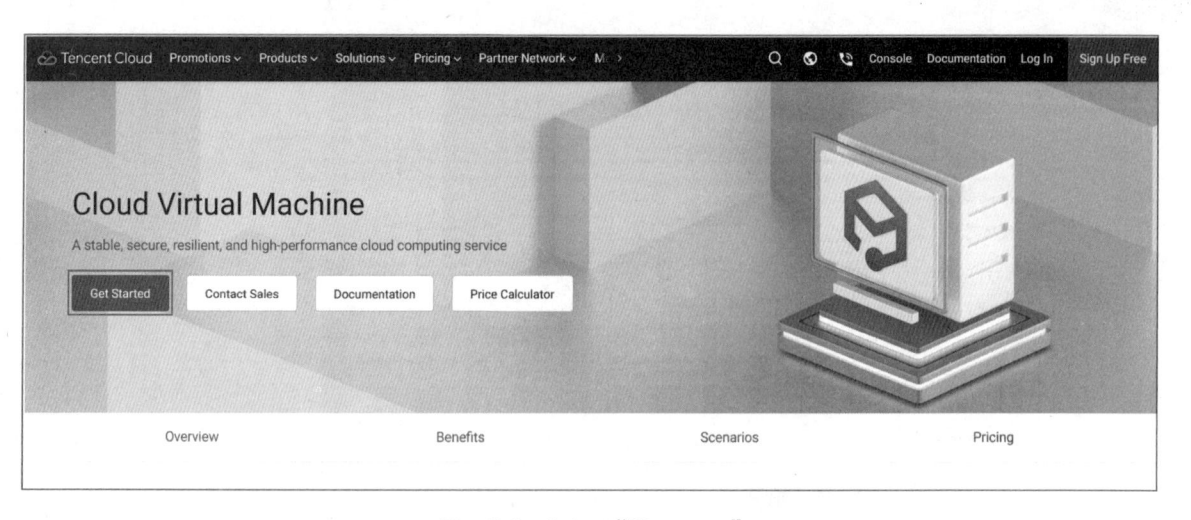

Fig. 2-4   Select "Shop now"

In the ECS configuration interface, select and purchase the corresponding services according to the business needs. This site needs to build a Web server in the Linux operating system and needs a public IP address for users' access. In this task, the demonstration steps are designed based on the experimental environment and the principle that the experiment can be completed at the lowest cost. Therefore, the computing resources and network are charged by usage, and the instance configuration is relatively low. In actual use, you can select yearly, monthly, and higher configurations according to specific needs, as shown in Fig. 2-5.

Set the host security groups, create a new security group, release the common port, and set the name and login password for the host, as shown in Fig. 2-6.

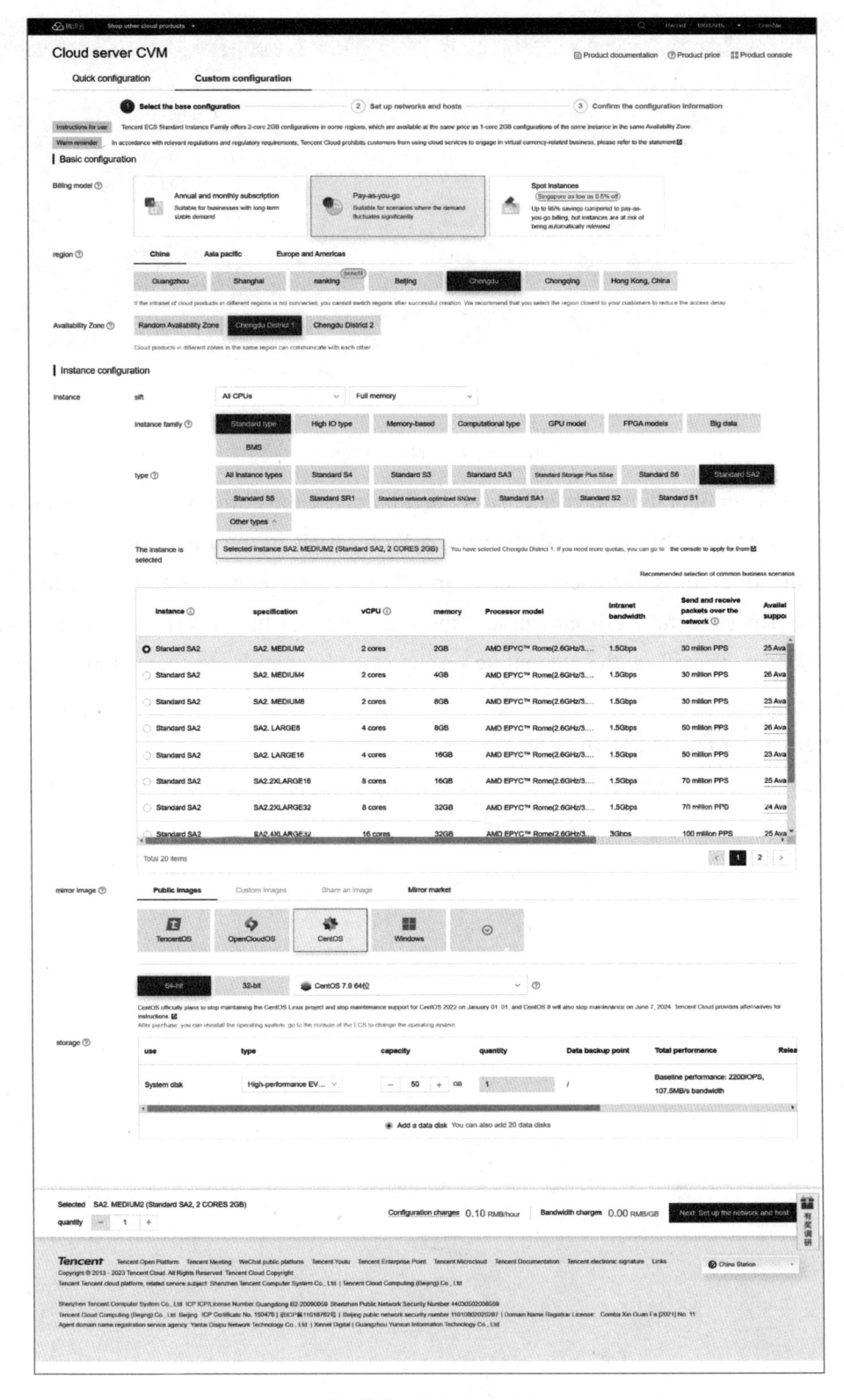

Fig. 2-5　Select model

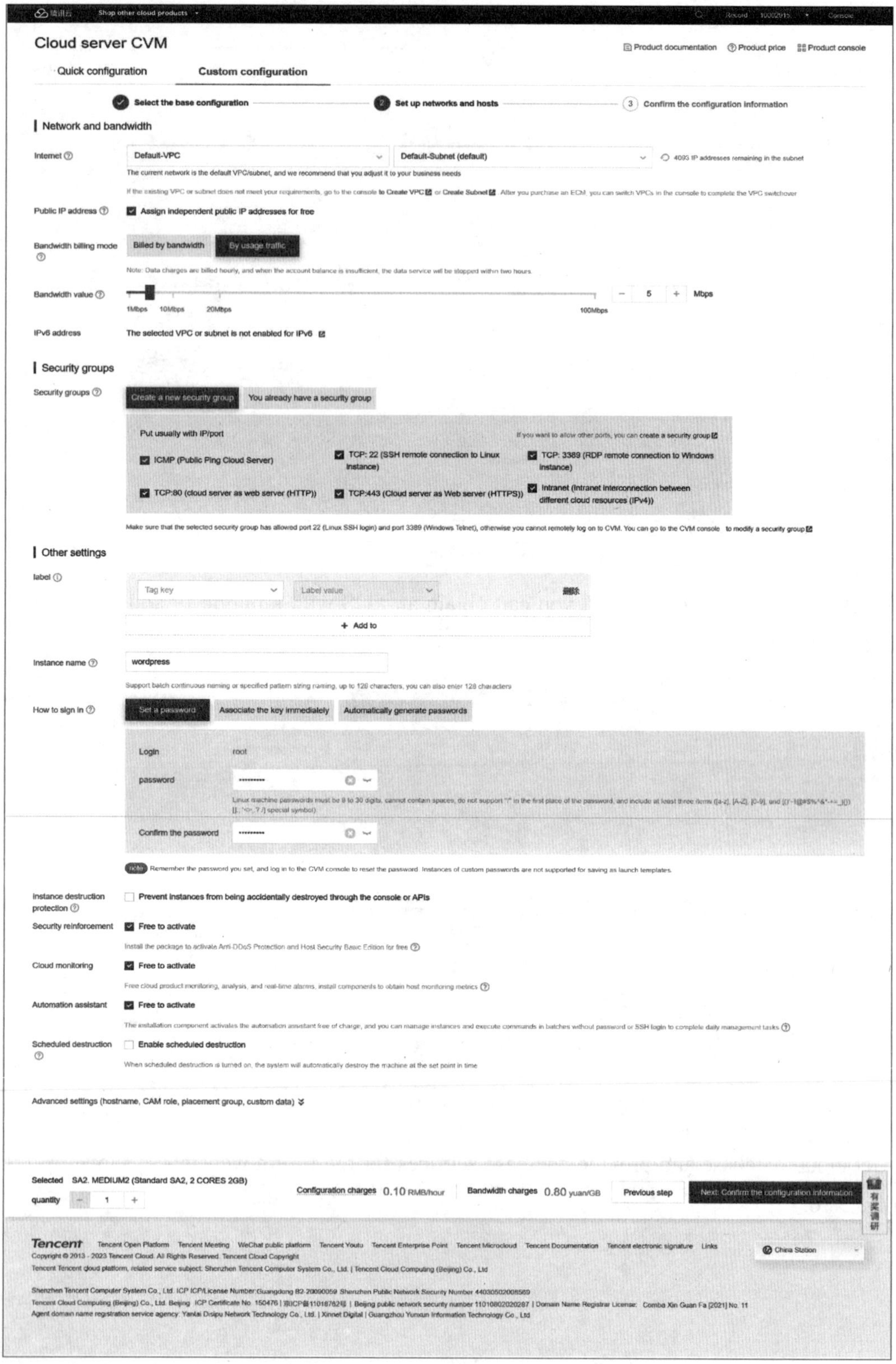

Fig. 2-6　Host settings

Confirm the configuration information and complete the purchase, as shown in Fig. 2-7.

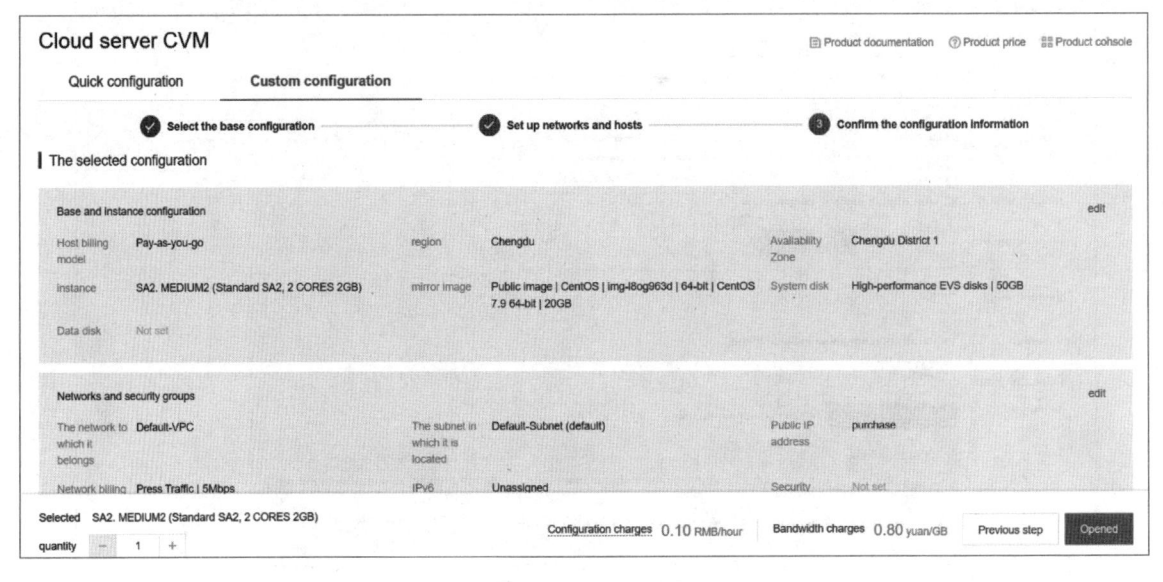

Fig. 2-7   Confirm configuration information

Click the "Opened" button and you can obtain an ECS with a public IP address, as shown in Fig. 2-8.

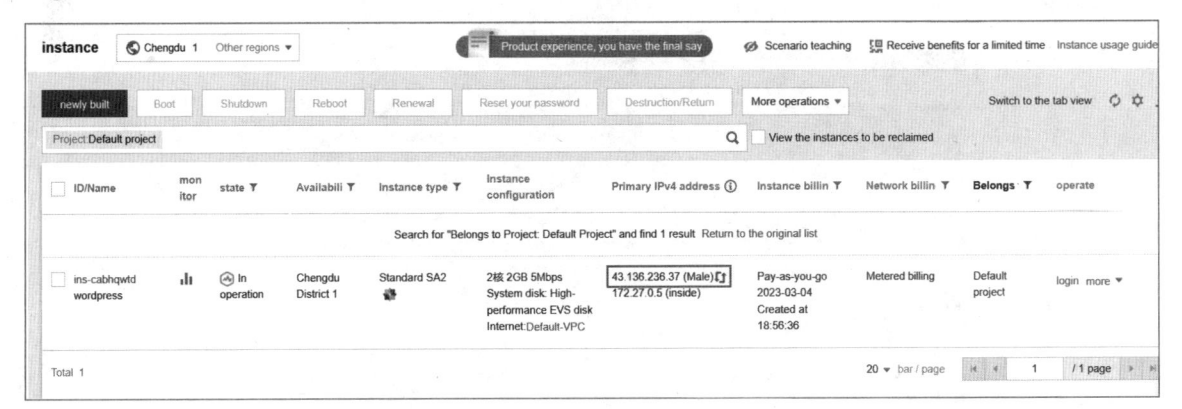

Fig. 2-8   Completion of ECS construction

Task 2   Get cloud database products

Get cloud database and provide database services for the forum website, as shown in Fig. 2-9 and Fig. 2-10.

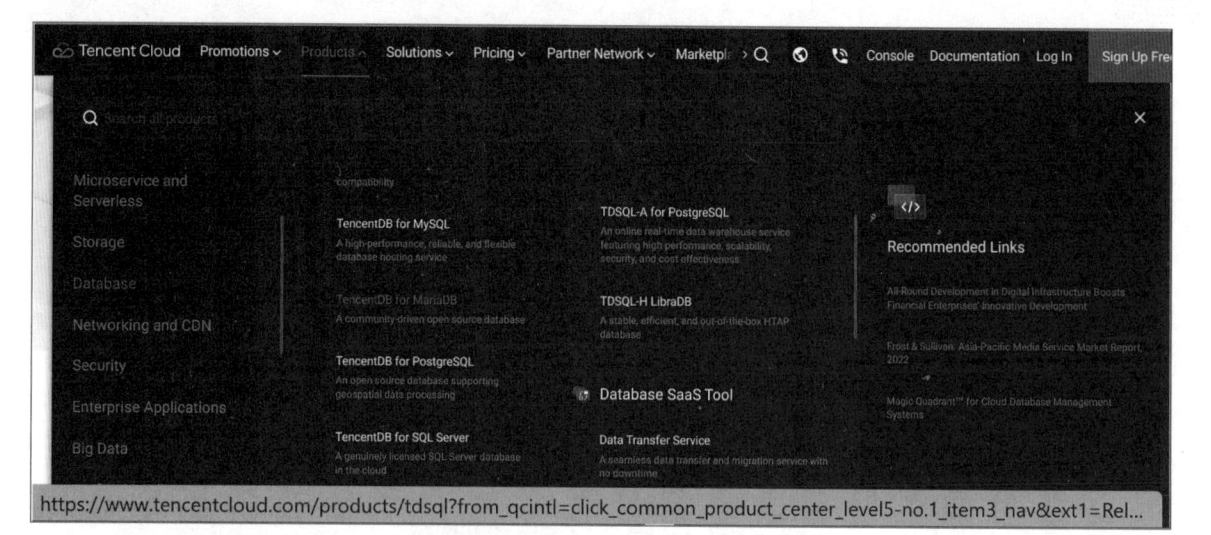

Fig. 2-9　Select database services

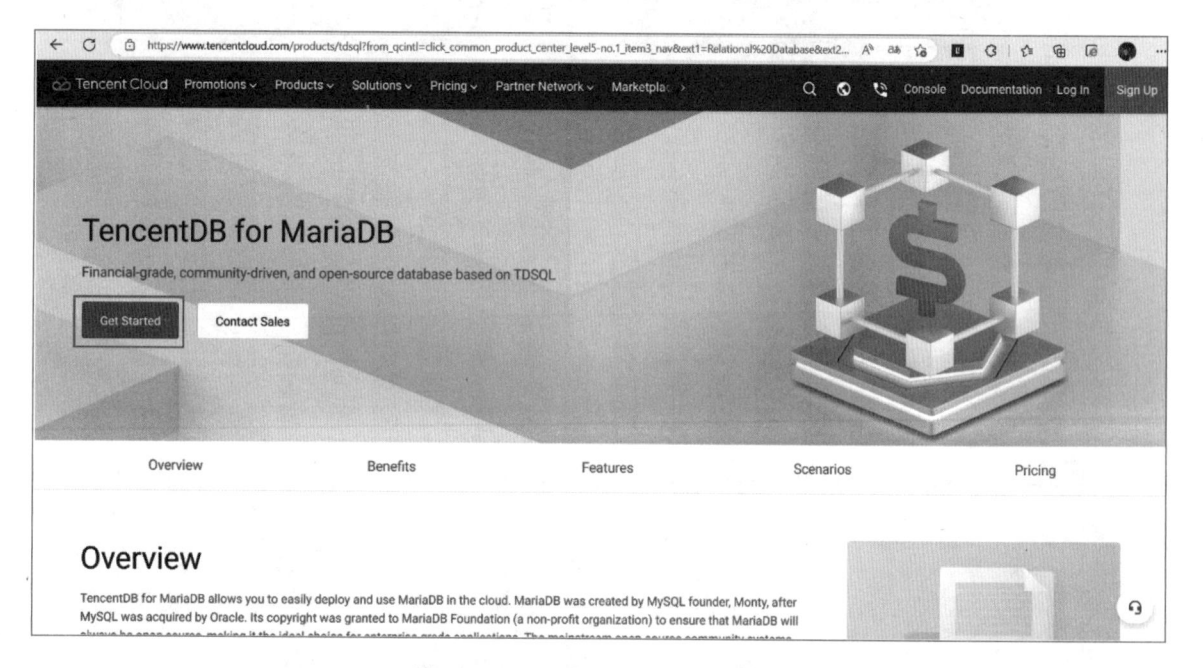

Fig. 2-10　Purchase cloud database

In the subsequent process of site building, the ECS needs to connect to the database, so when purchasing the database, select the same subnet in the same zone and the same private network as the ECS, as shown in Fig. 2-11 to Fig. 2-13.

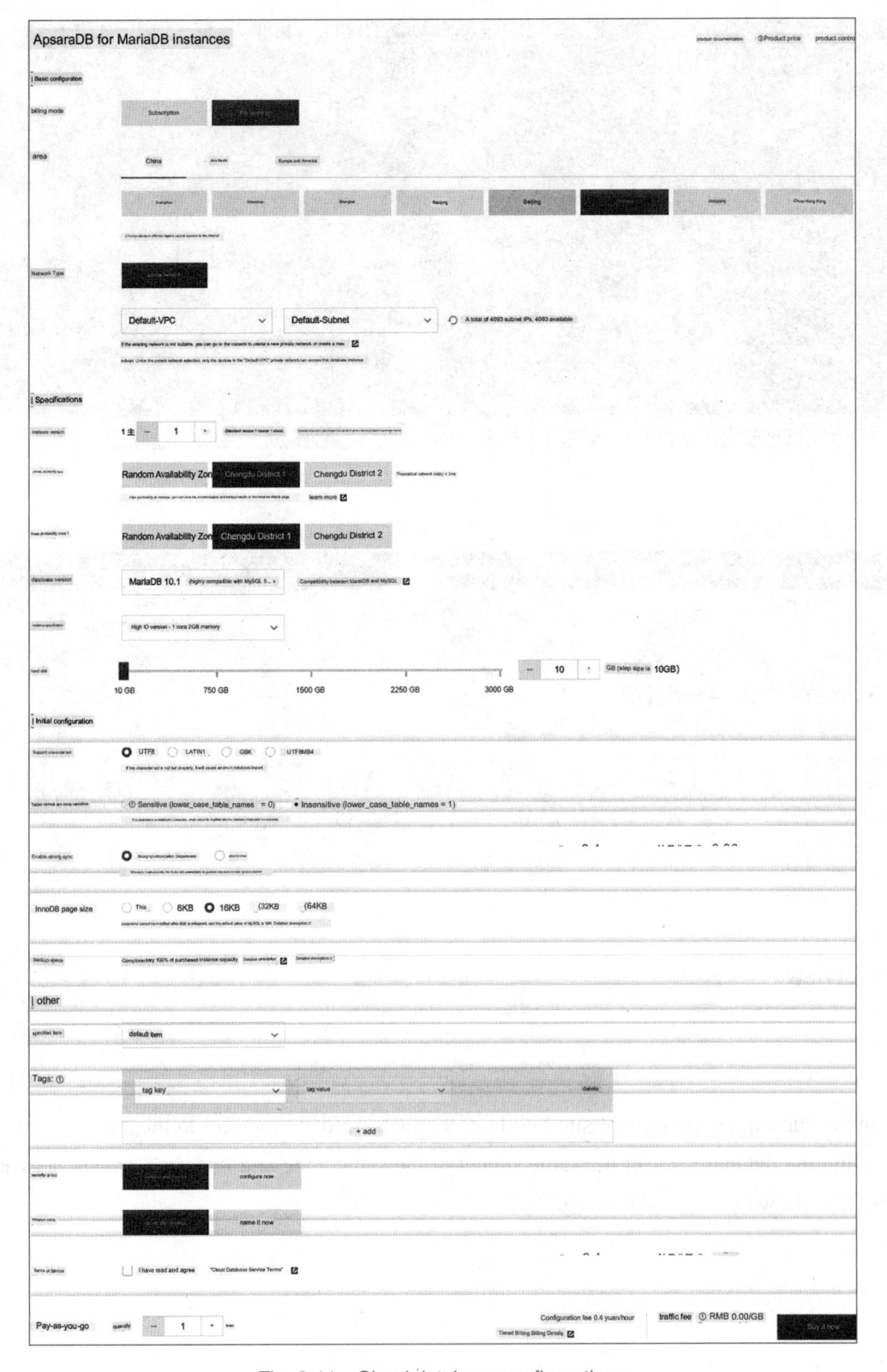

Fig. 2-11   Cloud database configurations

Fig. 2-12 Purchase confirmation instance

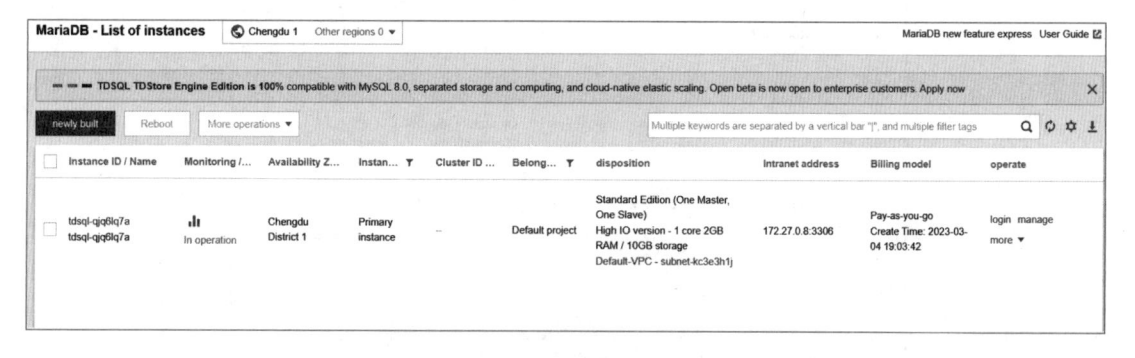

Fig. 2-13 View instance list

Create an account to connect to the database remotely, as shown in Fig. 2-14 and Fig. 2-15.

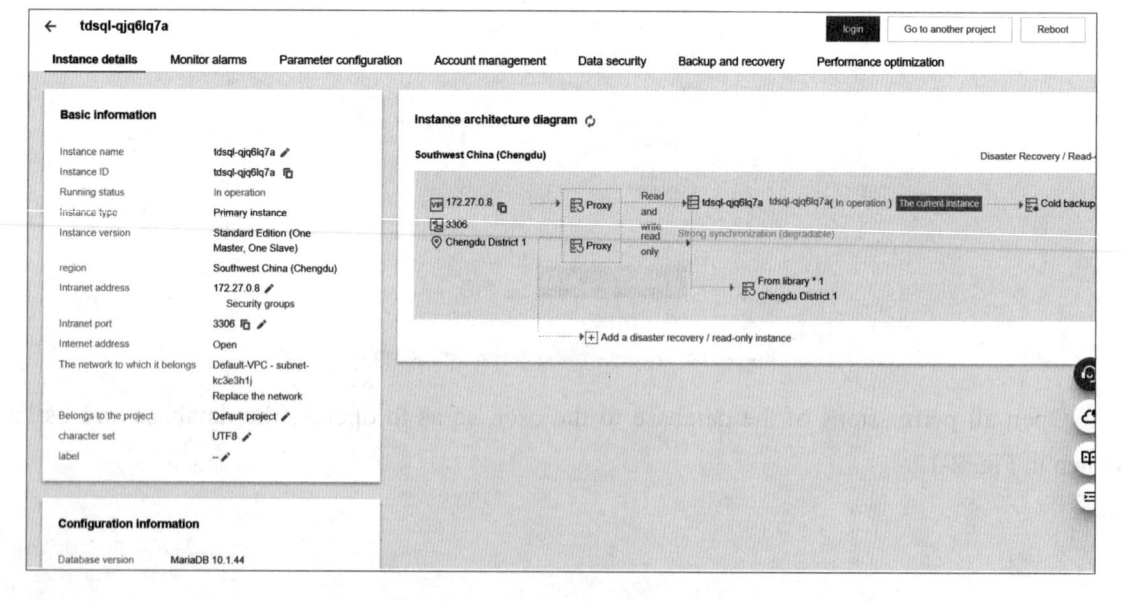

Fig. 2-14 Account management

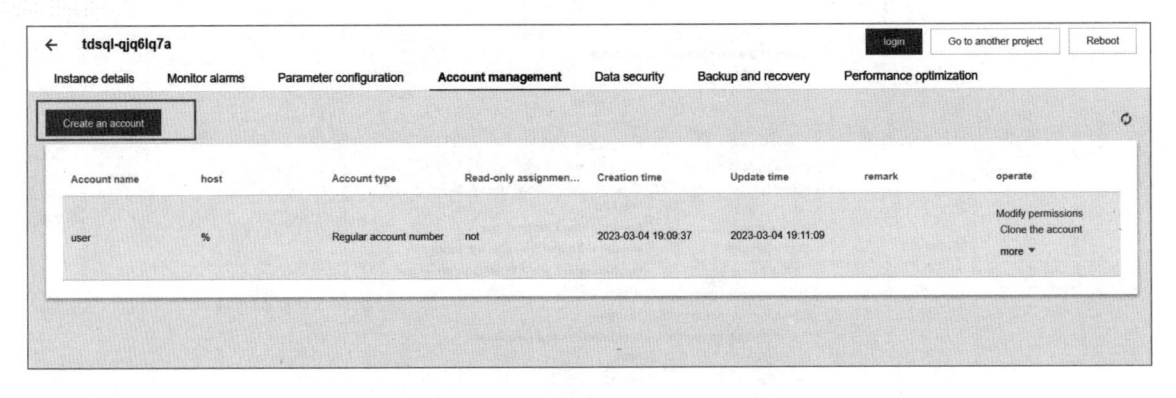

Fig. 2-15   New account

The host is configured as "%", which means that all IP addresses can use the user to access the database, and the access password is User123456 #, as shown in Fig. 2-16.

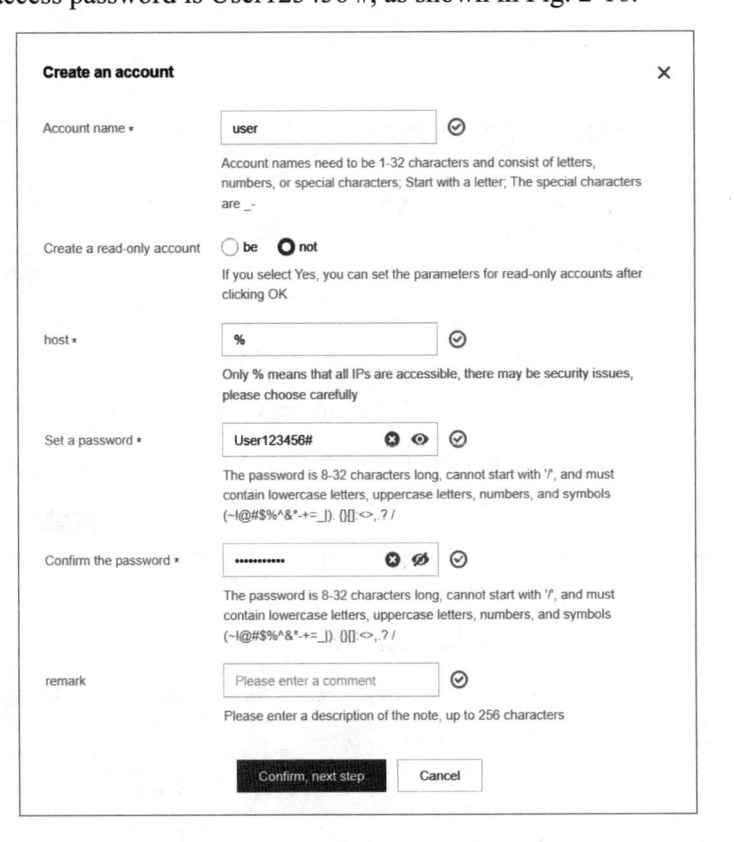

Fig. 2-16   Configure account information

Open all permissions of the database to the user, so as to operate the database and list, as shown in Fig. 2-17.

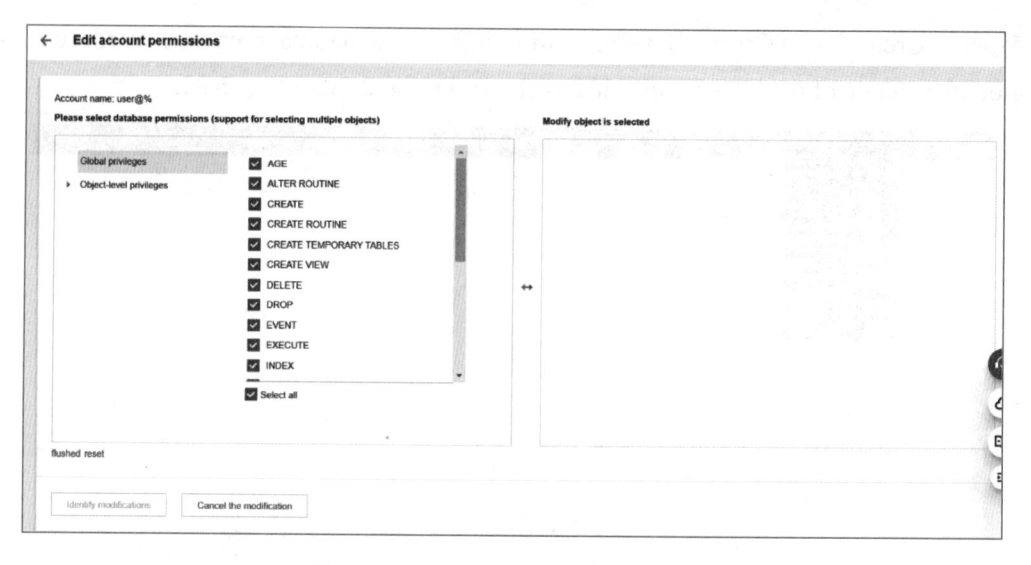

Fig. 2-17　Modify account permissions

After the account is created, use the user and corresponding password to log in to the database, as shown in Fig. 2-18 and Fig. 2-19.

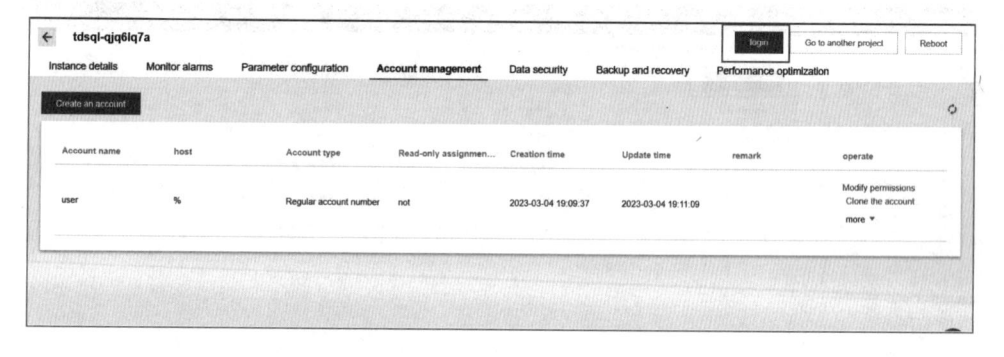

Fig. 2-18　Log in with account

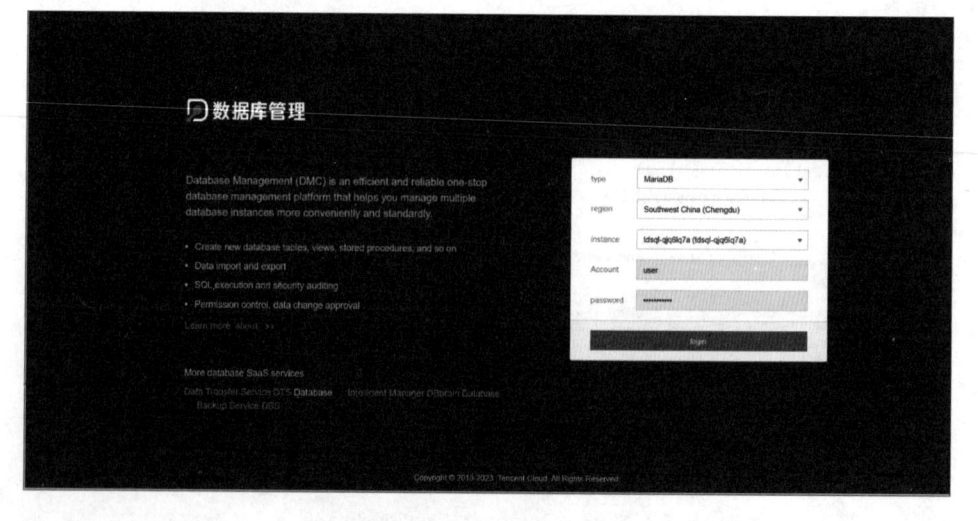

Fig. 2-19　Enter account information

Select "Create" → "Create Database" to create a new database named wordpress as the background database of the blog system site, as shown in Fig. 2-20 to Fig. 2-22.

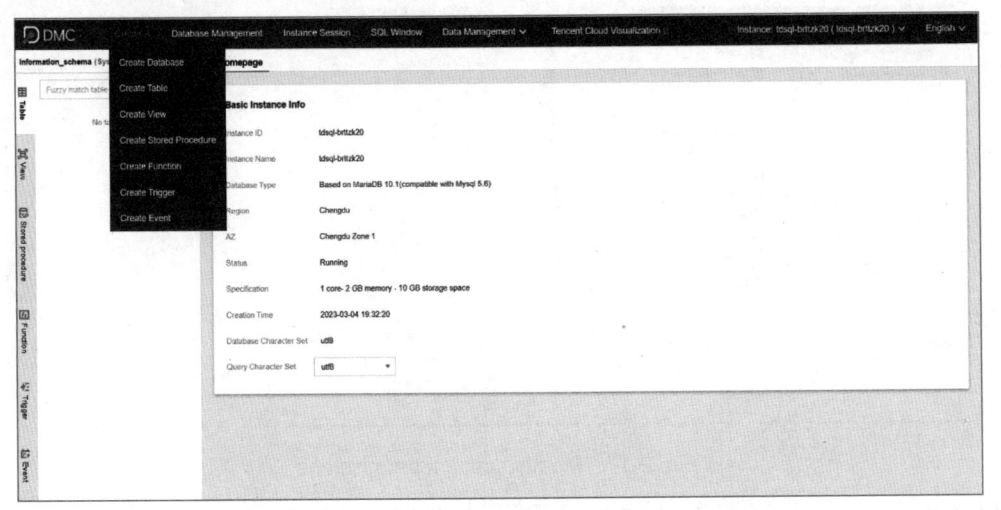

Fig. 2-20　Select "Create Database"

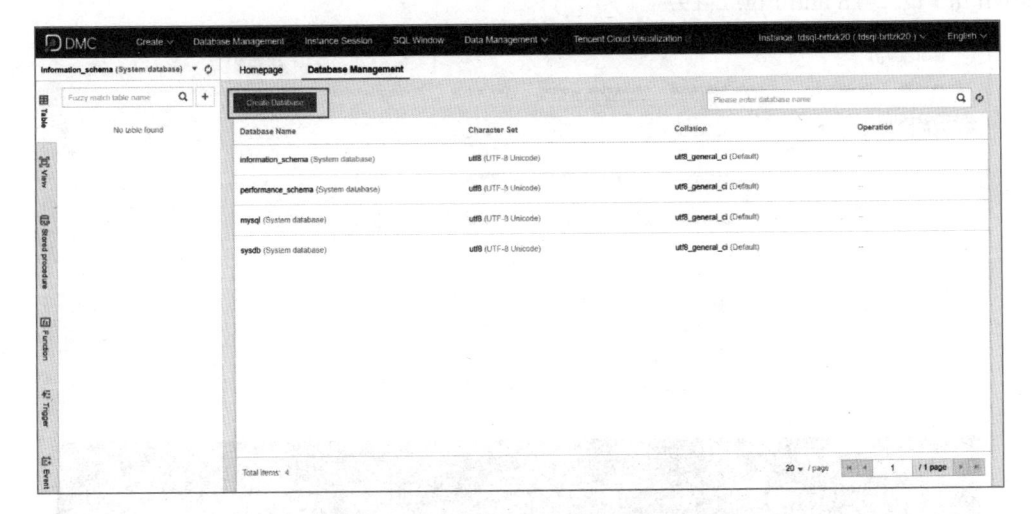

Fig. 2-21　Create a new database

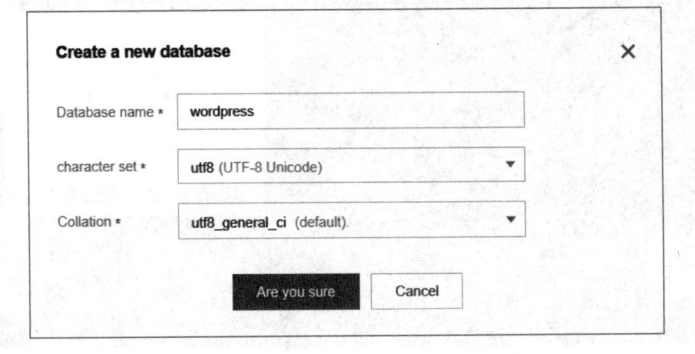

Fig. 2-22　Database Configurations

Task 3　Build a wordpress forum system

Log in to the ECS to build the Web server environment, as shown in Fig. 2-23.

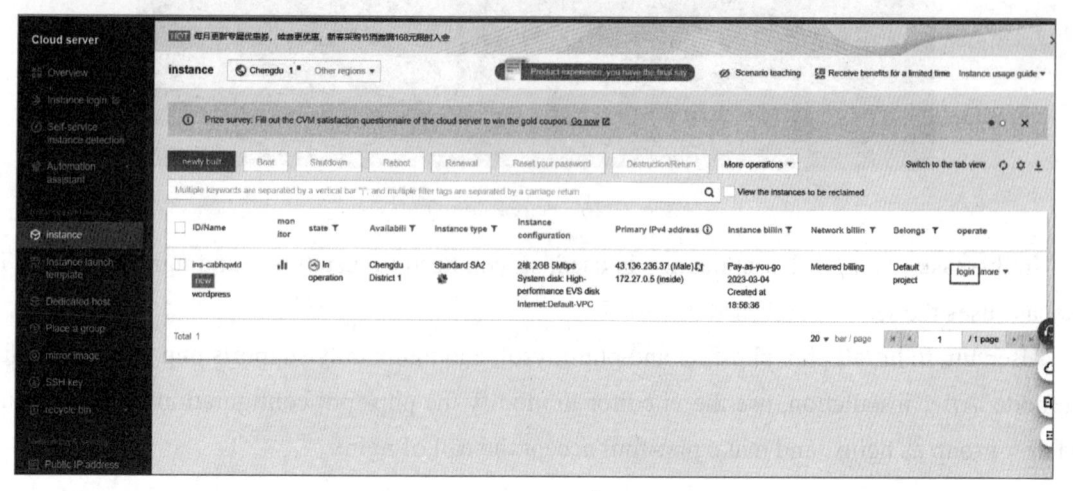

Fig. 2-23　Log in to ECS

Use the set user name and password to log in to the cloud host remotely via the ssh protocol, as shown in Fig. 2-24 and Fig. 2-25.

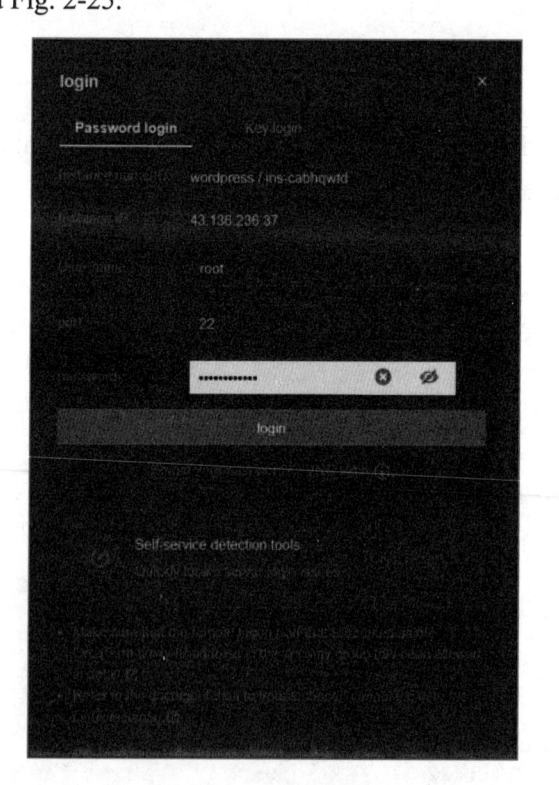

Fig. 2-24　Remote log in to cloud host via ssh protocol

Fig. 2-25   Server terminal

In this task, the LNMP architecture is used, so you need to install php and nginx services. The database uses the server purchased in Task 2.

Use yum to install php, php-fpm and php-mysql, and nginx only supports php work in the fast-cgi mode. After installation, use the vi editor to modify the php-fpm configuration file, set the user and user group as nginx, and make php-fpm accept the call of nginx.

```
[root@VM-0-6-centos ~]# yum -y install php php-fpm
[root@VM-0-6-centos ~]# vi /etc/php-fpm.d/www.conf
...
user = nginx
group = nginx
...
```

Start the php-fpm service working on the port 9000. Use the netstat command to view the port that the current system is working on and you can see that port 9000 is listening, which means that the php-fpm service is working normally.

```
[root@VM-0-6-centos ~]# systemctl start php-fpm          //start php service
[root@VM-0-6-centos ~]# systemctl enable php-fpm

[root@VM-0-6-centos ~]# netstat -nptl
Active Internet connections (only servers)
Proto Recv-Q Send-Q Local Address     Foreign Address   State    PID/Program name
tcp      0       0 127.0.0.1:25       0.0.0.0:*         LISTEN   1032/master
tcp      0       0 127.0.0.1:9000     0.0.0.0:*         LISTEN   15455/php-fpm: mast
tcp      0       0 0.0.0.0:3306       0.0.0.0:*         LISTEN   15300/mysqld
tcp      0       0 0.0.0.0:80         0.0.0.0:*         LISTEN   2248/nginx: master
tcp      0       0 0.0.0.0:22         0.0.0.0:*         LISTEN   867/sshd
tcp6     0       0 ::1:25             :::*             LISTEN   1032/master
tcp6     0       0 :::22              :::*             LISTEN   867/sshd
```

Install nginx. Configure the yum repository for nginx and use yum to install nginx services.

Log in to the nginx official website, select the RHEL/CentOS system, and write the yum configuration file /etc/yum.repos.d/nginx.repo, as shown in Fig. 2-26 to Fig. 2-28.

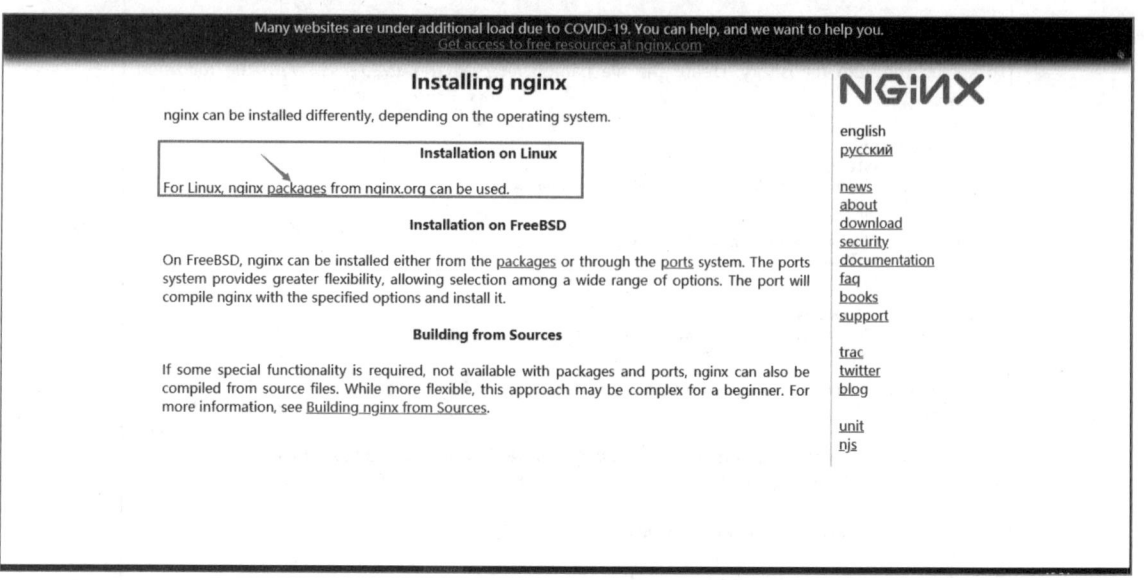

Fig. 2-26　Select software packages to install nginx

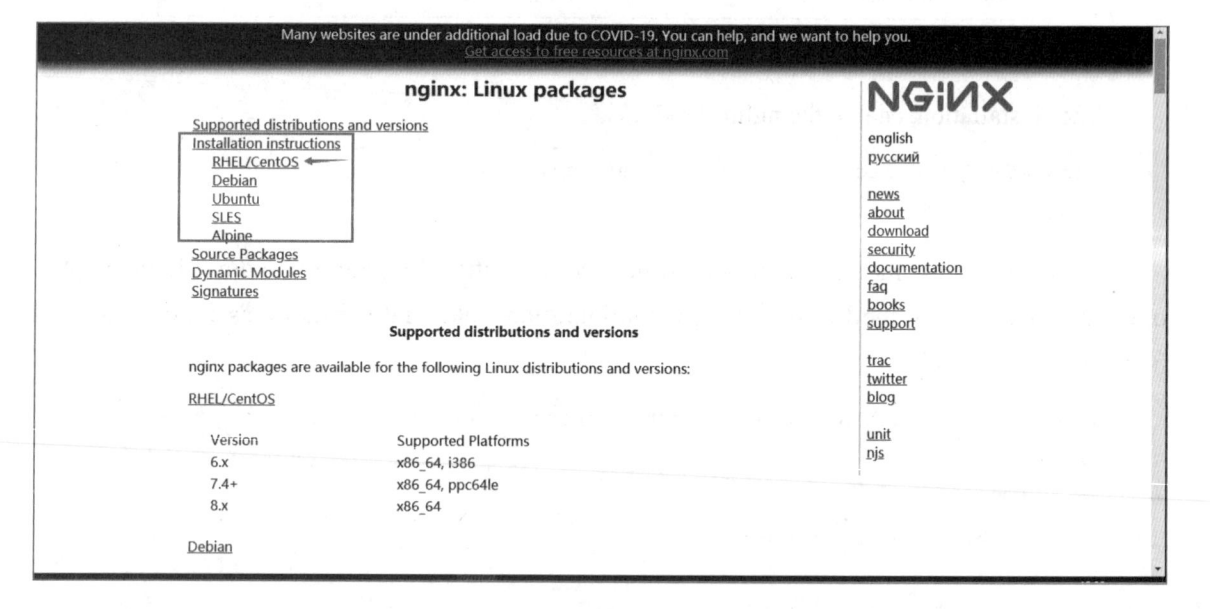

Fig. 2-27　Select the corresponding operating system

```
                                    RHEL/CentOS
Install the prerequisites:

     sudo yum install yum-utils

To set up the yum repository, create the file named /etc/yum.repos.d/nginx.repo with the following
contents:

     [nginx-stable]
     name=nginx stable repo
     baseurl=http://nginx.org/packages/centos/$releasever/$basearch/
     gpgcheck=1
     enabled=1
     gpgkey=https://nginx.org/keys/nginx_signing.key
     module_hotfixes=true

     [nginx-mainline]
     name=nginx mainline repo
     baseurl=http://nginx.org/packages/mainline/centos/$releasever/$basearch/
     gpgcheck=1
     enabled=0
     gpgkey=https://nginx.org/keys/nginx_signing.key
     module_hotfixes=true

By default, the repository for stable nginx packages is used. If you would like to use mainline nginx
packages, run the following command:

     sudo yum-config-manager --enable nginx-mainline

To install nginx, run the following command:

     sudo yum install nginx
```

Fig. 2-28   Configure yum source and install nginx

After installation, enable the nginx service.

```
[root@VM-0-6-centos ~]# systemctl start nginx
[root@ VM-0-6-centos ~]# systemctl enable nginx
```

Configure nginx, modify nginx virtual host configuration file, add the index file index.php, open the comments in the location~\.php$ configuration field, and configure as follows. Reload nginx after modification.

```
[root@VM-0-6-centos ~]# vi /etc/nginx/conf.d/default.conf
server {
    listen       80;
    server_name  localhost;

    #charset koi8-r;
    #access_log  /var/log/nginx/host.access.log  main;

    location / {
        root   /usr/share/nginx/html;
        index  index.php index.html index.htm;
    }
```

```
    #error_page   404                    /404.html;

    # redirect server error pages to the static page /50x.html
    #
    error_page    500 502 503 504  /50x.html;
    location = /50x.html {
        root    /usr/share/nginx/html;
    }

    # proxy the PHP scripts to Apache listening on 127.0.0.1:80
    #
    #location ~ \.php$ {
    #    proxy_pass   http://127.0.0.1;
    #}

    # pass the PHP scripts to FastCGI server listening on 127.0.0.1:9000
    #
    location ~ \.php$ {
        root            /usr/share/nginx/html;
        fastcgi_pass    127.0.0.1:9000;
        fastcgi_index   index.php;
        fastcgi_param   SCRIPT_FILENAME   $document_root$fastcgi_script_name;
        include         fastcgi_params;
    }

    # deny access to .htaccess files, if Apache's document root
    # concurs with nginx's one
    #
    #location ~ /\.ht {
    #    deny   all;
    #}
}


[root@VM-0-6-centos ~]# nginx -s reload
```

Deploy wordpress. Download the wordpress site code and extract the file.

```
    [root@VM-0-6-centos ~]# wget https://cn.wordpress.org/wordpress-5.0.3-zh_
CN.tar.gz--2022-04-25 23:20:22-- https://cn.wordpress.org/wordpress-5.0.3-zh_CN.tar.
gz
```

```
Resolving cn.wordpress.org (cn.wordpress.org)... 198.143.164.252
Connecting to cn.wordpress.org (cn.wordpress.org)|[198.143.164.252|:443... con-
nected.HTTP request sent,awaiting response... 200 OK
Length: 11098483 (11M) [application/octet-stream]
Saving to: "wordpress-5.0.3-zh_CM.tar.gz"
[root@VM-0-6-centos ~]# tar -zxvf latest-zh_CN.tar.gz
[root@VM-0-6-centos ~]# ll
total 19012
-rw-r--r-- 1 root root 19463922 Apr6 21:00 latest-zh_CN.tar.gz
drwxr-xr-x 5 1006 10064096 Apr6 21:00 wordpress
```

Delete the default file in the nginx working directory, copy the site code to the working directory, and set permissions.

```
[root@VM-0-6-centos ~]# rm -rf /usr/share/nginx/html/*
[root@VM-0-6-centos ~]# cp -r wordpress/* /usr/share/nginx/html/
[root@VM-0-6-centos ~]# chmod -R 777 /usr/share/nginx/html
```

Use the browser to access the public IP address of the cloud host. If there is a domain name available, you can also bind it to the cloud server and access it through the domain name to initialize the site, as shown in Fig. 2-29.

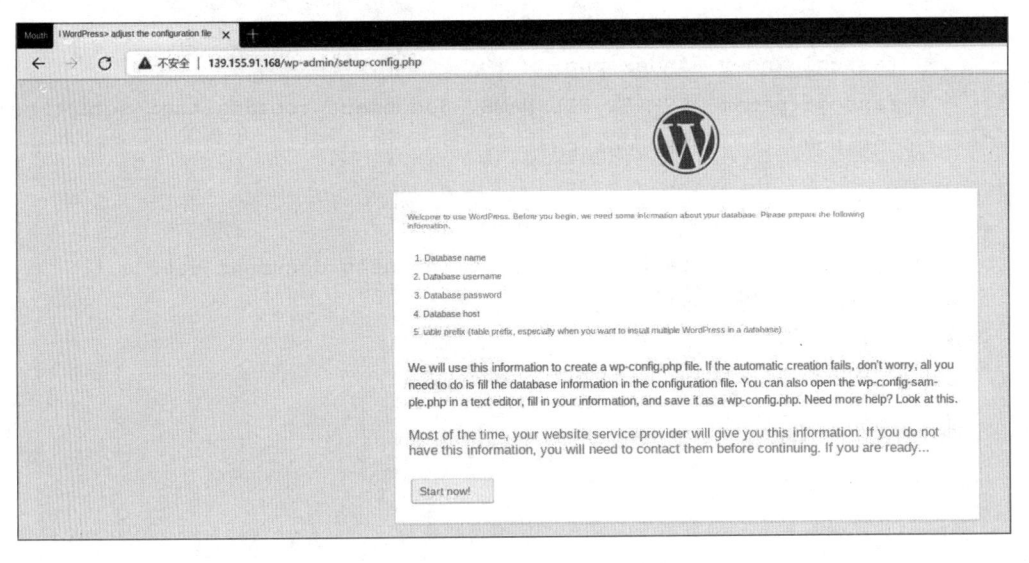

Fig. 2-29　Wordpress installer interface

Configure the database connection information and the database name is wordpress created in Task 2 of this unit; the user name and password are the remote access accounts created for the database in Task 2 of this unit; the database host is the intranet IP address of the database server purchased in the same subnet as the ECS in Task 2 of this unit, as shown in Fig. 2-30 and Fig. 2-31.

Fig. 2-30   Configure environment parameters

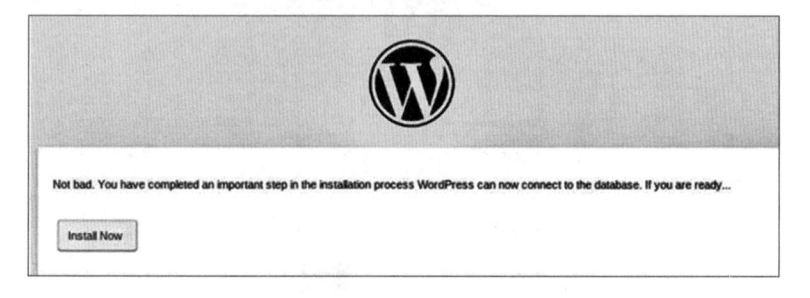

Fig. 2-31   Start to install

Set the site administrator information, as shown in Fig. 2-32 and Fig. 2-33.

After the installation, you can log in as an administrator and access the blog system site, as shown in Fig. 2-34 and Fig. 2-35.

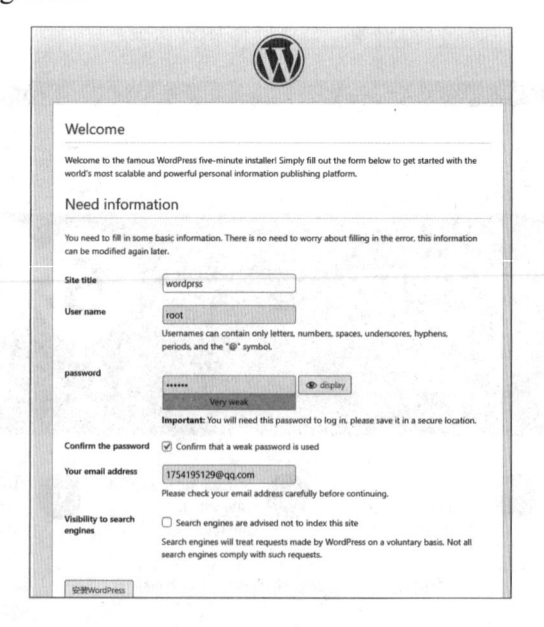

Fig. 2-32   Configure site administrator information

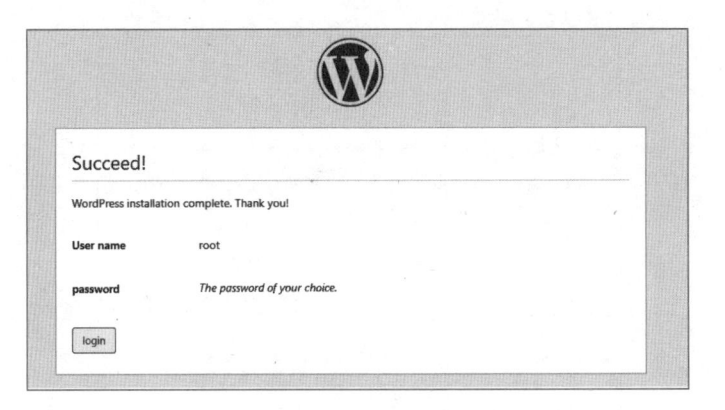

Fig. 2-33   Installation completed

Fig. 2-34   Log in to wordpress

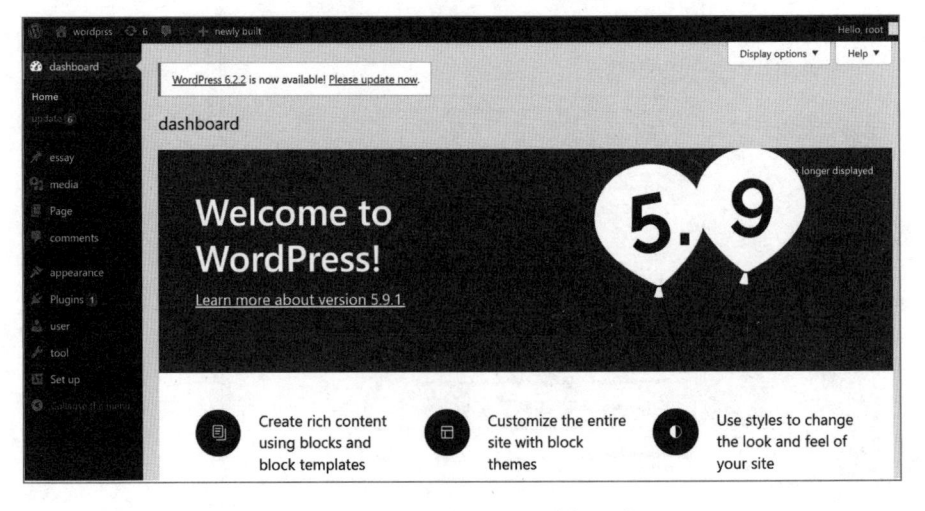

Fig. 2-35   wordpress dashboard

Unit 3

Private cloud technology —OpenStack

3.1 OpenStack cloud operating system

OpenStack is a source-open cloud operating system, which was first established by Rackspace and NASA (National Aeronautics and Space Administration). OpenStack can uniformly manage the computing resources, storage resources and network resources in large resource pools (including Bare Metal, Virtual Machines and Container). All these resources are managed and configured through the API of the unified authentication mechanism (OpenStack SDK). In addition, OpenStack also provides a dashboard to enable administrators to manage the cloud platform through the Web interface (Horizon Web UI), and at the same time gives users the right to view the corresponding resources. In addition to the standard infrastructure as a service function, third-party service components can also be deployed, such as providing Terraform, fault management and service management (such as Kubernetes and CloudFoundry for managing dockers), and other services to ensure the high availability of user applications. Fig.3-1 shows the OpenStack schematic diagram.

See Fig. 3-2 for the interface of the official website of OpenStack open source unit.

OpenStack is updated every half a year on average. Its version number starts from A, followed by B, C, D... Currently, version Y is used while version Z is under development, as shown in Fig. 3-3.

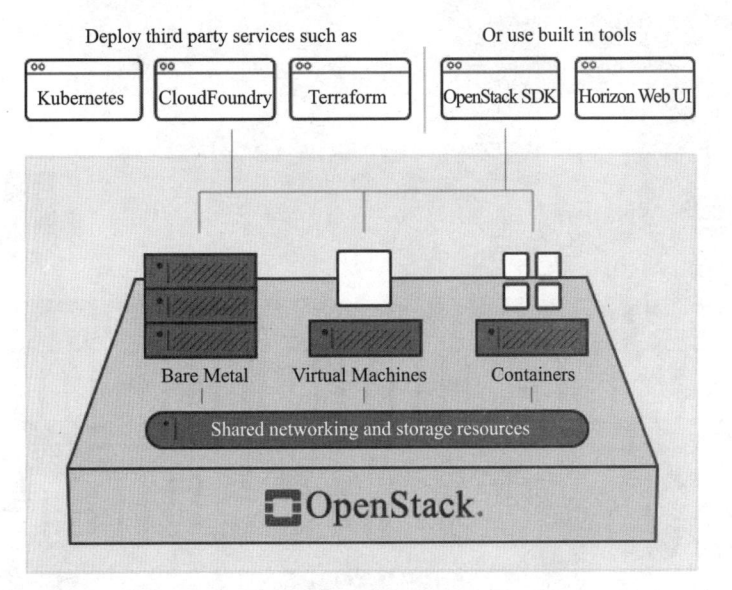

Fig. 3-1 Schematic diagram of OpenStack

Fig.3-2 OpenStack official website interface

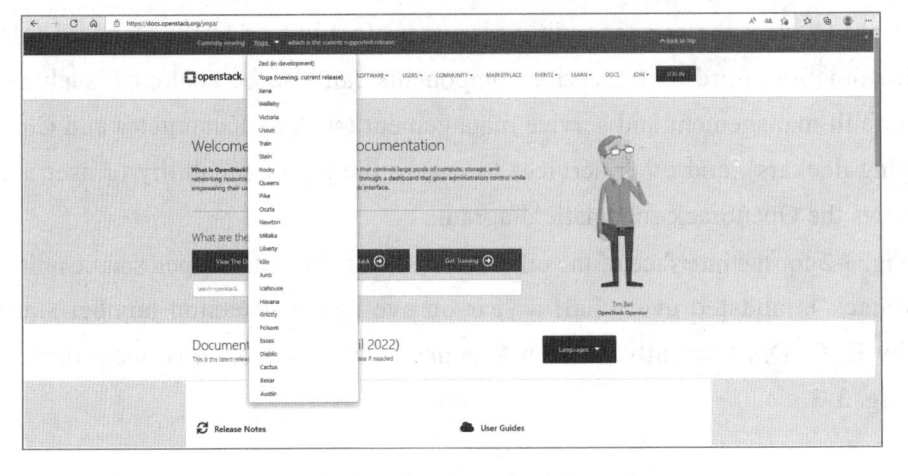

Fig. 3-3 OpenStack version information

3.2　OpenStack architecture

OpenStack's flexible scheduling of resources depends on various components in the entire unit, or services. These components can be plug-and-play based on application scenarios and needs. Fig. 3-4 shows the components and functions of OpenStack.

OpenStack's architecture ensures its high flexibility and its distributed deployment architecture to see where these services are applicable and how they work together.

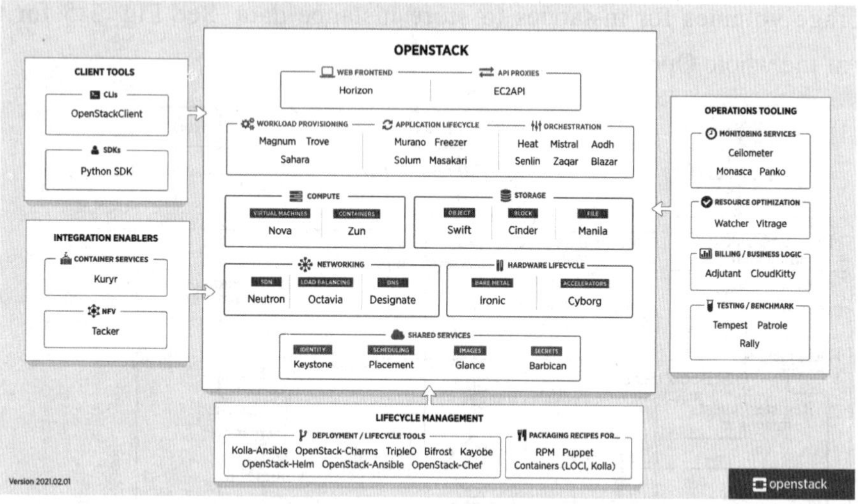

Fig. 3-4　OpenStack module schematic

Among many modules, the following six core modules are required:

① Nova, computing service.It provides large-scale scalable and on-demand self-service access to computing resources (including Bare Metal, Virtual Machines and Container).

② Glance, image service.The image service is responsible for discovering, registering and searching the images of virtual machines.

③ Cinder, block storage service.It provides virtualization management of block storage devices and self-service APIs, through which users can request and use storage resources.

④ Swift, object storage service.It provides object storage services with built-in redundancy and a high error tolerance mechanism, which can be used as backup storage for Cinder and provide storage services for Glance.

⑤ Keystone, identity authentication service.It is responsible for client identity authentication, service discovery and token management, etc.

⑥ Neutron, network service.It provides network virtualization services, provides users with API interfaces, customizes networks, subnets, routes, etc., configures DHCP, DNS, load balancing and other services, and supports VLAN, FLAT, GRE, VXLAN and other network modes.

In addition to the required core modules, there is also a common module, Horizon dashboard, which provides a dashboard management interface for the platform. OpenStack can be operated and maintained in the browser visualization environment. See Fig. 3-4 for the schematic diagram of OpenStack modules.

The most important work of OpenStack is to manage the life cycle of the virtual machine, with the virtual machine as the center, and other modules provide corresponding services for the virtual machine. Glance provides images for the virtual machine; Nova manages the life cycle of instances; Neutron provides network connection services for instances; Cinder provides storage volumes for instances to store instance data. See Fig. 3-5 for the module architecture of the whole OpenStack.

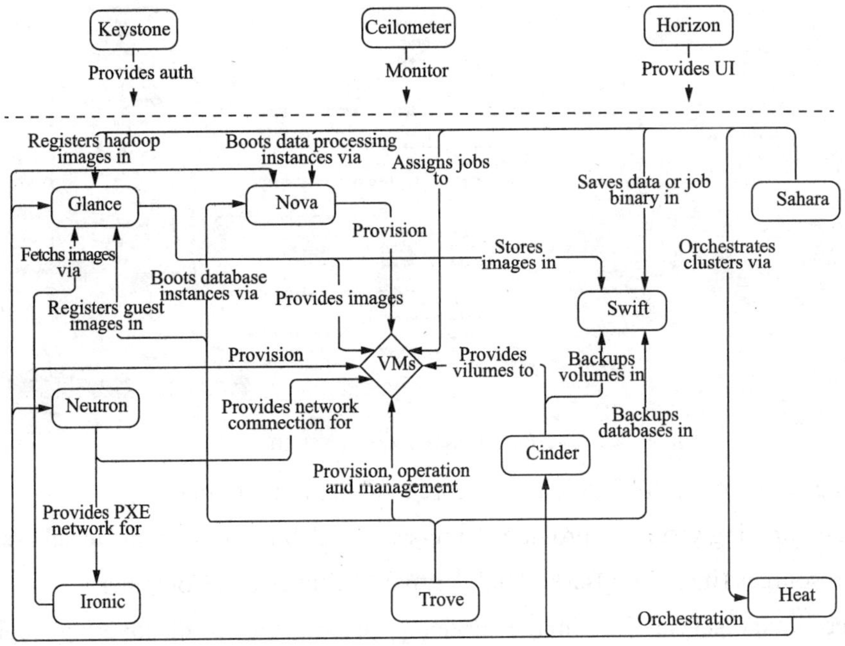

Fig. 3-5   Module architecture of the whole OpenStack

Each module provides services which are working together. Fig. 3-6 shows the logical architecture between services.

Here, the Nova module is taken as an example to illustrate how the services within the module work together. While Nova is managing the instance lifecycle, the following services are mainly involved.

Nova-api, responsible for receiving and responding to customer API calls. In addition to its own API, it is also compatible with the Amazon EC2 API. In other words, you can call the Amazon EC2 instance API in OpenStack.

Nova-scheduler, responsible for implementing the scheduling service and deciding which computing node to schedule the virtual machine to run on.

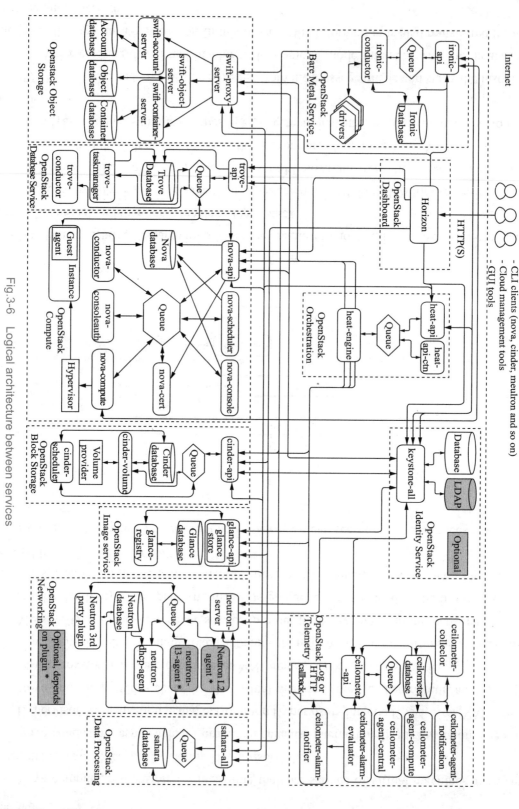

Fig.3-6 Logical architecture between services

Nova-compute, responsible for managing the core service of virtual machines and managing the virtual machine lifecycle by calling Hypervisor API.

Nova-conductor, responsible for providing database interaction for the nova-compute service.

Hypervisor, virtualization manager.Common Hypervisors include KVM, Xen, VMware, etc.

Nova-console provides access to the virtual machine console. The default access method is nova-novncproxy, which is a browser-based VNC remote desktop access service.

Nova-consoleauth, providing Token authentication for access to the virtual machine console.

Message Queue, message queue service.In order to reduce the coupling between services and improve the flexibility of the platform, the message queue is introduced as the information transfer station responsible for queue management of message data generated by services. OpenStack uses the RabbitMQ service by default.

Here, from the process of creating an instance, we can see how Nova services work together, as shown in Fig. 3-7.

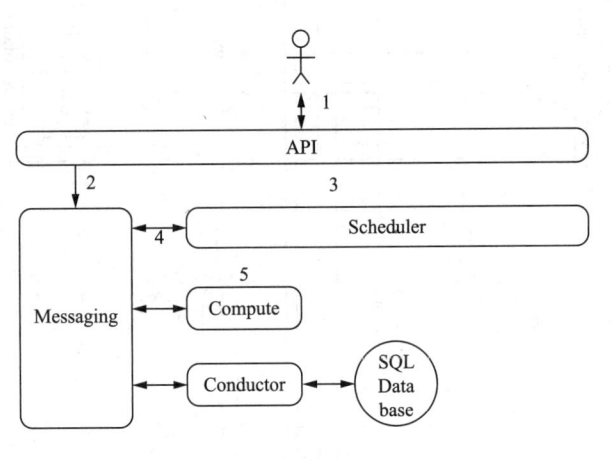

Fig. 3-7   Relationship of Nova services

① The user sends a "create virtual machine" request to nova-api. The user here can be the end user of OpenStack or other components of OpenStack.

② Nova-api submits the message to the message queue (RabbitMQ) for processing.

③ Nova-scheduler obtains the messages sent by nova-api from the message queue, and then executes the scheduling algorithm to select the appropriate computing node A.

④ Nova-scheduler submits the message "Create an instance on compute node A" to the Message queue.

⑤ The compute node A gets this message from the message queue and starts the instance.

⑥ While creating the virtual machine, the compute node interacts with the database "SQL Database" through nova-constructor. In this process, the information is also transferred through the message queue service.

So far, an instance has been created successfully.

3.3   OpenStack deployment sample architecture

See Fig. 3-8 for the typical deployment example architecture of OpenStack. The typical deployment architecture includes "Controller Node", "Compute Node", "Block Storage Node" and "Object Storage Node". Inclusively, the "Controller Node" and "Compute Node" are core components while others are optional components.

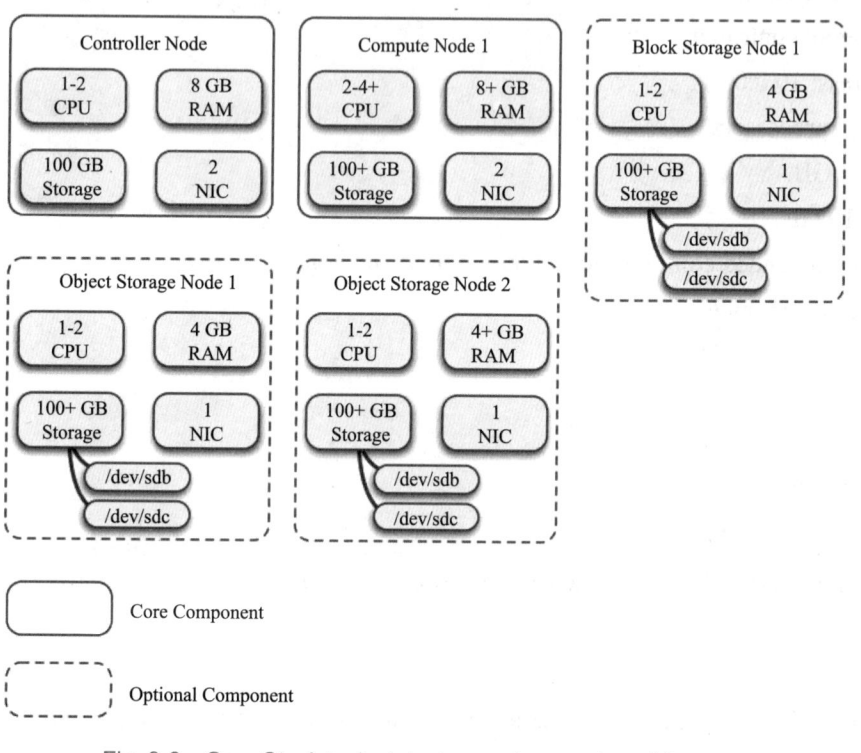

Fig. 3-8   OpenStack typical deployment example architecture

3.4   Project practice—build cloud computing OpenStack system platform

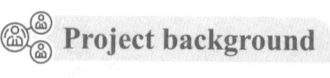 **Project background**

In order to improve the utilization rate of the existing hardware resources, a company decided to use the existing resources to build a cloud computing platform, uniformly manage the computing resources, storage resources and network resources in the resource pool, realize the reasonable allocation and flexible scheduling of resources, and realize the management and configuration of resources through the unified authentication mechanism API.

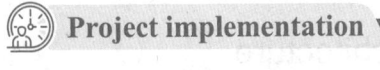 **Project implementation**

Task 1   Basic environment construction

1. Hardware and network configuration

Prepare two hosts as the controller node and compute node respectively in the Open Stack cluster.

Inclusively, the recommended configuration of the control node is as follows:

Host name: controller.

IP address: 10.0.0.11.

CPU：2 processors

RAM: 4 GB.

HDD: 10 GB.

The recommended configuration of the compute node is as follows:

Host name: compute.

IP address: 10.0.0.31.

CPU：2 processors.

RAM: 2 GB.

HDD: 20 GB.

2. Software warehouse

Install the following yum source software warehouse on all nodes of the cluster.

(1) Aliyun centos yum source

(2) Aliyun openstack yum source

Configure the software warehouse according to the content of the repo file in the following sample code.

```
[root@controller yum.repos.d]# cat CentOS-OpenStack-queens.repo
[centos-openstack-queens]
name=openstack-queens
baseurl=https://mirrors.aliyun.com/centos-vault/7.5.1804/cloud/x86_64/openstack-queens/
enabled=1
gpgcheck=0
[qume-kvm]
name=qemu-kvm
baseurl= https://mirrors.aliyun.com/centos-vault/7.5.1804/virt/x86_64/kvm-common/
enabled=1
gpgcheck=0
[root@controller yum.repos.d]# yum repolist
```

```
......
repo id                          repo name                                  status
base/7/x86_64                    CentOS-7-Base-mirrors.aliyun.com           10,072
centos-openstack-queens          openstack-queens                            2,449
extras/7/x86_64                  CentOS-7 - Extras - mirrors.aliyun.com      500
qume-kvm                         qemu-kvm                                     63
updates/7/x86_64                 CentOS-7 - Updates - mirrors.aliyun.com    3,252
repolist: 16,336
```

3. Time service

The time between nodes in the cluster requires strict synchronization. Install and configure the chrony service on the controller node and the compute node respectively to achieve time synchronization between nodes.

```
controller:
[root@controller yum.repos.d]# yum -y install chrony
Loaded plugins: fastestmirror
Loading mirror speeds from cached hostfile
 * base: mirrors.aliyun.com
 * extras: mirrors.aliyun.com
 * updates: mirrors.aliyun.com
centos-openstack-queens | 2.9 kB        00:00
qume-kvm                | 2.9 kB        00:00
Resolving Dependencies
--> Running transaction check
---> Package chrony.x86_64 0:3.4-1.el7 will be installed
--> Processing Dependency: libseccomp.so.2()(64 bit) for package: chro-
ny-3.4-1.el7.x86_64
--> Running transaction check
---> Package libseccomp.x86_64 0:2.3.1-4.el7 will be installed
--> Finished Dependency Resolution

Dependencies Resolved

================================================
 Package      Arch      Version       Repository
                                           Size
================================================
Installing:
 chrony      x86_64    3.4-1.el7      base   251 k
Installing for dependencies:
```

```
   libseccomp   x86_64   2.3.1-4.el7   base   56 k


Transaction Summary
==================================================
Install   1 Package (+1 Dependent package)


Total download size: 307 k
Installed size: 788 k
Downloading packages:
(1/2): libseccomp-2.3.1-4 |   56 kB   00:04
(2/2): chrony-3.4-1.el7.x | 251 kB   00:04
--------------------------------------------------
Total                 64 kB/s | 307 kB   00:04
Running transaction check
Running transaction test
Transaction test succeeded
Running transaction
   Installing : libseccomp-2.3.1-4.el7.x8     1/2
   Installing : chrony-3.4-1.el7.x86_64       2/2
   Verifying  : libseccomp-2.3.1-4.el7.x8     1/2
   Verifying  : chrony-3.4-1.el7.x86_64       2/2


Installed:
   chrony.x86_64 0:3.4-1.el7


Dependency Installed:
   libseccomp.x86_64 0:2.3.1-4.el7


Complete!
[root@controller yum.repos.d]# cat -n /etc/chrony.conf
   1 # Use public servers from the pool.ntp.org project.
   2 # Please consider joining the pool (http://www.pool.ntp.org/join.html).
   3 server ntp1.aliyun.com iburst
   4
   5 # Record the rate at which the system clock gains/losses time.
   6 driftfile /var/lib/chrony/drift
   7
   8 # Allow the system clock to be stepped in the first three updates
   9 # if its offset is larger than 1 second.
  10 makestep 1.0 3
  11
```

```
12 # Enable kernel synchronization of the real-time clock (RTC).
13 rtcsync
14
15 # Enable hardware timestamping on all interfaces that support it.
16 #hwtimestamp *
17
18 # Increase the minimum number of selectable sources required to adjust
19 # the system clock.
20 #minsources 2
21
22 # Allow NTP client access from local network.
23 allow 10.0.0.0/24
24
25 # Serve time even if not synchronized to a time source.
26 #local stratum 10
27
28 # Specify file containing keys for NTP authentication.
29 #keyfile /etc/chrony.keys
30
31 # Specify directory for log files.
32 logdir /var/log/chrony
33
34 # Select which information is logged.
35 #log measurements statistics tracking
```

```
[root@compute yum.repos.d]# systemctl restart chronyd
[root@compute yum.repos.d]# systemctl enable chronyd
compute:
[root@controller yum.repos.d]# yum -y install chrony
[root@compute yum.repos.d]# cat -n /etc/chrony.conf
 1 # Use public servers from the pool.ntp.org project.
 2 # Please consider joining the pool (http://www.pool.ntp.org/join.html).
 3 server controller iburst
 4 # Record the rate at which the system clock gains/losses time.
 5 driftfile /var/lib/chrony/drift
[root@compute yum.repos.d]# systemctl restart chronyd
[root@compute yum.repos.d]# systemctl enable chronyd
```

Synchronize with the time source and verify that the time has been synchronized.

```
[root@controller yum.repos.d]# chronyc sources
210 Number of sources = 1
```

```
MS Name/IP address          Stratum Poll Reach LastRx Last sample
===============================================================================
^* 120.25.115.20               2    6    377    11    -79us[ -326us] +/-    17ms
[root@controller yum.repos.d]# date
Fri Dec 24 22:30:35 CST 2021
[root@compute yum.repos.d]# chronyc sources
210 Number of sources = 1
MS Name/IP address          Stratum Poll Reach LastRx Last sample
===============================================================================
^* controller               3    6    77    49    +89us[ +956us] +/- 22ms
[root@compute yum.repos.d]# date
Fri Dec 24 22:30:35 CST 2021
```

4. Install OpenStack software

Install python-openstackclient and openstack-selinux on all nodes in the cluster.

```
[root@controller yum.repos.d]# yum-y install python-openstackclient
[root@controller yum.repos.d]# yum-y install openstack-selinux
```

5. Database service

Install the database service on the controller node, and perform initialization and permission settings.

```
[root@controller yum.repos.d]# yum -y install mariadb mariadb-server
[root@controller yum.repos.d]# vi /etc/my.cnf.d/openstack.cnf
[mysqld]
bind-address = 10.0.0.11

default-storage-engine = innodb
innodb_file_per_table = on
max_connections = 4096
collation-server = utf8_general_ci
character-set-server = utf8
[root@controller yum.repos.d]# systemctl restart mariadb
[root@controller yum.repos.d]# systemctl enable mariadb
Created symlink from /etc/systemd/system/multi-user.target.wants/mariadb.
service to /usr/lib/systemd/system/mariadb.service.
[root@controller yum.repos.d]# mysql_secure_installation

NOTE: RUNNING ALL PARTS OF THIS SCRIPT IS RECOMMENDED FOR ALL MariaDB
      SERVERS IN PRODUCTION USE!  PLEASE READ EACH STEP CAREFULLY!

In order to log into MariaDB to secure it, we'll need the current
```

```
password for the root user.  If you've just installed MariaDB, and
you haven't set the root password yet, the password will be blank,
so you should just press enter here.

Enter current password for root (enter for none):
OK, successfully used password, moving on...

Setting the root password ensures that nobody can log into the MariaDB root
```
user without the proper authorisation.
```

Set root password? [y/n] y
New password:
Re-enter new password:
Password updated successfully!
Reloading privilege tables.
 ... Success!
```

By default, a MariaDB installation has an anonymous user, allowing anyone to
log into MariaDB without having to have a user account created for them. This
is intended only for testing, and to make the installation go a bit smoother.
You should remove them before moving into a production environment.
```

Remove anonymous users? [y/n] y
 ... Success!
```

Normally, root should only be allowed to connect from 'localhost'. This ensures that someone cannot guess at the root password from the network.
```

Disallow root login remotely? [y/n] y
 ... Success!
```

By default, MariaDB comes with a database named 'test' that anyone can access. This is also intended only for testing, and should be removed before moving into a production environment.
```

Remove test database and access to it? [y/n] y
 - Dropping test database...
 ... Success!
 - Removing privileges on test database...
 ... Success!
```

```
    Reloading the privilege tables will ensure that all changes made so far will
take effect immediately.

    Reload privilege tables now? [y/n] y
     ... Success!

Cleaning up...

All done!  If you've completed all of the above steps, your MariaDB instal-
lation should now be secure.

Thanks for using MariaDB!
```

6. Message queuing service

Install rabbitmq-server on the controller node, add corresponding users to the database, and grant permissions accordingly.

```
    [root@controller yum.repos.d]# yum -y install rabbitmq-server
    [root@controller yum.repos.d]# systemctl start rabbitmq-server
    [root@controller yum.repos.d]# systemctl enable rabbitmq-server
    Created symlink from /etc/systemd/system/multi-user.target.wants/rabbit-
mq-server.service to /usr/lib/systemd/system/rabbitmq-server.service.
    [root@controller yum.repos.d]# rabbitmqctl add_user openstack RABBIT_PASS
    Creating user "openstack" ...
    [root@controller yum.repos.d]# rabbitmqctl set_permissions openstack "." ".*" ".*"
    Setting permissions for user "openstack" in vhost "/" ...
```

Monitoring plugins (web interface monitoring tools) can be installed.

```
    [root@controller yum.repos.d]# rabbitmq-plugins enable rabbitmq_management
    The following plugins have been enabled:
     mochiweb
     webmachine
     rabbitmq_web_dispatch
     amqp_client
     rabbitmq_management_agent
     rabbitmq_management

    Applying plugin configuration to rabbit@controller... started 6 plugins.
    [root@controller yum.repos.d]# netstat -ntpl
    Active Internet connections (only servers)
```

```
Proto Recv-Q Send-Q Local Address      Foreign Address   State       PID/Program name
tcp    0      0 0.0.0.0:25672          0.0.0.0:*         LISTEN      2621/beam.smp
tcp    0      0 10.0.0.11:3306         0.0.0.0:*         LISTEN      2420/mysqld
tcp    0      0 0.0.0.0:4369           0.0.0.0:*         LISTEN      1/systemd
tcp    0      0 10.0.0.11:53           0.0.0.0:*         LISTEN      794/named
tcp    0      0 127.0.0.1:53           0.0.0.0:*         LISTEN      794/named
tcp    0      0 0.0.0.0:22             0.0.0.0:*         LISTEN      774/sshd
tcp    0      0 0.0.0.0:15672          0.0.0.0:*         LISTEN      2621/beam.smp
tcp    0      0 127.0.0.1:25           0.0.0.0:*         LISTEN      905/master
tcp    0      0 127.0.0.1:953          0.0.0.0:*         LISTEN      794/named
tcp    0      0 :::5672                :::*              LISTEN      2621/beam.smp
tcp6   0      0 ::1:53                 :::*              LISTEN      794/named
tcp6   0      0 :::22                  :::*              LISTEN      774/sshd
tcp6   0      0 ::1:25                 :::*              LISTEN      905/master
tcp6   0      0 ::1:953                :::*              LISTEN      794/named
```

Monitor 5672 (for client), 25672 (for communication between cluster nodes) and 15672 (for monitoring web interface). Use the browser for access to http://10.0.0.11:15672, as shown in Fig. 3-9 and Fig. 3-10.

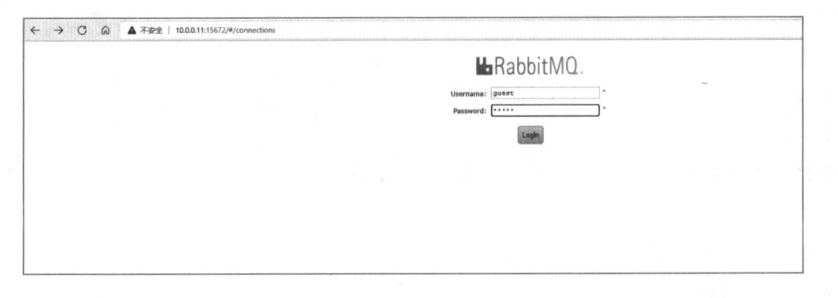

Fig. 3-9　RabbitMQ login interface

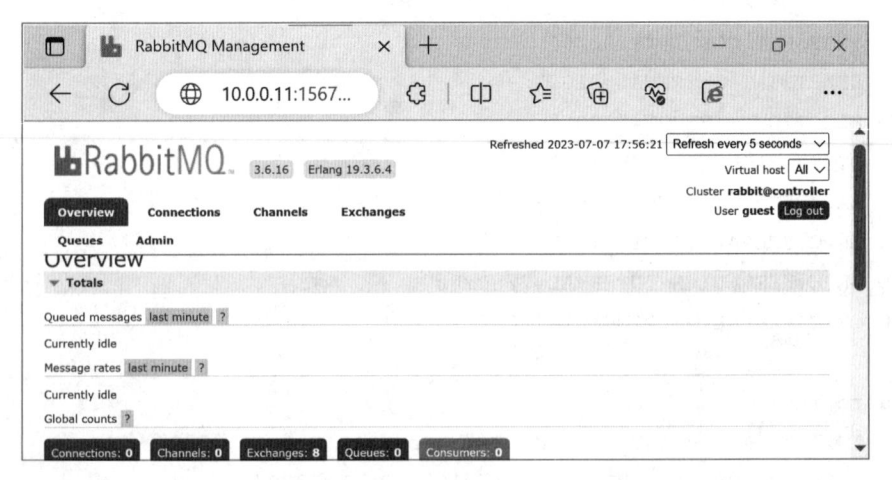

Fig. 3-10　RabbitMQ management interface

7. Cache service

Install the memcached service on the controller node.

```
[root@controller yum.repos.d]# yum -y install memcached python-memcached
[root@controller yum.repos.d]# cat /etc/sysconfig/memcached
PORT="11211"
USER="memcached"
MAXCONN="1024"
CACHESIZE="64"
OPTIONS="-l 127.0.0.1,::1,controller"
[root@controller yum.repos.d]# systemctl enable memcached.service
Created symlink from /etc/systemd/system/multi-user.target.wants/memcached.
service to /usr/lib/systemd/system/memcached.service.
[root@controller yum.repos.d]# systemctl start memcached.service
```

8. etcd

Install etcd on the controller node, modify the configuration file and set the client url address.

```
[root@controller yum.repos.d]# yum -y install etcd
[root@controller yum.repos.d]# vi /etc/etcd/etcd.conf
#[Member]
#ETCD_CORS=""
ETCD_DATA_DIR="/var/lib/etcd/default.etcd"
#ETCD_WAL_DIR=""
#ETCD_LISTEN_PEER_URLS="http://10.0.0.11:2380"
ETCD_LISTEN_CLIENT_URLS="http://10.0.0.11:2379"
#ETCD_MAX_SNAPSHOTS="5"
#ETCD_MAX_WALS="5"
ETCD_NAME="default"
#ETCD_SNAPSHOT_COUNT="100000"
#ETCD_HEARTBEAT_INTERVAL="100"
#ETCD_ELECTION_TIMEOUT="1000"
#ETCD_QUOTA_BACKEND_BYTES="0"
#ETCD_MAX_REQUEST_BYTES="1572864"
#ETCD_GRPC_KEEPALIVE_MIN_TIME="5s"
#ETCD_GRPC_KEEPALIVE_INTERVAL="2h0m0s"
#ETCD_GRPC_KEEPALIVE_TIMEOUT="20s"
#
#[Clustering]
#ETCD_INITIAL_ADVERTISE_PEER_URLS="http://10.0.0.11:2380"
ETCD_ADVERTISE_CLIENT_URLS="http://10.0.0.11:2379"
```

```
#ETCD_DISCOVERY=""
#ETCD_DISCOVERY_FALLBACK="proxy"
#ETCD_DISCOVERY_PROXY=""
#ETCD_DISCOVERY_SRV=""
#ETCD_INITIAL_CLUSTER="default=http://10.0.0.11:2380"
#ETCD_INITIAL_CLUSTER_TOKEN="etcd-cluster"
#ETCD_INITIAL_CLUSTER_STATE="new"
#ETCD_STRICT_RECONFIG_CHECK="true"
#ETCD_ENABLE_V2="true"
#
#[Proxy]
#ETCD_PROXY="off"
#ETCD_PROXY_FAILURE_WAIT="5000"
#ETCD_PROXY_REFRESH_INTERVAL="30000"
#ETCD_PROXY_DIAL_TIMEOUT="1000"
#ETCD_PROXY_WRITE_TIMEOUT="5000"
#ETCD_PROXY_READ_TIMEOUT="0"
#
#[Security]
#ETCD_CERT_FILE=""
#ETCD_KEY_FILE=""
#ETCD_CLIENT_CERT_AUTH="false"
:1,$s/localhost/10.0.0.11/g
[root@controller yum.repos.d]# systemctl restart etcd
[root@controller yum.repos.d]# systemctl enable etcd
Created symlink from /etc/systemd/system/multi-user.target.wants/etcd.ser-
vice to /usr/lib/systemd/system/etcd.service.
```

Task 2　Deploy OpenStack core modules

1. Keystone module deployment

Edit the script file and run the script openstack-keystone-install.sh to install keystone.

```
[root@controller ~]# cat openstack-keystone-install.sh
#!/bin/bash
################################################
# install keystone for Openstack on controller
# Author:Ann   Date:2022-1-5
################################################
```

```
    mysql -uroot -p000000 -e "create database IF NOT EXISTS keystone ;"
    mysql -uroot -p000000 -e "GRANT ALL PRIVILEGES ON keystone.* TO 'key-
stone'@'localhost' IDENTIFIED BY 'KEYSTONE_DBPASS';"
    mysql -uroot -p000000 -e "GRANT ALL PRIVILEGES ON keystone.* TO 'key-
stone'@'%' IDENTIFIED BY 'KEYSTONE_DBPASS';"


    yum -y install openstack-keystone httpd mod_wsgi
    yum -y install openstack-utils

    #Configure database connection
    openstack-config --set /etc/keystone/keystone.conf database connection
mysql+pymysql://keystone:KEYSTONE_DBPASS@controller/keystone
    #Use fernet to produce tokens
    openstack-config --set /etc/keystone/keystone.conf token provider  fernet
    #Synchronize database
    su -s /bin/sh -c "keystone-manage db_sync" keystone
    #Initialize fernet
    keystone-manage fernet_setup --keystone-user keystone --keystone-group keystone
    keystone-manage credential_setup --keystone-user keystone --keystone-group keystone
    #Register keystone
    keystone-manage bootstrap --bootstrap-password ADMIN_PASS \
    --bootstrap-admin-url http://controller:5000/v3/ \
    --bootstrap-internal-url http://controller:5000/v3/ \
    --bootstrap-public-url http://controller:5000/v3/ \
    --bootstrap-region-id RegionOne

    #Modify apache configurations
    sed -i "s/#ServerName www.example.com:80/ServerName controller/g" /etc/httpd/
conf/httpd.conf
    ln -s /usr/share/keystone/wsgi-keystone.conf /etc/httpd/conf.d/
    systemctl restart httpd
    systemctl enable httpd

    #Configure a temporary management account
    export OS_USERNAME=admin
    export OS_PASSWORD=ADMIN_PASS
    export OS_PROJECT_NAME=admin
    export OS_USER_DOMAIN_NAME=Default
    export OS_PROJECT_DOMAIN_NAME=Default
```

```
export OS_AUTH_URL=http://controller:5000/v3
export OS_IDENTITY_API_VERSION=3


#Create domains,projects,users and roles
openstack domain create --description "An Example Domain" example
openstack project create --domain default --description "Service Project" service
openstack project create --domain default --description "Demo Project" demo
openstack user create --domain default --password DEMO_PASS demo
openstack role create user
openstack role add --project demo --user demo user


unset OS_AUTH_URL OS_PASSWORD

cat > ~/admin-openrc.sh <<-EOF
export OS_PROJECT_DOMAIN_NAME=Default
export OS_USER_DOMAIN_NAME=Default
export OS_PROJECT_NAME=admin
export OS_USERNAME=admin
export OS_PASSWORD=ADMIN_PASS
export OS_AUTH_URL=http://controller:5000/v3
export OS_IDENTITY_API_VERSION=3
export OS_IMAGE_API_VERSION=2
EOF


cat > ~/demo-openrc.sh <<-EOF
export OS_PROJECT_DOMAIN_NAME=Default
export OS_USER_DOMAIN_NAME=Default
export OS_PROJECT_NAME=demo
export OS_USERNAME=demo
export OS_PASSWORD=DEMO_PASS
export OS_AUTH_URL=http://controller:5000/v3
export OS_IDENTITY_API_VERSION=3
export OS_IMAGE_API_VERSION=2
EOF


echo "source ~/admin-openrc.sh" >> ~/.bashrc
bash
[root@controller ~]# ./openstack-keystone-install.sh
```

Cite the environment variable in admin-openrc.sh and use the openstack token issue command to obtain the token to verify whether the keystone service is working properly.

```
[root@controller ~]# source admin-openrc.sh
[root@controller ~]# openstack token issue


+------------+-----------------------------------------------------------
----------------------------------------------------------------------------
-----------------------------------------------+
| Field | Value|
+------------+-----------------------------------------------------------
----------------------------------------------------------------------------
-----------------------------------------------+
| expires | 2021-12-25T08:21:49+0000        |
| id | gAAAAABhxsaN1kzGK_ggpWGbM1W4OW87Krf3HPO1DQ26AD8Xy5A99uOiVnp2je5Ad-
mxih-HJvTVcZQuEEXuEuIrBpRT9oJReC-Tfvlchw3IStrR_Mhfdjx0hVxRSY_79JBDuoQx-
ETQ7SCkKSI448GDKp7Xboyc06r8ECvf3qM7mPVIlzRzyd7NI |
| project_id | e3215238c9ac4193a42dbc88cab07150 |
| user_id    | a431c36941aa4b058cb254fdc52a8a5e |
+------------+-----------------------------------------------------------
----------------------------------------------------------------------------
-----------------------------------------------+
```

2. Glance module deployment

Edit the script file and run the script openstack-glance-install.sh to install glance.

```
[root@controller ~]# cat openstack-glance-install.sh
#!/bin/bash
##############################################
# install glance for Openstack on controller
# Author:Ann   Date:2022-1-5
##############################################


source  ~/admin-openrc.sh
mysql -uroot -p000000 -e "create database IF NOT EXISTS glance;"
mysql -uroot -p000000 -e "GRANT ALL PRIVILEGES ON glance.* TO 'glance'@'
localhost' IDENTIFIED BY 'GLANCE_DBPASS' ;"
mysql -uroot -p000000 -e "GRANT ALL PRIVILEGES ON glance.* TO 'glance'@'%'
IDENTIFIED BY 'GLANCE_DBPASS' ;"


openstack user create --domain default --password GLANCE_PASS glance
```

```
openstack role add --project service --user glance admin
openstack service create --name glance --description "OpenStack Image" image

openstack endpoint create --region RegionOne image public http://controller:9292
openstack endpoint create --region RegionOne image internal http://controller:9292
openstack endpoint create --region RegionOne image admin http://controller:9292

yum install -y openstack-glance

openstack-config --set /etc/glance/glance-api.conf database connection
mysql+pymysql://glance:GLANCE_DBPASS@controller/glance
openstack-config --set /etc/glance/glance-api.conf keystone_authtoken auth_
uri http://controller:5000
openstack-config --set /etc/glance/glance-api.conf keystone_authtoken auth_
url http://controller:5000
openstack-config --set /etc/glance/glance-api.conf keystone_authtoken mem-
cached_servers  controller:11211
openstack-config --set /etc/glance/glance-api.conf keystone_authtoken auth_
type password
openstack-config --set /etc/glance/glance-api.conf keystone_authtoken proj-
ect_domain_name default
openstack-config --set /etc/glance/glance-api.conf keystone_authtoken user_
domain_name default
openstack-config --set /etc/glance/glance-api.conf keystone_authtoken proj-
ect_name service
openstack-config --set /etc/glance/glance-api.conf keystone_authtoken username glance
openstack-config --set /etc/glance/glance-api.conf keystone_authtoken pass-
word GLANCE_PASS
openstack-config --set /etc/glance/glance-api.conf paste_deploy flavor keystone
openstack-config --set /etc/glance/glance-api.conf glance_store stores file,http
openstack-config --set /etc/glance/glance-api.conf glance_store default_store file
openstack-config --set /etc/glance/glance-api.conf glance_store filesystem_
store_datadir /var/lib/glance/images/

openstack-config --set /etc/glance/glance-registry.conf database connection
mysql+pymysql://glance:GLANCE_DBPASS@controller/glance
openstack-config --set /etc/glance/glance-registry.conf keystone_authtoken
auth_uri http://controller:5000
openstack-config --set /etc/glance/glance-registry.conf keystone_authtoken
auth_url http://controller:35357
```

```
    openstack-config --set /etc/glance/glance-registry.conf keystone_authtoken
memcached_servers controller:11211
    openstack-config --set /etc/glance/glance-registry.conf keystone_authtoken
auth_type password
    openstack-config --set /etc/glance/glance-registry.conf keystone_authtoken
project_domain_name default
    openstack-config --set /etc/glance/glance-registry.conf keystone_authtoken
user_domain_name default
    openstack-config --set /etc/glance/glance-registry.conf keystone_authtoken
project_name service
    openstack-config --set /etc/glance/glance-registry.conf keystone_authtoken
username glance
    openstack-config --set /etc/glance/glance-registry.conf keystone_authtoken
password GLANCE_PASS
    openstack-config --set /etc/glance/glance-registry.conf paste_deploy flavor keystone


    su -s /bin/sh -c "glance-manage db_sync" glance


    systemctl enable openstack-glance-api.service openstack-glance-registry.service
    systemctl restart openstack-glance-api.service openstack-glance-registry.service
    [root@controller ~]# ./openstack-glance-install.sh
```

Download the cirros image file and make an OpenStack image to verify whether the glance service is providing services normally.

```
    [root@controller ~]# wget http://download.cirros-cloud.net/0.4.0/cirros-
0.4.0-i386-disk.img
    ......
    Saving to: 'cirros-0.4.0-i386-disk.img'


    100%[=====================================================================
==========================================>] 12,094,976  8.76MB/s   in 1.3s


    2021-12-25 16:08:00 (8.76 MB/s) - 'cirros-0.4.0-i386-disk.img' saved
[12094976/12094976]


    [root@controller ~]# openstack image create "cirros" --file cirros-0.4.0-
i386-disk.img --disk-format qcow2 --container-format bare --public
    +----------------+------------------------------------------------------+
    | Field | Value |
```

```
+------------------+----------------------------------------------------------+
| checksum | b7d8ac291c698c3f1dc0705ce52a3b64 |
| container_format | bare |
| created_at | 2021-12-25T08:09:31Z |
| disk_format | qcow2 |
| file | /v2/images/1c27bb51-9d4a-47dc-be9d-13445f29580d/file |
| id | 1c27bb51-9d4a-47dc-be9d-13445f29580d |
| min_disk | 0 |
| min_ram | 0 |
| name | cirros |
| owner | 50467f099c214d8abba4e3f9cf8475bc |
| protected | False |
| schema | /v2/schemas/image |
| size | 12094976 |
| status | active |
| tags | |
| updated_at | 2021-12-25T08:09:31Z |
| virtual_size | None |
| visibility | public |
+------------------+----------------------------------------------------------+
[root@controller ~]# openstack image list
+--------------------------------------+--------+--------+
| ID | Name   | Status |
+--------------------------------------+--------+--------+
| 1c27bb51-9d4a-47dc-be9d-13445f29580d | cirros | active |
+--------------------------------------+--------+--------+
```

3. Nova module deployment

Edit the script file in the controller node and install nova by running the script.

```
[root@controller ~]# cat openstack-nova-install.sh
#!/bin/bash
###############################################
# install nova for Openstack on controller
# Author:Ann   Date:2022-1-5
###############################################


source ~/admin-openrc.sh

mysql -uroot -p000000 -e "create database IF NOT EXISTS nova;"
```

```
mysql -uroot -p000000 -e "create database IF NOT EXISTS nova_api;"
mysql -uroot -p000000 -e "create database IF NOT EXISTS nova_cell0 ;"

mysql -uroot -p000000 -e " GRANT ALL PRIVILEGES ON nova_api.* TO 'nova'@'
localhost' IDENTIFIED BY 'NOVA_DBPASS';"
mysql -uroot -p000000 -e " GRANT ALL PRIVILEGES ON nova_api.* TO 'nova'@'%'
IDENTIFIED BY 'NOVA_DBPASS';"
mysql -uroot -p000000 -e " GRANT ALL PRIVILEGES ON nova.* TO 'nova'@
'localhost' IDENTIFIED BY 'NOVA_DBPASS';"
mysql -uroot -p000000 -e " GRANT ALL PRIVILEGES ON nova.* TO 'nova'@'%'
IDENTIFIED BY 'NOVA_DBPASS';"
mysql -uroot -p000000 -e " GRANT ALL PRIVILEGES ON nova_cell0.* TO 'nova'
@'localhost' IDENTIFIED BY 'NOVA_DBPASS';"
mysql -uroot -p000000 -e " GRANT ALL PRIVILEGES ON nova_cell0.* TO
'nova'@'%' IDENTIFIED BY 'NOVA_DBPASS';"


openstack user create --domain default --password NOVA_PASS nova
openstack role add --project service --user nova admin
openstack service create --name nova --description "OpenStack Compute" compute
openstack endpoint create --region RegionOne compute public http://control-
ler:8774/v2.1
openstack endpoint create --region RegionOne compute internal http://con-
troller:8774/v2.1
openstack endpoint create --region RegionOne compute admin http://control-
ler:8774/v2.1


openstack user create --domain default --password PLACEMENT_PASS placement
openstack role add --project service --user placement admin
openstack service create --name placement --description "Placement API"
placement
openstack endpoint create --region RegionOne placement public http://con-
troller:8778
openstack endpoint create --region RegionOne placement internal http://con-
troller:8778
openstack endpoint create --region RegionOne placement admin http://controller:8778


yum install -y openstack-nova-api openstack-nova-conductor openstack-no-
va-console openstack-nova-novncproxy openstack-nova-scheduler openstack-no-
va-placement-api
```

```
    openstack-config --set /etc/nova/nova.conf database connection mysql+py-
mysql://nova:NOVA_DBPASS@controller/nova
    openstack-config --set /etc/nova/nova.conf DEFAULT  transport_url  rabbit://
openstack:RABBIT_PASS@controller
    openstack-config --set /etc/nova/nova.conf DEFAULT  enabled_apis osapi_com-
pute,metadata
    openstack-config --set /etc/nova/nova.conf DEFAULT  my_ip 10.0.0.11
    openstack-config --set /etc/nova/nova.conf DEFAULT  use_neutron True
    openstack-config --set /etc/nova/nova.conf DEFAULT  firewall_driver  nova.
virt.firewall.NoopFirewallDriver
    openstack-config --set /etc/nova/nova.conf api auth_strategy keystone
    openstack-config --set /etc/nova/nova.conf api_database connection mysql+py-
mysql://nova:NOVA_DBPASS@controller/nova_api
    openstack-config --set /etc/nova/nova.conf keystone_authtoken auth_url
http://controller:5000/v3
    openstack-config --set /etc/nova/nova.conf keystone_authtoken memcached_serv-
ers controller:11211
    openstack-config --set /etc/nova/nova.conf keystone_authtoken auth_type
password
    openstack-config --set /etc/nova/nova.conf keystone_authtoken project_domain_
name default
    openstack-config --set /etc/nova/nova.conf keystone_authtoken user_domain_name  default
    openstack-config --set /etc/nova/nova.conf keystone_authtoken project_name service
    openstack-config --set /etc/nova/nova.conf keystone_authtoken username nova
    openstack-config --set /etc/nova/nova.conf keystone_authtoken password NOVA_PASS
    openstack-config --set /etc/nova/nova.conf vnc enabled  true
    openstack-config --set /etc/nova/nova.conf vnc server_listen '$my_ip'
    openstack-config --set /etc/nova/nova.conf vnc server_proxyclient_address  '$my_ip'
    openstack-config --set /etc/nova/nova.conf glance api_servers http://controller:9292
    openstack-config --set /etc/nova/nova.conf oslo_concurrency lock_path  /var/
lib/nova/tmp
    openstack-config --set /etc/nova/nova.conf placement os_region_name  RegionOne
    openstack-config --set /etc/nova/nova.conf placement  project_domain_name
Default
    openstack-config --set /etc/nova/nova.conf placement  project_name  service
    openstack-config --set /etc/nova/nova.conf placement  auth_type  password
    openstack-config --set /etc/nova/nova.conf placement  user_domain_name  Default
    openstack-config --set /etc/nova/nova.conf placement  auth_url  http://con-
troller:5000/v3
```

```
openstack-config --set /etc/nova/nova.conf placement  username  placement
openstack-config --set /etc/nova/nova.conf placement  password  PLACEMENT_PASS

cat > /etc/httpd/conf.d/00-nova-placement-api.conf <<-EOF
Listen 8778

<VirtualHost *:8778>
  WSGIProcessGroup nova-placement-api
  WSGIApplicationGroup %{GLOBAL}
  WSGIPassAuthorization On
  WSGIDaemonProcess nova-placement-api processes=3 threads=1 user=nova group=nova
  WSGIScriptAlias / /usr/bin/nova-placement-api
  <IfVersion >= 2.4>
    ErrorLogFormat "%M"
  </IfVersion>
  ErrorLog /var/log/nova/nova-placement-api.log
  <Directory /usr/bin>
   <IfVersion >= 2.4>
     Require all granted
   </IfVersion>
   <IfVersion < 2.4>
     Order allow,deny
     Allow from all
   </IfVersion>
  </Directory>

  #SSLEngine On
  #SSLCertificateFile ...
  #SSLCertificateKeyFile ...
</VirtualHost>

Alias /nova-placement-api /usr/bin/nova-placement-api
<Location /nova-placement-api>
  SetHandler wsgi-script
  Options +ExecCGI
  WSGIProcessGroup nova-placement-api
  WSGIApplicationGroup %{GLOBAL}
  WSGIPassAuthorization On
</Location>
```

```
<Directory /usr/bin>
   <IfVersion >= 2.4>
      Require all granted
   </IfVersion>
   <IfVersion < 2.4>
      Order allow,deny
      Allow from all
   </IfVersion>
</Directory>
EOF
systemctl restart httpd

su -s /bin/sh -c "nova-manage api_db sync" nova
su -s /bin/sh -c "nova-manage cell_v2 map_cell0" nova
su -s /bin/sh -c "nova-manage cell_v2 create_cell --name=cell1" nova
su -s /bin/sh -c "nova-manage db sync" nova

systemctl enable openstack-nova-api.service openstack-nova-consoleauth.
service openstack-nova-scheduler.service openstack-nova-conductor.service open-
stack-nova-novncproxy.service
   systemctl start openstack-nova-api.service openstack-nova-consoleauth.ser-
vice openstack-nova-scheduler.service openstack-nova-conductor.service open-
stack-nova-novncproxy.service
   nova service-list
[root@controller ~]# ./openstack-nova-install-controller.sh
```

Edit the script file "openstack-dashboard-install.sh" on the controller node and run the script file to install the service of nova on the compute node.

```
[root@compute ~]# vi openstack-nova-install-compute.sh
#!/bin/bash
################################################
# install nova for Openstack on compute
# Author:Ann   Date:2022-1-6
################################################


yum install -y openstack-utils
yum install -y openstack-nova-compute
```

```
    openstack-config --set /etc/nova/nova.conf DEFAULT enabled_apis  osapi_compute,meta-
data
    openstack-config --set /etc/nova/nova.conf DEFAULT transport_url  rabbit://
openstack:RABBIT_PASS@controller
    openstack-config --set /etc/nova/nova.conf DEFAULT my_ip  10.0.0.31
    openstack-config --set /etc/nova/nova.conf DEFAULT use_neutron  True
    openstack-config --set /etc/nova/nova.conf DEFAULT firewall_driver  nova.virt.
firewall.NoopFirewallDriver
    openstack-config --set /etc/nova/nova.conf api auth_strategy  keystone
    openstack-config --set /etc/nova/nova.conf keystone_authtoken auth_url
http://controller:5000/v3
    openstack-config --set /etc/nova/nova.conf keystone_authtoken memcached_servers  con-
troller:11211
    openstack-config --set /etc/nova/nova.conf keystone_authtoken auth_type  password
    openstack-config --set /etc/nova/nova.conf keystone_authtoken project_domain_name  default
    openstack-config --set /etc/nova/nova.conf keystone_authtoken user_domain_name  default
    openstack-config --set /etc/nova/nova.conf keystone_authtoken project_name  service
    openstack-config --set /etc/nova/nova.conf keystone_authtoken username  nova
    openstack-config --set /etc/nova/nova.conf keystone_authtoken password  NOVA_PASS
    openstack-config --set /etc/nova/nova.conf vnc enabled  True
    openstack-config --set /etc/nova/nova.conf vnc server_listen  0.0.0.0
    openstack-config --set /etc/nova/nova.conf vnc server_proxyclient_address  '$my_ip'
    openstack-config --set /etc/nova/nova.conf vnc novncproxy_base_url  http://control-
ler:6080/vnc_auto.html
    openstack-config --set /etc/nova/nova.conf glance api_servers http://controller:9292
    openstack-config --set /etc/nova/nova.conf oslo_concurrency lock_path  /var/lib/nova/tmp
    openstack-config --set /etc/nova/nova.conf placement os_region_name  RegionOne
    openstack-config --set /etc/nova/nova.conf placement project_domain_name  Default
    openstack-config --set /etc/nova/nova.conf placement project_name  service
    openstack-config --set /etc/nova/nova.conf placement auth_type  password
    openstack-config --set /etc/nova/nova.conf placement user_domain_name  Default
    openstack-config --set /etc/nova/nova.conf placement auth_url  http://control-
ler:5000/v3
    openstack-config --set /etc/nova/nova.conf placement username  placement
    openstack-config --set /etc/nova/nova.conf placement password  PLACEMENT_PASS
    systemctl enable libvirtd.service openstack-nova-compute.service
    systemctl start libvirtd.service openstack-nova-compute.service


    [root@compute ~]# ./openstack-nova-install-compute.sh
```

Operate on the control node to add the compute node to the database.

```
[root@controller ~]# openstack compute service list --service nova-compute
+----+--------------+---------+------+---------+-------+--------------------
---------+
| ID | Binary       | Host    | Zone | Status  | State | Updated At |
+----+--------------+---------+------+---------+-------+--------------------
---------+
|  8 | nova-compute | compute | nova | enabled | up    | 2021-12-27T01:17:26.000000 |
+----+--------------+---------+------+---------+-------+--------------------
---------+
[root@controller ~]# su -s /bin/sh -c "nova-manage cell_v2 discover_hosts
--verbose" nova
/usr/lib/python2.7/site-packages/oslo_db/sqlalchemy/enginefacade.py:332:
NotSupportedWarning: Configuration option(s) ['use_tpool'] not supported
  exception.NotSupportedWarning
Found 2 cell mappings.
Skipping cell0 since it does not contain hosts.
Getting computes from cell 'cell1': a4a04e93-ca56-44df-8262-f5ac902d0ca6
Checking host mapping for compute host 'compute': 9a4ac031-6276-48a6-b02e-589b9005c284
Creating host mapping for compute host 'compute': 9a4ac031-6276-48a6-b02e-589b9005c284
Found 1 unmapped computes in cell: a4a04e93-ca56-44df-8262-f5ac902d0ca6
[root@controller ~]# openstack compute service list
+--------------------------------------+------------------+------------+---------
-+---------+-------+--------------------------+------------------+--------------+
| Id | Binary | Host | Zone | Status | State | Updated_at | Disabled Reason
| Forced down |
+--------------------------------------+------------------+------------+---------
-+---------+-------+--------------------------+------------------+--------------+
| b1984dce-b9d0-40ad-9c34-d9f28380fdc1 | nova-consoleauth | controller | inter-
nal | enabled | up | 2021-12-27T01:19:36.000000 | - | False |
| dc112131-d714-4464-8208-d7357fbcd4f0 | nova-conductor | controller | internal |
enabled | up | 2021-12-27T01:19:39.000000 | - | False |
| ad83513a-0c2a-45de-a269-c7abb62a2a7d | nova-scheduler | controller | internal
| enabled | up | 2021-12-27T01:19:31.000000 | - | False |
| 7a6c8b70-0990-4427-8076-a8ce55bc8e72 | nova-compute | compute | nova | enabled
| up | 2021-12-27T01:19:36.000000 | - | False |
+--------------------------------------+------------------+------------+---------
-+---------+-------+--------------------------+------------------+--------------+
```

4. Neutron module deployment

Edit the script file in the controller node and install neutron by running the script.

```
[root@controller ~]# cat openstack-neutron-install.sh
#!/bin/bash
################################################
# install neutron for Openstack on controller
# Author:Ann  Date:2022-1-6
################################################


source  ~/admin-openrc.sh


mysql -uroot -p000000 -e "create database IF NOT EXISTS neutron;"
mysql -uroot -p000000 -e "GRANT ALL PRIVILEGES ON neutron.* TO 'neutron'
@'localhost' IDENTIFIED BY 'NEUTRON_DBPASS';"
mysql -uroot -p000000 -e "GRANT ALL PRIVILEGES ON neutron.* TO 'neutron'@'%'
IDENTIFIED BY 'NEUTRON_DBPASS';"


openstack user create --domain default --password NEUTRON_PASS neutron
openstack role add --project service --user neutron admin
openstack service create --name neutron --description "OpenStack Networking" network
openstack endpoint create --region RegionOne network public http://control-
ler:9696
openstack endpoint create --region RegionOne network internal http://con-
troller:9696
openstack endpoint create --region RegionOne network admin http://controller:9696


yum -y install openstack-neutron openstack-neutron-ml2 openstack-neutron-li-
nuxbridge ebtables


openstack-config --set /etc/neutron/neutron.conf database connection
mysql://neutron:NEUTRON_DBPASS@controller/neutron
openstack-config --set /etc/neutron/neutron.conf DEFAULT core_plugin  ml2
openstack-config --set /etc/neutron/neutron.conf DEFAULT service_plugins
openstack-config --set /etc/neutron/neutron.conf DEFAULT transport_url  rab-
bit://openstack:RABBIT_PASS@controller
openstack-config --set /etc/neutron/neutron.conf DEFAULT auth_strategy  key-
stone
openstack-config --set /etc/neutron/neutron.conf keystone_authtoken auth_uri
http://controller:5000
```

```
openstack-config --set /etc/neutron/neutron.conf keystone_authtoken auth_url
http://controller:35357
    openstack-config --set /etc/neutron/neutron.conf keystone_authtoken mem-
cached_servers  controller:11211
    openstack-config --set /etc/neutron/neutron.conf keystone_authtoken auth_type  password
    openstack-config --set /etc/neutron/neutron.conf keystone_authtoken project_
domain_name  default
    openstack-config --set /etc/neutron/neutron.conf keystone_authtoken user_do-
main_name default
    openstack-config --set /etc/neutron/neutron.conf keystone_authtoken project_
name  service
    openstack-config --set /etc/neutron/neutron.conf keystone_authtoken username
neutron
    openstack-config --set /etc/neutron/neutron.conf keystone_authtoken password
NEUTRON_PASS
    openstack-config --set /etc/neutron/neutron.conf DEFAULT notify_nova_on_port_
status_changes  true
    openstack-config --set /etc/neutron/neutron.conf DEFAULT notify_nova_on_port_
data_changes  true
    openstack-config --set /etc/neutron/neutron.conf  nova auth_url  http://con-
troller:35357
    openstack-config --set /etc/neutron/neutron.conf  nova auth_type  password
    openstack-config --set /etc/neutron/neutron.conf  nova project_domain_name  default
    openstack-config --set /etc/neutron/neutron.conf  nova user_domain_name  default
    openstack-config --set /etc/neutron/neutron.conf  nova region_name  RegionOne
    openstack-config --set /etc/neutron/neutron.conf  nova project_name  service
    openstack-config --set /etc/neutron/neutron.conf  nova username  nova
    openstack-config --set /etc/neutron/neutron.conf  nova password  NOVA_PASS
    openstack-config --set /etc/neutron/neutron.conf oslo_concurrency lock_path /var
/lib/neutron/tmp


    openstack-config --set /etc/neutron/plugins/ml2/ml2_conf.ini ml2 type_drivers
flat,vlan
    openstack-config --set /etc/neutron/plugins/ml2/ml2_conf.ini ml2 tenant_net-
work_types
    openstack-config --set /etc/neutron/plugins/ml2/ml2_conf.ini ml2 mechanism_
drivers  linuxbridge
    openstack-config --set /etc/neutron/plugins/ml2/ml2_conf.ini ml2 extension_
drivers  port_security
```

```
    openstack-config --set  /etc/neutron/plugins/ml2/ml2_conf.ini ml2_type_flat
flat_networks  provider
    openstack-config --set  /etc/neutron/plugins/ml2/ml2_conf.ini securitygroup
enable_ipset  true


    openstack-config --set /etc/neutron/plugins/ml2/linuxbridge_agent.ini linux_
bridge physical_interface_mappings provider:ens34
    openstack-config --set /etc/neutron/plugins/ml2/linuxbridge_agent.ini vxlan
enable_vxlan false
    openstack-config --set /etc/neutron/plugins/ml2/linuxbridge_agent.ini securi-
tygroup enable_security_group true
    openstack-config --set /etc/neutron/plugins/ml2/linuxbridge_agent.ini securi-
tygroup firewall_driver neutron.agent.linux.iptables_firewall.IptablesFirewallDriver


    openstack-config --set /etc/neutron/dhcp_agent.ini DEFAULT interface_driver
linuxbridge
    openstack-config --set /etc/neutron/dhcp_agent.ini DEFAULT dhcp_driver neu-
tron.agent.linux.dhcp.Dnsmasq
    openstack-config --set /etc/neutron/dhcp_agent.ini DEFAULT enable_isolated_
metadata true


    openstack-config --set /etc/neutron/metadata_agent.ini DEFAULT nova_metadata_
host  controller
    openstack-config --set /etc/neutron/metadata_agent.ini DEFAULT metadata_
proxy_shared_secret METADATA_SECRET


    openstack-config --set /etc/nova/nova.conf  neutron url  http://controller:9696
    openstack-config --set /etc/nova/nova.conf  neutron auth_url  http://controller:35357
    openstack-config --set /etc/nova/nova.conf  neutron auth_type  password
    openstack-config --set /etc/nova/nova.conf  neutron project_domain_name  default
    openstack-config --set /etc/nova/nova.conf  neutron user_domain_name  default
    openstack-config --set /etc/nova/nova.conf  neutron region_name  RegionOne
    openstack-config --set /etc/nova/nova.conf  neutron project_name  service
    openstack-config --set /etc/nova/nova.conf  neutron username  neutron
    openstack-config --set /etc/nova/nova.conf  neutron password  NEUTRON_PASS
    openstack-config --set /etc/nova/nova.conf  neutron service_metadata_proxy  true
    openstack-config --set /etc/nova/nova.conf  neutron metadata_proxy_shared_se-
cret  METADATA_SECRET


    ln -s /etc/neutron/plugins/ml2/ml2_conf.ini /etc/neutron/plugin.ini
```

```
    su -s /bin/sh -c "neutron-db-manage --config-file /etc/neutron/neutron.conf
--config-file /etc/neutron/plugins/ml2/ml2_conf.ini upgrade head" neutron
    systemctl restart openstack-nova-api.service
    systemctl enable neutron-server.service neutron-linuxbridge-agent.service
neutron-dhcp-agent.service neutron-metadata-agent.service
    systemctl start neutron-server.service neutron-linuxbridge-agent.service
neutron-dhcp-agent.service neutron-metadata-agent.service
    openstack network agent list
    [root@controller ~]# ./openstack-neutron-install-controller.sh
```

Edit the script file "openstack-neutron-install-compute.sh" on the compute node and run the script file to install the service of neutron on the compute node.

```
    [root@compute ~]# vi openstack-nova-install-compute.sh
    #!/bin/bash
    ################################################
    # install nova for Openstack on controller
    # Author:Ann   Date:2022-1-6
    ################################################

    source  ~/admin-openrc.sh

    yum install -y openstack-neutron-linuxbridge ebtables ipset

    openstack-config --set /etc/neutron/neutron.conf DEFAULT transport_url rab-
bit://openstack:RABBIT_PASS@controller
    openstack-config --set /etc/neutron/neutron.conf DEFAULT auth_strategy  keystone
    openstack-config --set /etc/neutron/neutron.conf keystone_authtoken auth_uri
http://controller:5000
    openstack-config --set /etc/neutron/neutron.conf keystone_authtoken auth_url
http://controller:35357
    openstack-config --set /etc/neutron/neutron.conf keystone_authtoken mem-
cached_servers  controller:11211
    openstack-config --set /etc/neutron/neutron.conf keystone_authtoken auth_type  password
    openstack-config --set /etc/neutron/neutron.conf keystone_authtoken project_
domain_name  default
    openstack-config --set /etc/neutron/neutron.conf keystone_authtoken user_do-
main_name  default
    openstack-config --set /etc/neutron/neutron.conf keystone_authtoken project_
name  service
```

```
    openstack-config --set /etc/neutron/neutron.conf keystone_authtoken username
neutron
    openstack-config --set /etc/neutron/neutron.conf keystone_authtoken password
NEUTRON_PASS
    openstack-config --set /etc/neutron/neutron.conf oslo_concurrency lock_path
/var/lib/neutron/tmp
    openstack-config --set /etc/neutron/neutron.conf linux_bridge physical_inter-
face_mappings  provider:ens34
    openstack-config --set /etc/neutron/neutron.conf vxlan enable_vxlan  false
    openstack-config --set /etc/neutron/neutron.conf securitygroup enable_securi-
ty_group  true
    openstack-config --set /etc/neutron/neutron.conf securitygroup firewall_driver
neutron.agent.linux.iptables_firewall.IptablesFirewallDriver

    openstack-config --set /etc/nova/nova.conf neutron url  http://control-
ler:9696
    openstack-config --set /etc/nova/nova.conf neutron auth_url  http://control-
ler:35357
    openstack-config --set /etc/nova/nova.conf neutron auth_type  password
    openstack-config --set /etc/nova/nova.conf neutron project_domain_name  de-
fault
    openstack-config --set /etc/nova/nova.conf neutron user_domain_name  default
    openstack-config --set /etc/nova/nova.conf neutron region_name  RegionOne
    openstack-config --set /etc/nova/nova.conf neutron project_name  service
    openstack-config --set /etc/nova/nova.conf neutron username  neutron
    openstack-config --set /etc/nova/nova.conf neutron password  NEUTRON_PASS

    systemctl restart openstack-nova-compute.service
    systemctl enable neutron-linuxbridge-agent.service
    systemctl start neutron-linuxbridge-agent.service

    [root@compute ~]# ./openstack-neutron-install-compute.sh
    neutron
```

Task 3　Install the dashboard management interface

Edit the script file and install the dashboard by running the script. Edit the script file openstack-dashboard-install.sh on the control node.

```
    [root@controller ~]# cat openstack-dashboard-install.sh
    #!/bin/bash
```

```
###############################################
# install dashboard for Openstack on controller
# Author:Ann   Date:2022-1-6
###############################################


source  ~/admin-openrc.sh
yum -y install openstack-dashboard
cat > /etc/openstack-dashboard/local_settings <<-EOF
import os
from django.utils.translation import ugettext_lazy as _
from openstack_dashboard.settings import HORIZON_CONFIG
DEBUG = False
WEBROOT = '/dashboard/'
ALLOWED_HOSTS = ['*',]
SESSION_ENGINE = 'django.contrib.sessions.backends.cache'
OPENSTACK_API_VERSIONS = {
    "identity": 3,
    "image": 2,
    "volume": 2,
}
OPENSTACK_KEYSTONE_MULTIDOMAIN_SUPPORT = True
OPENSTACK_KEYSTONE_DEFAULT_DOMAIN = 'Default'
LOCAL_PATH = '/tmp'
SECRET_KEY='a98fb726ae49aeefb5ab'
CACHES = {
  'default': {
    'BACKEND': 'django.core.cache.backends.memcached.MemcachedCache',
    'LOCATION': 'controller:11211',
  },
}
EMAIL_BACKEND = 'django.core.mail.backends.console.EmailBackend'
OPENSTACK_HOST = "controller"
OPENSTACK_KEYSTONE_URL = "http://%s:5000/v3" % OPENSTACK_HOST
OPENSTACK_KEYSTONE_DEFAULT_ROLE = "user"
OPENSTACK_KEYSTONE_BACKEND = {
  'name': 'native',
  'can_edit_user': True,
  'can_edit_group': True,
  'can_edit_project': True,
  'can_edit_domain': True,
```

```
        'can_edit_role': True,
    }
    OPENSTACK_HYPERVISOR_FEATURES = {
        'can_set_mount_point': False,
        'can_set_password': False,
        'requires_keypair': False,
        'enable_quotas': True
    }
    OPENSTACK_CINDER_FEATURES = {
        'enable_backup': False,
    }
    OPENSTACK_NEUTRON_NETWORK = {
        'enable_router': False,
        'enable_quotas': False,
        'enable_distributed_router': False,
        'enable_ha_router': False,
        'enable_lb': False,
        'enable_firewall': False,
        'enable_vpn': False,
        'enable_fip_topology_check': False,
        'supported_vnic_types': ['*'],
        'physical_networks': [],
    }
    OPENSTACK_HEAT_STACK = {
        'enable_user_pass': True,
    }
    IMAGE_CUSTOM_PROPERTY_TITLES = {
        "architecture": _("Architecture"),
        "kernel_id": _("Kernel ID"),
        "ramdisk_id": _("Ramdisk ID"),
        "image_state": _("Euca2ools state"),
        "project_id": _("Project ID"),
        "image_type": _("Image Type"),
    }
    IMAGE_RESERVED_CUSTOM_PROPERTIES = []
    API_RESULT_LIMIT = 1000
    API_RESULT_PAGE_SIZE = 20
    SWIFT_FILE_TRANSFER_CHUNK_SIZE = 512 * 1024
    INSTANCE_LOG_LENGTH = 35
    DROPDOWN_MAX_ITEMS = 30
```

```
TIME_ZONE = "Asia/Shanghai"
POLICY_FILES_PATH = '/etc/openstack-dashboard'
LOGGING = {
  'version': 1,
  'disable_existing_loggers': False,
  'formatters': {
    'console': {
        'format': '%(levelname)s %(name)s %(message)s'
    },
    'operation': {
      'format': '%(message)s'
    },
  },
  'handlers': {
    'null': {
      'level': 'DEBUG',
      'class': 'logging.NullHandler',
    },
    'console': {
      'level': 'INFO',
      'class': 'logging.StreamHandler',
      'formatter': 'console',
    },
    'operation': {
      'level': 'INFO',
      'class': 'logging.StreamHandler',
      'formatter': 'operation',
    },
  },
  'loggers': {
    'horizon': {
      'handlers': ['console'],
      'level': 'DEBUG',
      'propagate': False,
    },
    'horizon.operation_log': {
      'handlers': ['operation'],
      'level': 'INFO',
      'propagate': False,
    },
```

```
'openstack_dashboard': {
  'handlers': ['console'],
  'level': 'DEBUG',
  'propagate': False,
},
'novaclient': {
  'handlers': ['console'],
  'level': 'DEBUG',
  'propagate': False,
},
'cinderclient': {
  'handlers': ['console'],
  'level': 'DEBUG',
  'propagate': False,
},
'keystoneauth': {
  'handlers': ['console'],
  'level': 'DEBUG',
  'propagate': False,
},
'keystoneclient': {
  'handlers': ['console'],
  'level': 'DEBUG',
  'propagate': False,
},
'glanceclient': {
  'handlers': ['console'],
  'level': 'DEBUG',
  'propagate': False,
},
'neutronclient': {
  'handlers': ['console'],
  'level': 'DEBUG',
  'propagate': False,
},
'swiftclient': {
  'handlers': ['console'],
  'level': 'DEBUG',
  'propagate': False,
},
```

```
    'oslo_policy': {
      'handlers': ['console'],
      'level': 'DEBUG',
      'propagate': False,
    },
    'openstack_auth': {
      'handlers': ['console'],
      'level': 'DEBUG',
      'propagate': False,
    },
    'nose.plugins.manager': {
      'handlers': ['console'],
      'level': 'DEBUG',
      'propagate': False,
    },
    'django': {
      'handlers': ['console'],
      'level': 'DEBUG',
      'propagate': False,
    },
    'django.db.backends': {
      'handlers': ['null'],
      'propagate': False,
    },
    'requests': {
      'handlers': ['null'],
      'propagate': False,
    },
    'urllib3': {
      'handlers': ['null'],
      'propagate': False,
    },
    'chardet.charsetprober': {
      'handlers': ['null'],
      'propagate': False,
    },
    'iso8601': {
      'handlers': ['null'],
      'propagate': False,
    },
```

```
        'scss': {
          'handlers': ['null'],
          'propagate': False,
        },
      },
    }
    SECURITY_GROUP_RULES = {
      'all_tcp': {
        'name': _('All TCP'),
        'ip_protocol': 'tcp',
        'from_port': '1',
        'to_port': '65535',
      },
      'all_udp': {
        'name': _('All UDP'),
        'ip_protocol': 'udp',
        'from_port': '1',
        'to_port': '65535',
      },
      'all_icmp': {
        'name': _('All ICMP'),
        'ip_protocol': 'icmp',
        'from_port': '-1',
        'to_port': '-1',
      },
      'ssh': {
        'name': 'SSH',
        'ip_protocol': 'tcp',
        'from_port': '22',
        'to_port': '22',
      },
      'smtp': {
        'name': 'SMTP',
        'ip_protocol': 'tcp',
        'from_port': '25',
        'to_port': '25',
      },
      'dns': {
        'name': 'DNS',
        'ip_protocol': 'tcp',
```

```
        'from_port': '53',
        'to_port': '53',
    },
    'http': {
        'name': 'HTTP',
        'ip_protocol': 'tcp',
        'from_port': '80',
        'to_port': '80',
    },
    'pop3': {
        'name': 'POP3',
        'ip_protocol': 'tcp',
        'from_port': '110',
        'to_port': '110',
    },
    'imap': {
        'name': 'IMAP',
        'ip_protocol': 'tcp',
        'from_port': '143',
        'to_port': '143',
    },
    'ldap': {
        'name': 'LDAP',
        'ip_protocol': 'tcp',
        'from_port': '389',
        'to_port': '389',
    },
    'https': {
        'name': 'HTTPS',
        'ip_protocol': 'tcp',
        'from_port': '443',
        'to_port': '443',
    },
    'smtps': {
        'name': 'SMTPS',
        'ip_protocol': 'tcp',
        'from_port': '465',
        'to_port': '465',
    },
    'imaps': {
```

```
        'name': 'IMAPS',
        'ip_protocol': 'tcp',
        'from_port': '993',
        'to_port': '993',
    },
    'pop3s': {
        'name': 'POP3S',
        'ip_protocol': 'tcp',
        'from_port': '995',
        'to_port': '995',
    },
    'ms_sql': {
        'name': 'MS SQL',
        'ip_protocol': 'tcp',
        'from_port': '1433',
        'to_port': '1433',
    },
    'mysql': {
        'name': 'MYSQL',
        'ip_protocol': 'tcp',
        'from_port': '3306',
        'to_port': '3306',
    },
    'rdp': {
        'name': 'RDP',
        'ip_protocol': 'tcp',
        'from_port': '3389',
        'to_port': '3389',
    },
}
REST_API_REQUIRED_SETTINGS = ['OPENSTACK_HYPERVISOR_FEATURES',
                              'LAUNCH_INSTANCE_DEFAULTS',
                              'OPENSTACK_IMAGE_FORMATS',
                              'OPENSTACK_KEYSTONE_DEFAULT_DOMAIN',
                              'CREATE_IMAGE_DEFAULTS',
                              'ENFORCE_PASSWORD_CHECK']
ALLOWED_PRIVATE_SUBNET_CIDR = {'ipv4': [], 'ipv6': []}
EOF
```

```
      sed -i '1i\WSGIApplicationGroup %{GLOBAL}' /etc/httpd/conf.d/openstack-dashboard.
conf
      systemctl restart httpd.service memcached.service
      [root@controller ~]# ./openstack-keystone-install.sh
```

Verify dashboard access.

Access 10.0.0 11/dashboard via browser to verify access, where the domain is "default", the user name is "admin" and the password is "ADMIN_PASS". After successful login, you can see the relevant information of the entire cloud platform in the dashboard interface, as shown in Fig. 3-11 and Fig. 3-12.

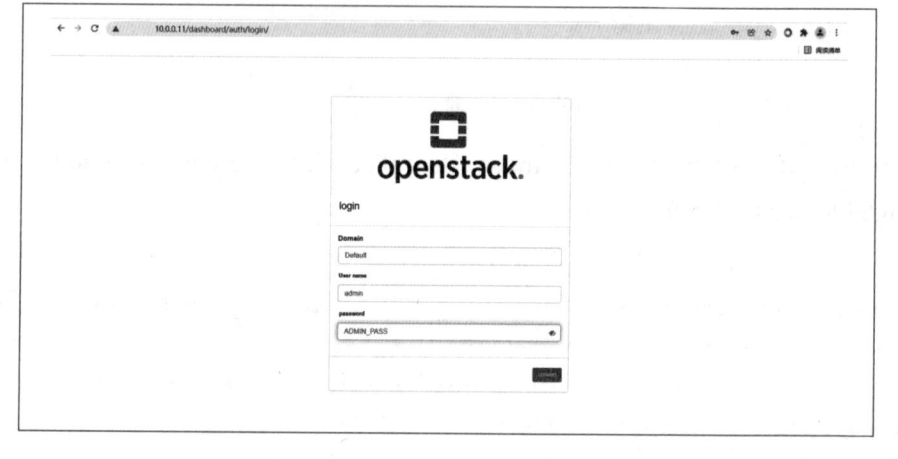

Fig. 3-11 Openstack dashboard login interface

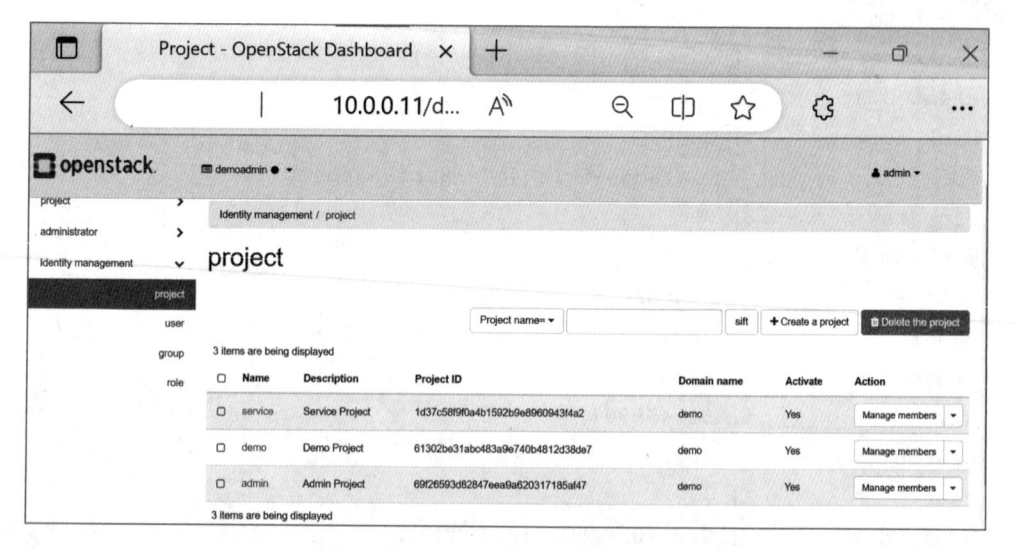

Fig. 3-12 Openstack dashboard management interface

3.5 Project practice—use the dashboard web interface to manage virtual machines

 **Project background**

In order to facilitate the daily operation and maintenance management of the OpenStack platform, the company has provided a dashboard web graphical management interface to more intuitively manage the operation and maintenance of the cloud host life cycle, image and virtual network, etc. via the browser.

 **Project implementation**

Task 1   Create an image

Prepare a local CentOS7.5 qcow2 image file, upload it to the controller node, and create a glance image file CentOS7.5 for use by ECS.

```
[root@controller ~]# glance image-create --name "CentOS7.5" --disk-format
qcow2  --container-format bare --progress < /opt/CentOS_7.5_x86_64_XD.qcow2
[==============================>] 100%
+------------------+--------------------------------------+
| Property         | Value                                |
+------------------+--------------------------------------+
| checksum         | 3d3e9c954351a4b6953fd156f0c29f5c     |
| container_format | bare                                 |
| created_at       | 2022-03-19T15:22:26Z                 |
| disk_format      | qcow2                                |
| id               | 91af8dd4-581e-4c8a-8bb5-8a4d4f09a4ab |
| min_disk         | 0                                    |
| min_ram          | 0                                    |
| name             | CentOS7.5                            |
| owner            | 2b2476cd9c1a4429b3d1f6f5ad29e686     |
| protected        | False                                |
| size             | 510459904                            |
| status           | active                               |
| tags             | []                                   |
| updated_at       | 2022-03-19T15:22:38Z                 |
| virtual_size     | None                                 |
| visibility       | shared                               |
+------------------+--------------------------------------+
```

You can also upload images on the "Administrator" → "Compute" → "Image" interface in the dashboard, as shown in Fig. 3-13 and Fig. 3-14.

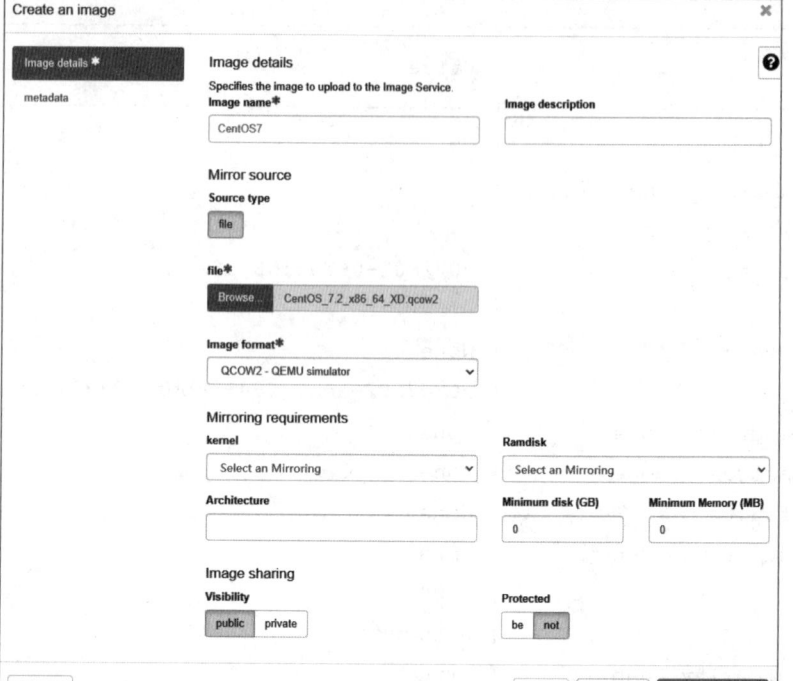

Fig. 3-13   Image Upload

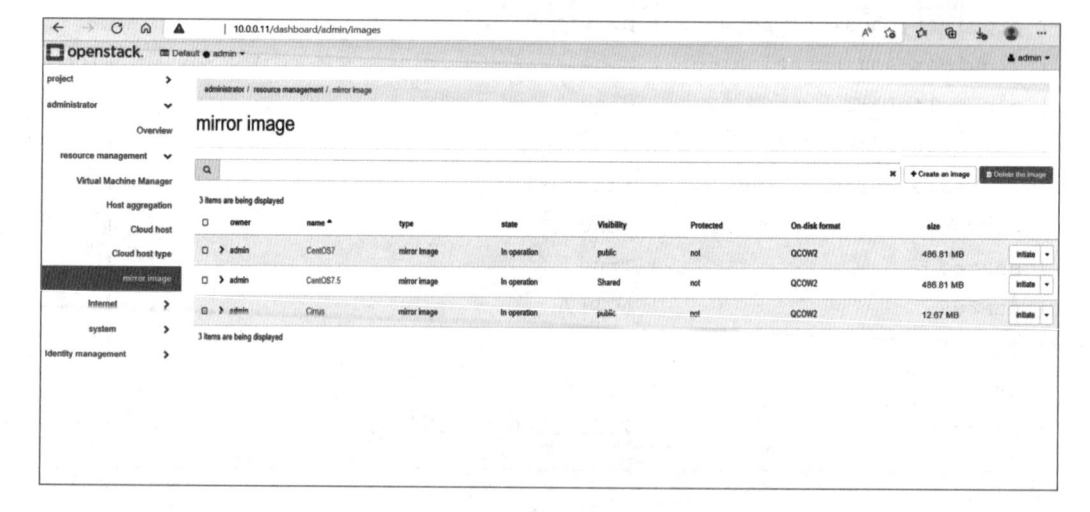

Fig. 3-14   View the image list

Task 2   Create a virtual network

Create a flat virtual network with "provider" as its corresponding physical network card.

```
[root@controller ~]# openstack network create --share --external \
>    --provider-physical-network provider \
>    --provider-network-type flat network
+--------------------------+--------------------------------------+
| Field                    | Value                                |
+--------------------------+--------------------------------------+
| admin_state_up           | UP                                   |
| availability_zone_hints  |                                      |
| availability_zones       |                                      |
| created_at               | 2022-01-06T05:56:35Z                 |
| description              |                                      |
| dns_domain               | None                                 |
| id                       | c08d7e52-aca3-4946-8bbf-808787b36fe1 |
| ipv4_address_scope       | None                                 |
| ipv6_address_scope       | None                                 |
| is_default               | None                                 |
| is_vlan_transparent      | None                                 |
| mtu                      | 1500                                 |
| name                     | network                              |
| port_security_enabled    | True                                 |
| project_id               | 928b3512a31c47e59b4fc1287db1ad7b     |
| provider:network_type    | flat                                 |
| provider:physical_network| provider                             |
| provider:segmentation_id | None                                 |
| qos_policy_id            | None                                 |
| revision_number          | 4                                    |
| router:external          | External                             |
| segments                 | None                                 |
| shared                   | True                                 |
| status                   | ACTIVE                               |
| subnets                  |                                      |
| tags                     |                                      |
| updated_at               | 2022-01-06T05:56:35Z                 |
+--------------------------+--------------------------------------+
```

Create a subnet in the virtual network. In the flat network, the virtual network is bridged with the corresponding physical network card "provide", and the subnet segment is set to 192.168.0.0/24, and the range of the address pool is defined as 192.168.0.2~192.168.0.254. The gateway is usually the first address available in the subnet, namely 192.168.0.1.

```
[root@controller ~]# openstack subnet create --network provider    --allocation
-pool start=192.168.0.2,end=192.168.0.254    --dns-nameserver 192.168.0.1 --gateway
192.168.0.1 --subnet-range 192.168.0.0/24 provider
+-------------------+---------------------------------------+
| Field             | Value                                 |
+-------------------+---------------------------------------+
| allocation_pools  | 192.168.0.2-192.168.0.254             |
| cidr              | 192.168.0.0/24                        |
| created_at        | 2022-01-06T06:00:15Z                  |
| description       |                                       |
| dns_nameservers   | 192.168.0.1                           |
| enable_dhcp       | True                                  |
| gateway_ip        | 192.168.0.1                           |
| host_routes       |                                       |
| id                | ecf65ebe-91f0-4ea5-876b-dbba0ad870f8  |
| ip_version        | 4                                     |
| ipv6_address_mode | None                                  |
| ipv6_ra_mode      | None                                  |
| name              | provider                              |
| network_id        | c08d7e52-aca3-4946-8bbf-808787b36fe1  |
| project_id        | 928b3512a31c47e59b4fc1287db1ad7b      |
| revision_number   | 0                                     |
| segment_id        | None                                  |
| service_types     |                                       |
| subnetpool_id     | None                                  |
| tags              |                                       |
| updated_at        | 2022-01-06T06:00:15Z                  |
+-------------------+---------------------------------------+
```

View the network list and you can obtain the ID of the virtual network, and use the ID to further view the details of the network.

```
[root@controller ~]# openstack network list
+--------------------------------------+---------+---------------------------+
| ID                                   | Name    | Subnets                   |
+--------------------------------------+---------+---------------------------+
| 0d91e83b-3d1d-4ebe-9c37-7964bca3c2a3 | network | 65c7720c-ac26-4a02-b15e-
59910c0b4fd7 |
+--------------------------------------+---------+---------------------------+
```

```
[root@controller ~]# openstack network show 0d91e83b-3d1d-4ebe-9c37-7964bca-
3c2a3
+--------------------------+-----------------------------------------+
| Field                    | Value                                   |
+--------------------------+-----------------------------------------+
| admin_state_up           | UP                                      |
| availability_zone_hints  |                                         |
| availability_zones       | nova                                    |
| created_at               | 2022-03-16T07:18:22Z                    |
| description              |                                         |
| dns_domain               | None                                    |
| id                       | 0d91e83b-3d1d-4ebe-9c37-7964bca3c2a3    |
| ipv4_address_scope       | None                                    |
| ipv6_address_scope       | None                                    |
| is_default               | None                                    |
| is_vlan_transparent      | None                                    |
| mtu                      | 1500                                    |
| name                     | network                                 |
| port_security_enabled    | True                                    |
| project_id               | 2b2476cd9c1a4429b3d1f6f5ad29e686        |
| provider:network_type    | flat                                    |
| provider:physical_network| provider                                |
| provider:segmentation_id | None                                    |
| qos_policy_id            | None                                    |
| revision_number          | 5                                       |
| router:external          | External                                |
| segments                 | None                                    |
| shared                   | True                                    |
| status                   | ACTIVE                                  |
| subnets                  | 65c7720c-ac26-4a02-b15e-59910c0b4fd7    |
| tags                     |                                         |
| updated_at               | 2022-03-16T07:18:22Z                    |
+--------------------------+-----------------------------------------+
```

In the dashboard, you can create a virtual network and its subnet in the "administrator" → "Internet" → "Internet" interface, as shown in Fig. 3-15 to Fig. 3-18.

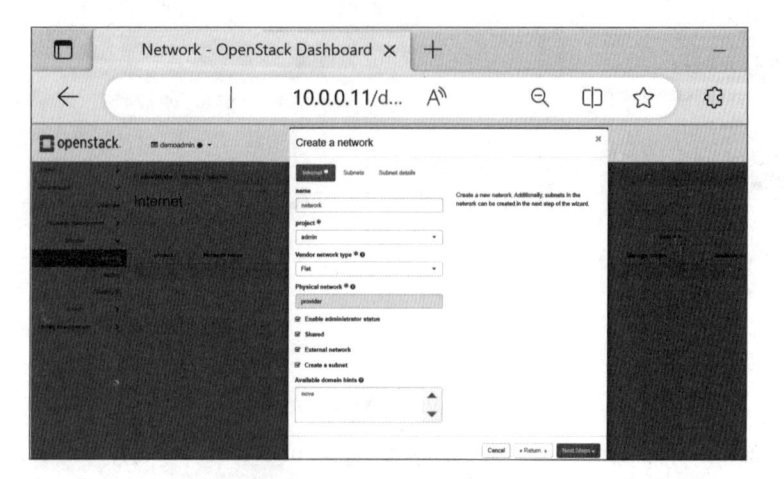

Fig. 3-15   Configure network information

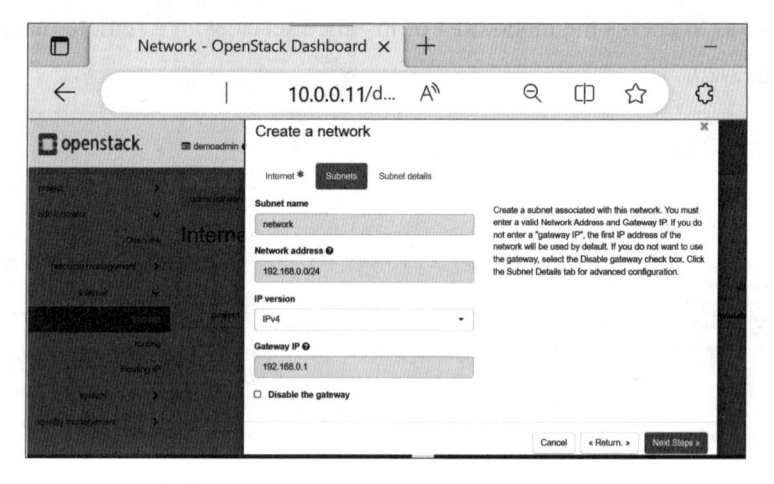

Fig. 3-16   Configure subnet information

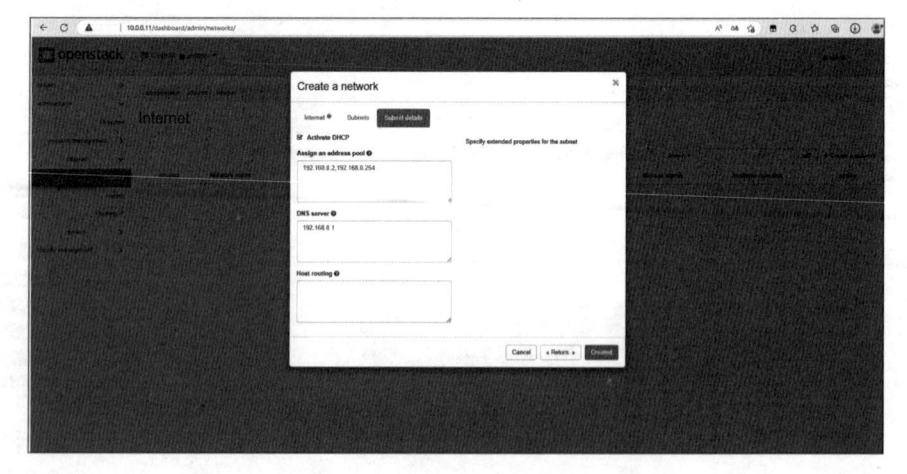

Fig. 3-17   Configure subnet details

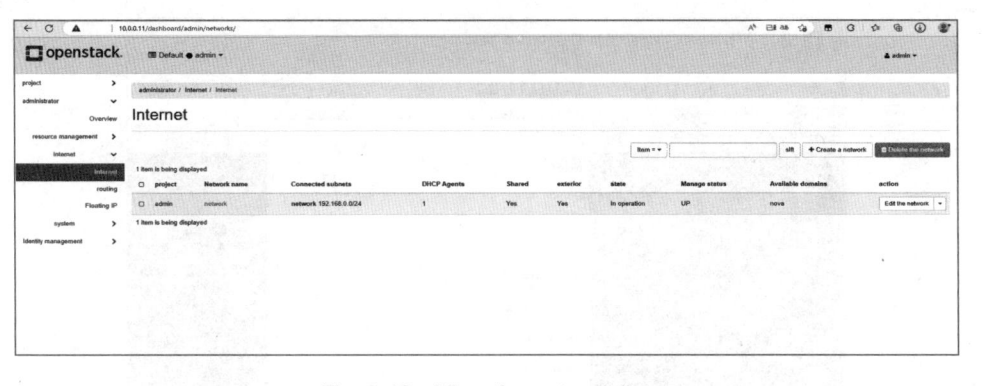

Fig. 3-18　View the network list

In the "Project" → "Internet" → "Security groups" interface, modify the firewall rules, delete the original rules, add new rules, and release the traffic of all directions and ports of ICMP, TCP and UDP protocols, so as to access the cloud host or instance services in a remote manner, as shown in Fig. 3-19 and Fig. 3-20.

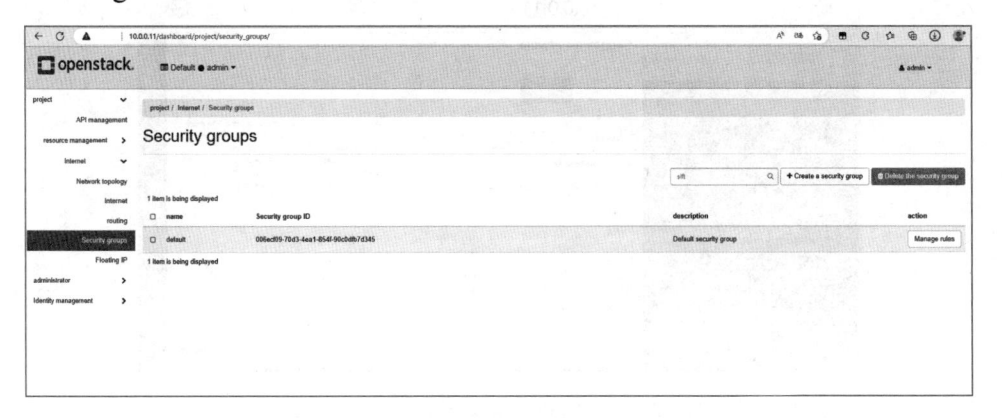

Fig. 3-19　Select security groups

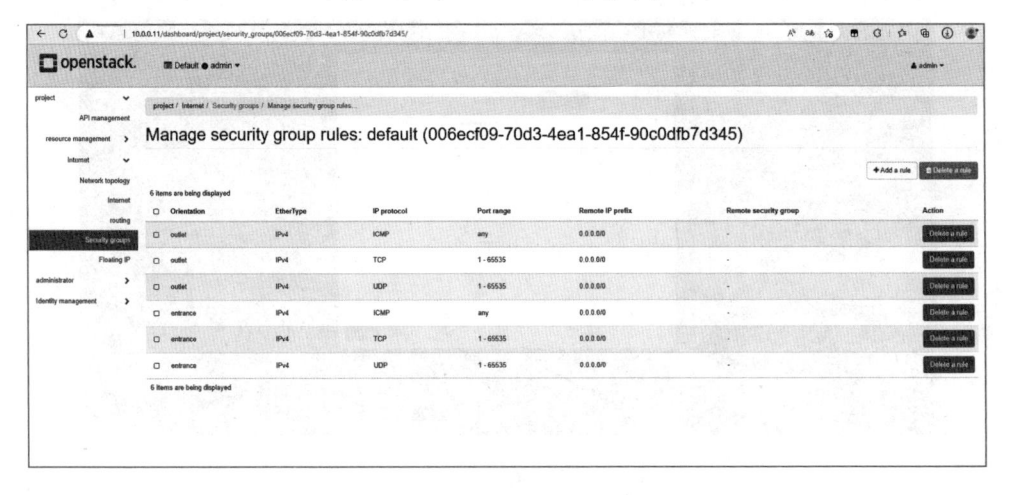

Fig. 3-20　Configure security group rules

Task 3　Start the cloud host instance

First, create a template named m1.small. The number of virtual CPUs is 1, the memory is 2 048 MB, and the disk capacity is 20 GB. View the template list and you can see the m1.small template created successfully.

```
[root@controller ~]# openstack flavor create --id 0 --vcpus 1 --ram 2048
--disk 20 m1.small
[root@controller ~]# openstack flavor list
+----+-----------+------+------+-----------+-------+-----------+
| ID | Name      | RAM  | Disk | Ephemeral | VCPUs | Is Public |
+----+-----------+------+------+-----------+-------+-----------+
| 0  | m1.small  | 2048 |  20  |         0 |     1 | True      |
+----+-----------+------+------+-----------+-------+-----------+
```

You can create instance types in "administrator" → "resource management" → "Cloud host type" in the dashboard interface, as shown in Fig. 3-21 and Fig. 3-22.

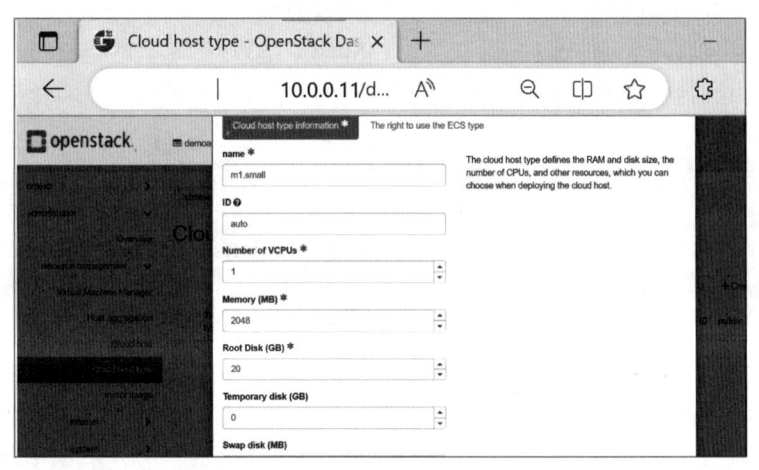

Fig. 3-21　Configure instance type information

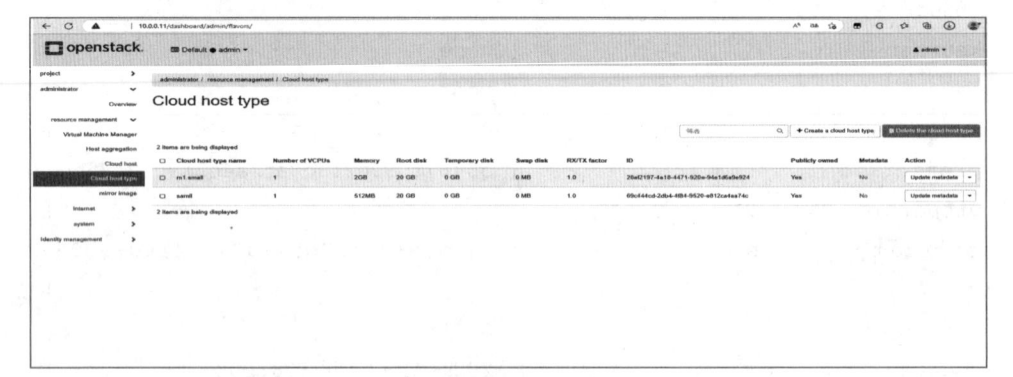

Fig. 3-22　View instance type list

Start the instance test, use the template m1.small, use CentOS7.5 as the image, let the network id as the virtual network created in Task 2, and select the default security group as the security group.

```
[root@controller ~]# openstack server create --flavor m1.small --image cir-
ros   --nic net-id=0d91e83b-3d1d-4ebe-9c37-7964bca3c2a3 --security-group default test
+------------------------------------+------------------------------------+
| Field                              | Value                              |
+------------------------------------+------------------------------------+
| OS-DCF:diskConfig                  | MANUAL                             |
| OS-EXT-AZ:availability_zone        |                                    |
| OS-EXT-SRV-ATTR:host               | None                               |
| OS-EXT-SRV-ATTR:hypervisor_hostname | None                              |
| OS-EXT-SRV-ATTR:instance_name      |                                    |
| OS-EXT-STS:power_state             | NOSTATE                            |
| OS-EXT-STS:task_state              | scheduling                         |
| OS-EXT-STS:vm_state                | building                           |
| OS-SRV-USG:launched_at             | None                               |
| OS-SRV-USG:terminated_at           | None                               |
| accessIPv4                         |                                    |
| accessIPv6                         |                                    |
| addresses                          |                                    |
| adminPass                          | bzq4S9SgeExs                       |
| config_drive                       |                                    |
| created                            | 2021-12-29T02:08:20Z               |
| flavor                             | m1.small                           |
| hostId                             |                                    |
| id                                 | e415d184-c0dc-4e14-bc3d-fb8a3f071ade|
| image                              | cirros (1c27bb51-9d4a-47dc-be9d-   |
13445f29580d)                                                             |
| key_name                           | none                               |
| name                               | test                               |
| progress                           | 0                                  |
| project_id                         | 50467f099c214d8abba4e3f9cf8475bc   |
| properties                         |                                    |
| security_groups                    | name='81b12d54-314f-4190-8d06-3ac9d-|
b21492e'                                                                  |
| status                             | BUILD                              |
| updated                            | 2021-12-29T02:08:20Z               |
| user_id                            | c346da06399c429f864eed7afff641e6   |
```

```
| volumes_attached                     |                                      |
+-------------------------------------+--------------------------------------+
[root@controller ~]# openstack server list
+-------------------------------------+--------+------+-------+--------+--------+
| ID                                  | Name  | Status | Networks
| Image  | Flavor                                                             |
+-------------------------------------+--------+------+-------+--------+--------+
| e415d184-c0dc-4e14-bc3d-fb8a3f071ade | test | ACTIVE | network=192.168.0.3
| cirros | m1.small                                                          |
+-------------------------------------+--------+------+-------+--------+--------+
[root@controller ~]# openstack console url show test
+--------+----------------------------------------------------------------------+
| Field | Value                                                                 |
+--------+----------------------------------------------------------------------+
| type  | novnc                                                                 |
| url   | http://controller:6080/vnc_auto.html?token=4c3287c5-520b-44e0-
91e1-289743f19296                                                               |
+--------+----------------------------------------------------------------------+
```

Use the browser to access the instance through VNC, as shown in Fig. 3-23.

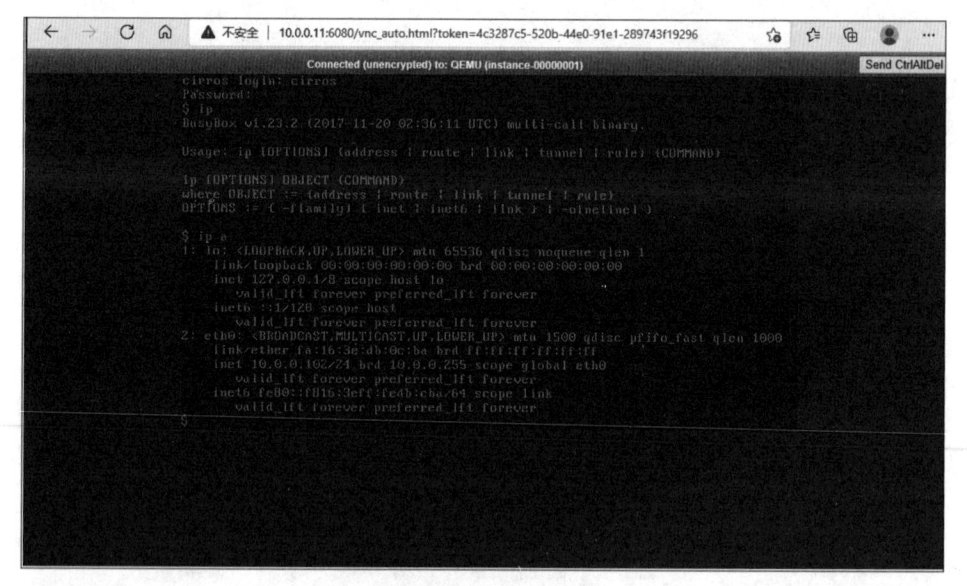

Fig. 3-23　Instance VNC access

In the "Project" → "resource management" → "Cloud host" interface of the dashboard, you can create instances through interface operations, as shown in Fig. 3-24 to Fig. 3-30.

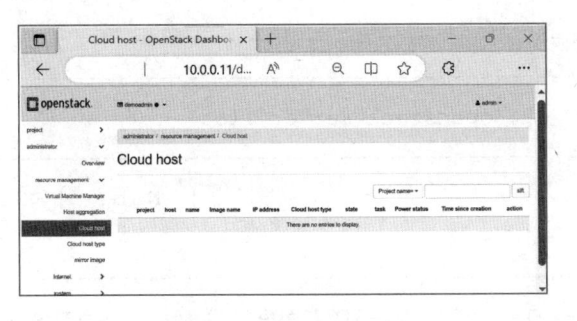

Fig. 3-24   New instance

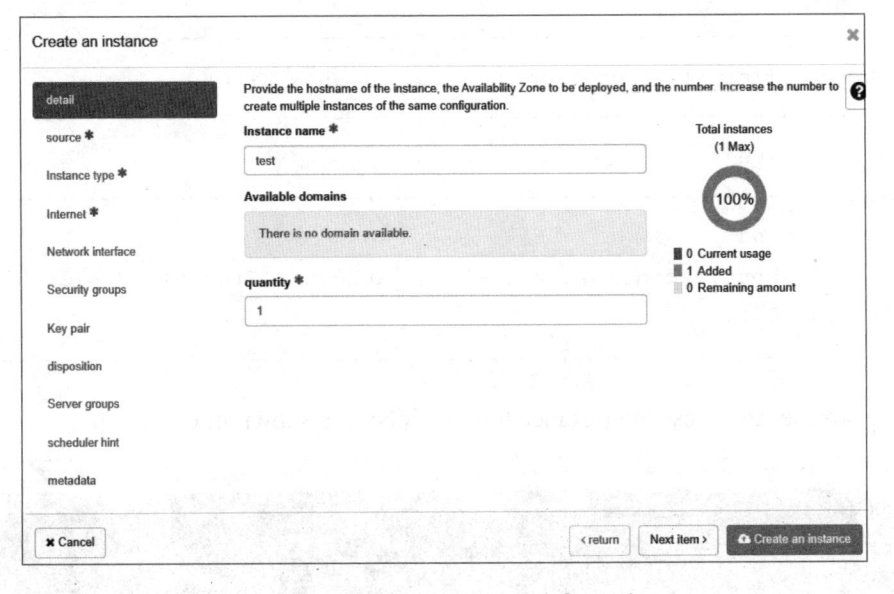

Fig. 3-25   Configure instance information

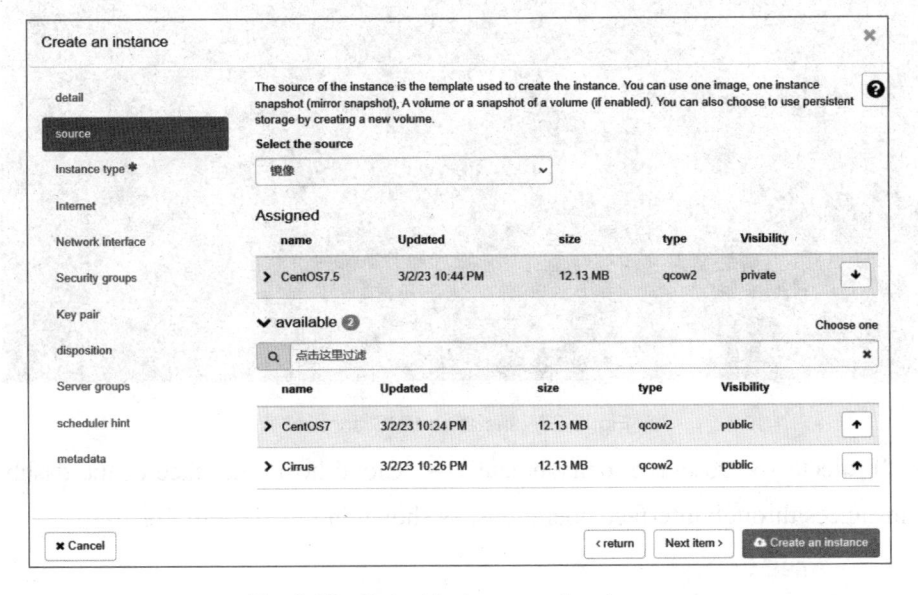

Fig. 3-26   Select instance system image

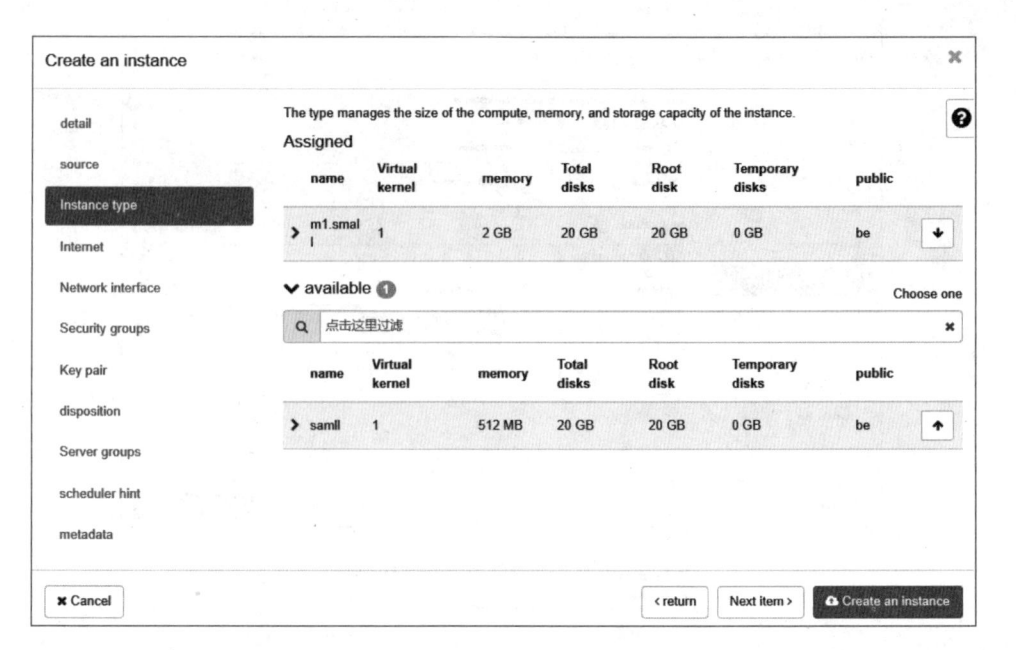

Fig. 3-27　Select instance type

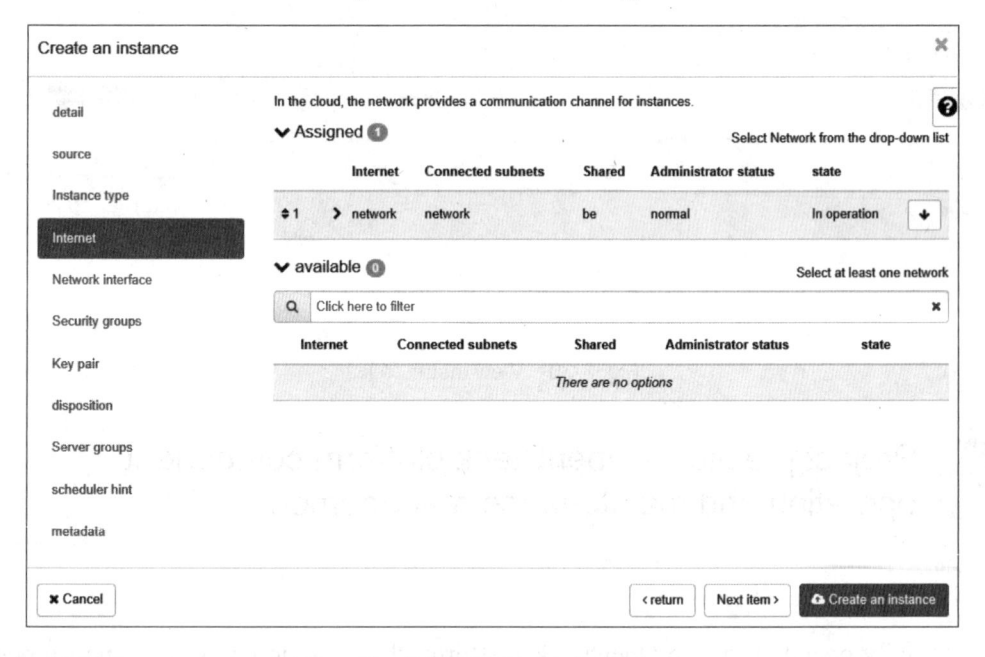

Fig. 3-28　Select network

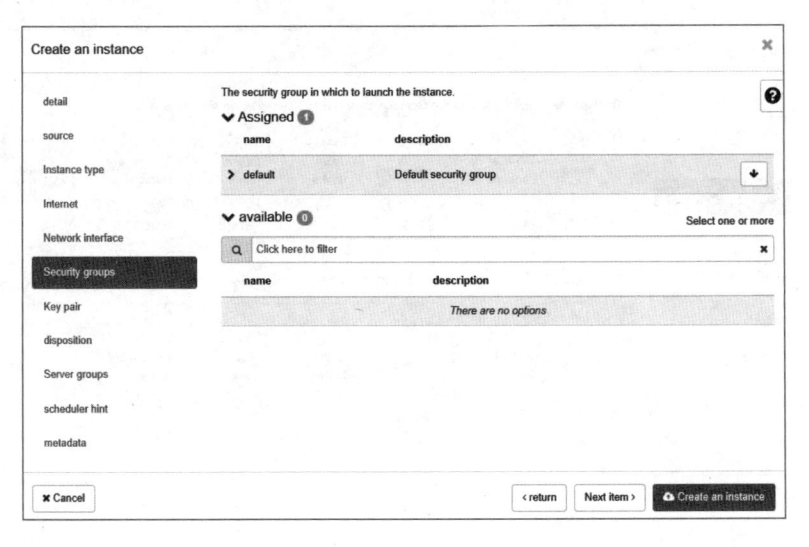

Fig. 3-29   Select security group

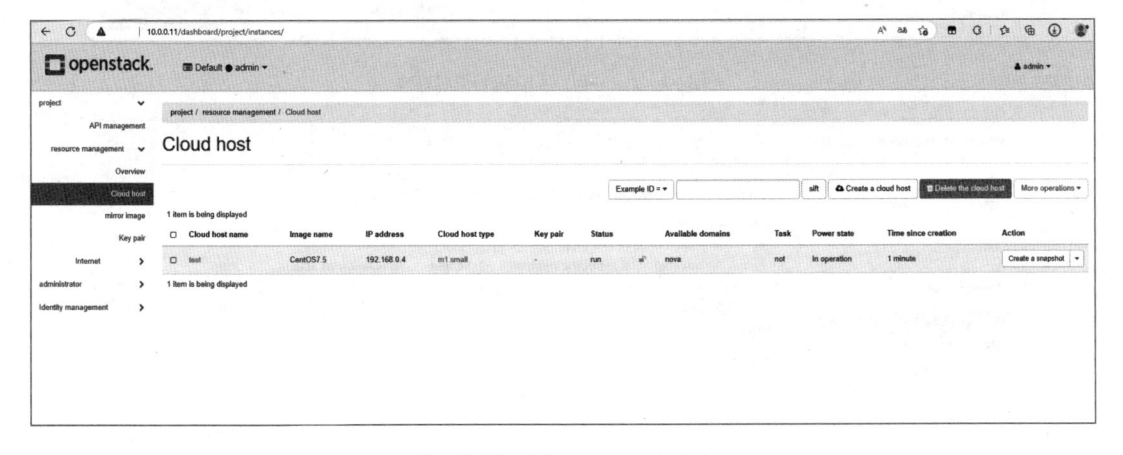

Fig. 3-30   View instance list

3.6   Project practice—OpenStack platform component operation and maintenance management

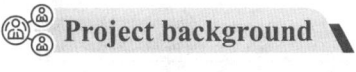 **Project background**

In the daily operation of the OpenStack platform, the operation and maintenance personnel need to undertake the operation and maintenance management on all components of the platform, including identity authentication, image, cloud host, network and other components.

Project implementation

Task 1   Keystone operation and maintenance

Keystone provides identity authentication services for the OpenStack platform and

performs identity authentication for platform users. First, create an account named "asus" with the password "mypassword123", and the mailbox is asus@example.com.

```
[root@container ~]# source/etc/keystone/admin-openrc.sh
[root@container ~]# openstack user create --password mypassword123 --email
asus@example.com --domain demo asus
+--------------------+------------------------------------+
| Field              | Value                              |
+--------------------+------------------------------------+
| domain_id          | b5fb5dbb3a0247f9b9d964457aab5d52   |
| email              | asus@example.com                   |
| enabled            | True                               |
| id                 | 0107f0547cd84c8db5fd8c7d98a6c430   |
| name               | asus                               |
| options            | {}                                 |
| password_expires_at | None                              |
+--------------------+------------------------------------+
```

When creating a user, you can also specify other parameters besides the above, as shown below:

```
[root@controller ~]# openstack user create
usage: openstack user create [-h]
                             [-f {html,json,json,shell,table,value,yaml,yaml}]
                             [-c COLUMN] [--max-width <integer>] [--noindent]
                             [--prefix PREFIX] [--domain <domain>]
                             [--project <project>]
                             [--project-domain <project-domain>]
                             [--password <password>] [--password-prompt]
                             [--email <email-address>]
                             [--description <description>]
                             [--enable | --disable] [--or-show]
                             <name>
```

Change the password of the "asus" account to "000000".

```
[root@container ~]# openstack user set --password 000000 asus
```

In OpenStack, all resources are divided by units. Usually, you can define a team or a unit in a project and create a unit named "myproject".

```
[root@conytainer ~]# openstack project create --domain demo  myproject
+-------------+------------------------------------+
| Field       | Value                              |
```

```
+-------------+-----------------------------------+
| description |                                   |
| domain_id   | b5fb5dbb3a0247f9b9d964457aab5d52  |
| enabled     | True                              |
| id          | c1bb4c90c7884e5c954300fcf009ade3  |
| is_domain   | False                             |
| name        | myproject                         |
| parent_id   | b5fb5dbb3a0247f9b9d964457aab5d52  |
| tags        | []                                |
+-------------+-----------------------------------+
```

When creating a unit, you can also specify other parameters besides the above, as shown below:

```
[root@controller ~]# openstack project create
usage: openstack project create [-h]
                                [-f {html,json,json,shell,table,value,yaml,yaml}]
                                [-c COLUMN] [--max-width <integer>]
                                [--noindent] [--prefix PREFIX]
                                [--domain <domain>] [--parent <project>]
                                [--description <description>]
                                [--enable | --disable]
                                [--property <key=value>] [--or-show]
                                <project-name>
```

Change the name of the unit "myproject" to "key_ project".

```
[root@container ~]# openstack project set --name key_project myproject
```

User's permission to use resources in the unit is determined by the associated roles. Binding different roles for users means that users have different permissions. Create a role compute-user.

```
[root@container ~]# openstack role create compute-user
+-----------+-----------------------------------+
| Field     | Value                             |
+-----------+-----------------------------------+
| domain_id | None                              |
| id        | 6c7bf68f8d1f439d8b420a49d8f725a5  |
| name      | compute-user                      |
+-----------+-----------------------------------+
```

When it is necessary to assign permissions to users, you can bind users with corresponding roles, for example, assign the compute-user role under the key_project unit to the user "asus"

subject to the following command:

```
[root@container ~]#   openstack role add --user alice --project key_project
compute-user
```

Task 2   Glance operation and maintenance

Create an image named after "cirros" in the disk format of qcow2. The image file uses cirros-0.4.0-x86_ 64-disk.img.

```
[root@container ~]# glance image-create --name cirros --disk-format qcow2
--container-format bare --progress <cirros-0.4.0-x86_64-disk.img
[==============================>] 100%
+-----------------+---------------------------------------+
| Property        | Value                                 |
+-----------------+---------------------------------------+
| checksum        | 443b7623e27ecf03dc9e01ee93f67afe      |
| container_format | bare                                 |
| created_at      | 2022-08-14T15:27:08Z                  |
| disk_format     | qcow2                                 |
| id              | c7f5e328-89a9-46ec-b5b9-c5a1c1040295 |
| min_disk        | 0                                     |
| min_ram         | 0                                     |
| name            | cirros                                |
| owner           | b9cd2a3e9a2345128def2d8236cdd6a5      |
| protected       | False                                 |
| size            | 12716032                              |
| status          | active                                |
| tags            | []                                    |
| updated_at      | 2022-08-14T15:27:09Z                  |
| virtual_size    | None                                  |
| visibility      | shared                                |
```

Query the list of the current images.

```
[root@container ~]# glance image-list
+---------------------------------------+--------+
| ID                                    | Name|
+---------------------------------------+--------+
| c7f5e328-89a9-46ec-b5b9-c5a1c1040295 | cirros |
+---------------------------------------+--------+
```

Use the list command and you can only see the list information about the image. To view the

details of the image, use the "glance image-show" command, where the parameters can be image ID or image name, as shown below:

```
[root@container ~]# glance image-show c7f5e328-89a9-46ec-b5b9-c5a1c1040295
```

The query results are as follows:

```
+------------------+------------------------------------+
| Property         | Value                              |
+------------------+------------------------------------+
| checksum         | 443b7623e27ecf03dc9e01ee93f67afe   |
| container_format | bare                               |
| created_at       | 2022-08-14T15:27:08Z               |
| disk_format      | qcow2                              |
| id               | c7f5e328-89a9-46ec-b5b9-c5a1c1040295 |
| min_disk         | 0                                  |
| min_ram          | 0                                  |
| name             | cirros                             |
| owner            | b9cd2a3e9a2345128def2d8236cdd6a5   |
| protected        | False                              |
| size             | 12716032                           |
| status           | active                             |
| tags             | []                                 |
| updated_at       | 2022-08-14T15:27:09Z               |
| virtual_size     | None                               |
| visibility       | shared                             |
+------------------+------------------------------------+
```

Delete the image "cirros".

```
[root@container ~]# glance image-delete c7f5e328-89a9-46ec-b5b9-c5a1c1040295
[root@container ~]# glance image-list
+----+------+
| ID | Name |
+----+------+
+----+------+
```

Task 3　Nova operation and maintenance

Nova is responsible for managing the entire life cycle of the cloud host.

Before creating a cloud host, you usually need to create a host type to define the resources allocated to the host. For example, create a cloud host type named "Fmin" (ID 1, RAM 2048 MB,

HDD 20 GB and 2 VCPUs).

```
[root@container ~]#  nova flavor-create Fmin 1 2048 20 2
+----+-------+-----------+------+-----------+------+-------+--------+-------------+-----------+
| ID | Name | Memory_MB | Disk | Ephemeral | Swap | VCPUs | RXTX_Factor | Is_Public |
+----+-------+-----------+------+-----------+------+-------+--------+-------------+-----------+
| 19999   | m1.flavor | 2048      | 20   | 0         |      | 2     | 1.0
| True      |
```

View the cloud host type.

```
[root@container ~]# nova flavor-show m1.flavor
+----------------------------+--------+
| Property                   | Value |
+----------------------------+--------+
| OS-FLV-DISABLED:disabled   | False |
| OS-FLV-EXT-DATA:ephemeral  | 0      |
| description                | -      |
| disk                       | 20     |
| extra_specs                | {}     |
| id                         | 1      |
| name                       | Fmin   |
| os-flavor-access:is_public | True   |
| ram                        | 2048   |
| rxtx_factor                | 1.0    |
| swap                       |        |
| vcpus                      | 2      |
+----------------------------+--------+
```

Use the created cloud host type to create a cloud host subject to the "cirros" image and the network of "public". The cloud host is named VM1.

```
# openstack server create VM1 -image cirros --flavor Fmin --network public
```

View the created cloud host.

```
[root@container ~]# nova list
+--------------------------------------+------+--------+------------+-------------+-------------+
| ID                                   | Name | Status | Task State | Power State | Networks    |
+--------------------------------------+------+--------+------------+-------------+-------------+
| 6924f0ac-81da-42a5-9cb2-4cf3ea5bf503 | VM1  | ACTIVE | -          | Running     |public=192.168.1.2 |
+--------------------------------------+------+--------+------------+-------------+-------------+
```

When creating a cloud host, you can specify other parameters besides the above parameters, as shown below:

```
[root@controller ~]# openstack server create
usage: openstack server create [-h]
                               [-f {html,json,json,shell,table,value,yaml,yaml}]
                               [-c COLUMN] [--max-width <integer>]
                               [--noindent] [--prefix PREFIX]
                               (--image <image> | --volume <volume>) --flavor
                               <flavor>
                               [--security-group <security-group-name>]
                               [--key-name <key-name>]
                               [--property <key=value>]
                               [--file <dest-filename=source-filename>]
                               [--user-data <user-data>]
                               [--availability-zone <zone-name>]
                               [--block-device-mapping <dev-name=mapping>]
                               [--nic <net-id=net-uuid,v4-fixed-ip=ip-addr,v6-
fixed-ip=ip-addr,port-id=port-uuid>]
                               [--hint <key=value>]
                               [--config-drive <config-drive-volume>|True]
                               [--min <count>] [--max <count>] [--wait]
                               <server-name>
```

Delete the virtual machine VM1.

```
[root@container ~]# openstack server delete VM1
```

Task 4 Neutron operation and maintenance

Neutron provides network services for the OpenStack platform and network access for the cloud host.

Create a network named "public".

```
[root@container ~]# neutron net-create public
neutron CLI is deprecated and will be removed in the future. Use openstack
CLI instead.
Created a new network:
+---------------------------+----------------------------------------+
| Field                     | Value                                  |
+---------------------------+----------------------------------------+
| admin_state_up            | True                                   |
```

```
| availability_zone_hints |                                      |
| availability_zones      |                                      |
| created_at              | 2022-08-15T09:19:39Z                 |
| description             |                                      |
| id                      | 0bbb8a16-f04c-4b08-b964-52802fa33b60 |
| ipv4_address_scope      |                                      |
| ipv6_address_scope      |                                      |
| is_default              | False                                |
| mtu                     | 1450                                 |
| name                    | public                               |
| port_security_enabled   | True                                 |
| project_id              | b9cd2a3e9a2345128def2d8236cdd6a5     |
| provider:network_type   | vxlan                                |
| provider:physical_network |                                    |
| provider:segmentation_id | 13                                  |
| revision_number         | 2                                    |
| router:external         | False                                |
| shared                  | False                                |
| status                  | ACTIVE                               |
| subnets                 |                                      |
| tags                    |                                      |
| tenant_id               | b9cd2a3e9a2345128def2d8236cdd6a5     |
| updated_at              | 2022-08-15T09:19:39Z                 |
+-------------------------+--------------------------------------+
```

Create a subnet "newnet", set DNS to 8.8.8.8, gateway to 192.168.1.1, and address pool range to 192.168.1.2-192.168.1.254.

```
    [root@container ~]# neutron subnet-create --name newnet --dns-nameserver
8.8.8.8 --gateway 192.168.1.1 public 192.168.1.0/24
    neutron CLI is deprecated and will be removed in the future. Use openstack
CLI instead.
    Created a new subnet:
    +------------------+-----------------------------------------------------+
    | Field            | Value                                               |
    +------------------+-----------------------------------------------------+
    | allocation_pools | {"start": "192.168.1.2", "end": "192.168.1.254"}    |
    | cidr             | 192.168.1.0/24                                      |
    | created_at       | 2022-08-15T09:19:54Z                               |
    | description      |                                                     |
    | dns_nameservers  | 8.8.8.8                                             |
```

```
| enable_dhcp         | True                                       |
| gateway_ip          | 192.168.1.1                                |
| host_routes         |                                            |
| id                  | e97d3ebd-6fec-43f3-82b2-bff97f3b71c2       |
| ip_version          | 4                                          |
| ipv6_address_mode   |                                            |
| ipv6_ra_mode        |                                            |
| name                | newnet                                     |
| network_id          | 0bbb8a16-f04c-4b08-b964-52802fa33b60       |
| project_id          | b9cd2a3e9a2345128def2d8236cdd6a5           |
| revision_number     | 0                                          |
| service_types       |                                            |
| subnetpool_id       |                                            |
| tags                |                                            |
| tenant_id           | b9cd2a3e9a2345128def2d8236cdd6a5           |
| updated_at          | 2022-08-15T09:19:54Z                       |
+---------------------+--------------------------------------------+
```

Query the network details.

```
[root@container ~]# neutron net-list
neutron CLI is deprecated and will be removed in the future. Use openstack
CLI instead.
+----------------------+--------+----------------------+---------------------+
| id                   | name   | tenant_id            | subnets             |--
+----------------------+--------+----------------------+---------------------+
  0bbb8a16-f04c-4b08-b964-52802fa33b60 | public | b9cd2a3e9a2345128def2d-
8236cdd6a5 |
  e97d3ebd-6fec-43f3-82b2-bff97f3b71c2 192.168.1.0/24 |
+----------------------+--------+----------------------+---------------------+
[root@container ~]# neutron subnet-list
+----------------------+--------+----------------------+---------------------+
| id                   | name   | tenant_id            | cidr    | allocation_pools  |
+----------------------+--------+----------------------+---------------------+
| e97d3ebd-6fec-43f3-82b2-bff97f3b71c2 | newnet | b9cd2a3e9a2345128def2d-
8236cdd6a5 | 192.168.1.0/24 | {"start": "192.168.1.2", "end": "192.168.1.254"} |
+----------------------+--------+----------------------+---------------------+
```

Query the relevant information of network services.

```
[root@controller ~]# neutron agent-list
+----------+----------+----------+----------+-------+----------+--------+
```

id	agent_type	host	availability_zone	alive	admin_state_up	binary
1ccc8d2d-b40a-44 06-b15b-tch-ab4b2b8e 8398	Open vSwitch agent	controller		:-)	True	neutron-openvswi tch-agent
245ae2f9-6220-4c 00-a36d-542a746a 3f9e	DHCP agent	controller	nova	:-)	True	neutron-dhcp-agent
26f38d05-3553-46 f6-8740-847f9774 1f89	L3 agent	controller	nova	:-)	True	neutron-l3-agent
5bd0c9e2-a2ef-48 90-a4ac-	Metadata agent	controller		:-)	True	neutron-metadata-agent

```
      | a4cb6bcc |              |          |                        |         |
|          |
      | f71b         |              |          |                        |         |
|          |
      | f1d1fc34 | Loadbalanc | controll |                        | :-)     | True
| neutron- |
      | -6e9d-46 | er agent    | er       |                        |         |
| lbaas-   |
      | b5-8370- |              |          |                        |         |
| agent    |
      | 0579f3b4 |              |          |                        |         |
|          |
      | 6b6f         |              |          |                        |         |
|          |
      +----------+----------+----------+------------+-------+-----------+---------+
```

Unit 4

Container cloud technology — Docker

4.1 | Docker engine

As a source-open container engine technology, Docker is used to package software into standardized units for developing, shipping and deploying software applications. Compared with traditional virtualization technology, Docker shares the host operating system nucleus, so the container is more lightweight and independent, greatly improving the server use efficiency.

Container is also a virtualization technology, but it is quite different from traditional virtualization technology. The virtual machine is to virtualize computing resources and build a virtual host on the basis of the physical host. Each virtual machine needs to install an operating system, deploy applications and complete business functions. Moreover, the virtual machine usually takes a long time to start up because it needs to boot the operating system. However, the container abstracts the application layer and packages the code and dependencies together. A variety of containers can run on the same computer and share the operating system nucleus with other containers. Each container runs as an isolated process in the user space. Compared with the virtual machine, the container is much smaller in size and usually has only tens of megabytes, while the virtual machine may need several gigabytes. Therefore, under the same conditions, the container technology can enable more applications with fewer virtual machines and OS resources.

See Fig. 4-1 for the model comparison between container and virtual machine.

The idea of container is "running anywhere once packaged", which determines the three advantages of Docker:

1. Standardization

The server can run any container as long as the standard runtime operating environment is well configured, which can greatly improve the efficiency of operation and maintenance. In the process of traditional product manufacturing, the development environment, test environment and production environment are often inconsistent, which makes all sectors more difficult to be connected. Moreover, the operation and maintenance personnel shall spare no effort to solve the problem of environment inconsistency. Once a container is available, you can repeatedly use the container in any environment as long as you package it once. You only need to configure the container running environment.

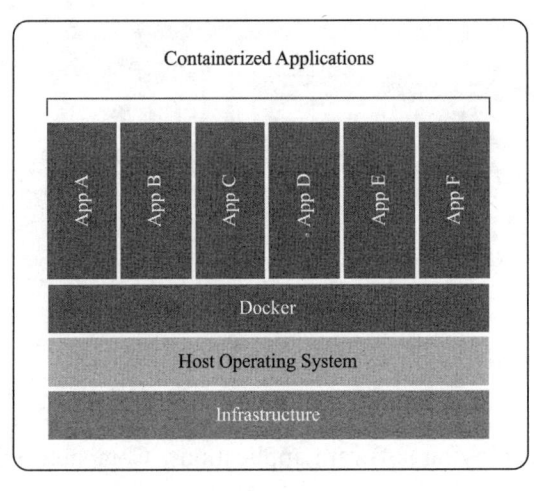
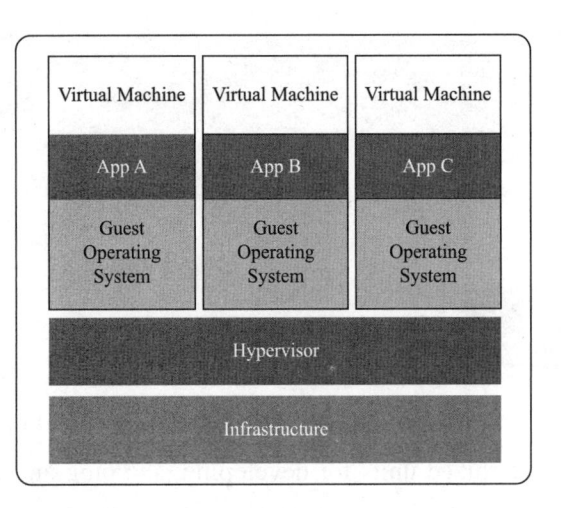

Fig. 4-1   Comparison between container and virtual machine

2. Lightweight

Container is called a "lightweight" virtualization technology. In traditional virtualization, whenever we need to build a business and run some programs, we need to start with creating a virtual machine, further install the operating system and deploy the corresponding environment. However, in some use scenarios, it may only be a simple program or just a process. If we still use the traditional virtualization method, it will make a fuss to some extent. However, this can be done through a container The container does not need to create a virtual machine, nor does it need to install additional operating system. It only needs one command to complete the operation of the container.

3. Safety

In traditional computers, applications are all running in the same operating system. However, when packaged in a container, the applications can run more safely in the container because it is strong in isolation.

See Fig. 4-2 for the official website interface of the Docker open source unit. Here the most

authoritative documents, operation instances and best practices are available.

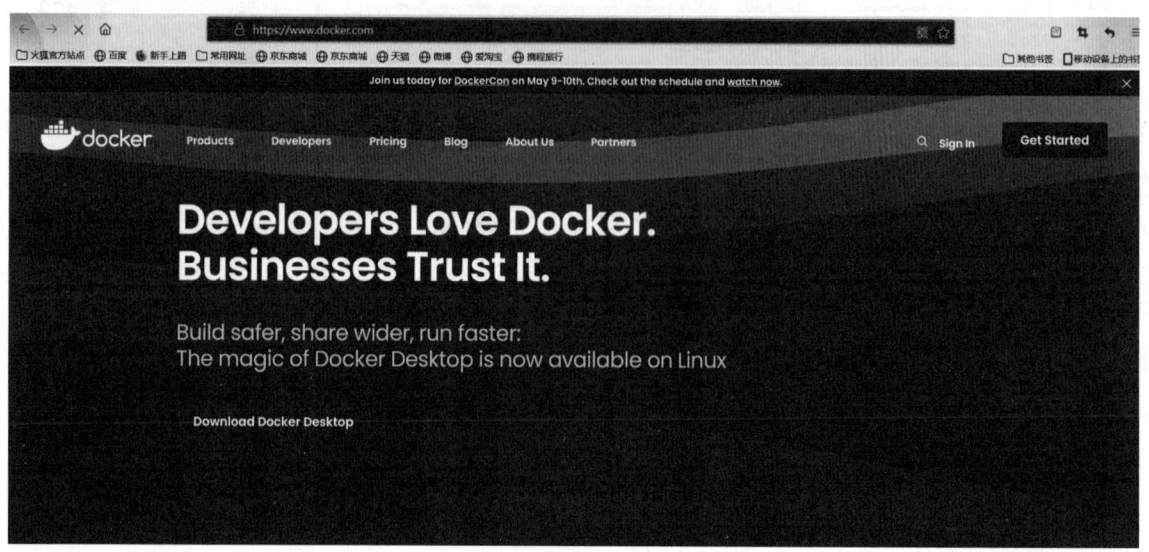

Fig. 4-2　Docker official website interface

4.2　Docker architecture

Docker is a C/S (Client/Server) structured system. The client sends requests to the Docker daemon, that is, its server. The daemon is responsible for building, running and distributing dockers. The core components of Docker include Docker client—Client, Docker server—Docker daemon, Docker image—Images, image warehouse—Registry, and Docker container—Containers. Fig. 4-3 shows the Docker architecture.

The whole life cycle of the container is completed by the cooperation of various components. To run a container, you must first pull the image from the warehouse registry to the local. Warehouses include public and private ones. Docker uses the Docker Hub warehouse by default. This is the image warehouse maintained by Docker company, in which tens of thousands of container images are maintained for users. After obtaining an image, instantiate the image, start and run it, and then obtain a container. In this process, such operations as image pulling and container startup are implemented through the Docker command tool on the client side. The client sends these operation requests to the server Docker daemon, and the server program on the host completes the creation, operation, and monitoring of the container. On the contrary, sometimes when a container is built and you want to keep the current state, package it and run it in other circumstances, you need to submit the container, build it as an image and push it to the warehouse so that it can be pulled and instantiated in other circumstances. See Fig. 4-4 for the above process.

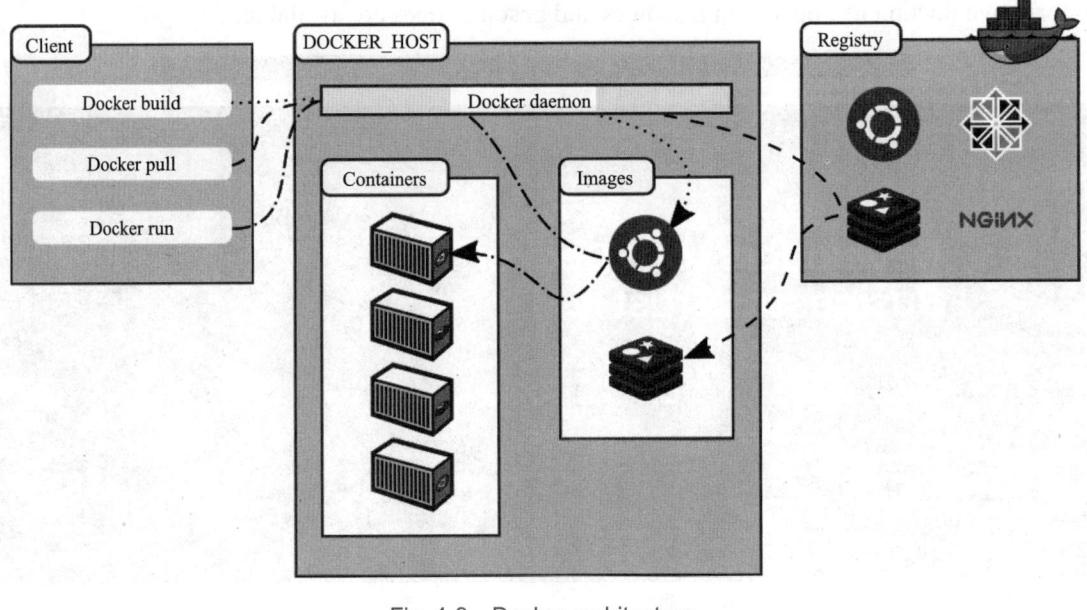

Fig. 4-3   Docker architecture

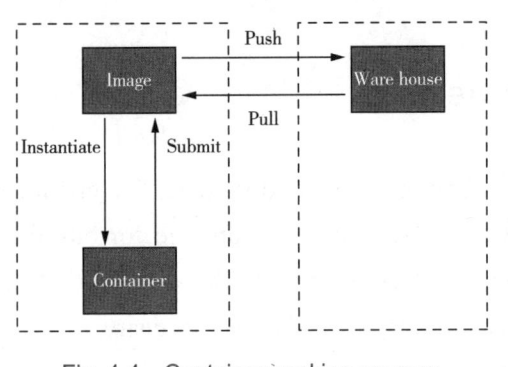

Fig. 4-4   Container working process

The entire life cycle of a container starts with an image. Firstly, check whether the required image exists locally. If not, pull it from the warehouse; then, use the pulled local image to start the container and allocate a file system, that is, a container writable layer shall be added on the top layer based on the original image and this is a layer where users can write new data; next, the host bridge assigns a virtual interface to the container, and assigns an available IP address to the container from the address pool, so that users can access the container through the network; execute user-specified instructions in the container; finally, the container is terminated after business execution is completed. The life cycle process of the entire container is shown in Fig. 4-5.

The container image adopts a hierarchical structure. The bottom layer is the shared kernel. On this basis, the first layer is the base image, and the top layer is the writable container layer. In the container hierarchy, each layer is software added on the basis of the lower layer, which is called the parent image of the upper layer. Below the container layer is the read-only layer. See Fig. 4-6

for the hierarchical structure of the container. This hierarchical structure of containers can fully realize resource sharing. The Host will only save one base image on the disk and be shared by all containers. Each container will add its own writable layer accordingly.

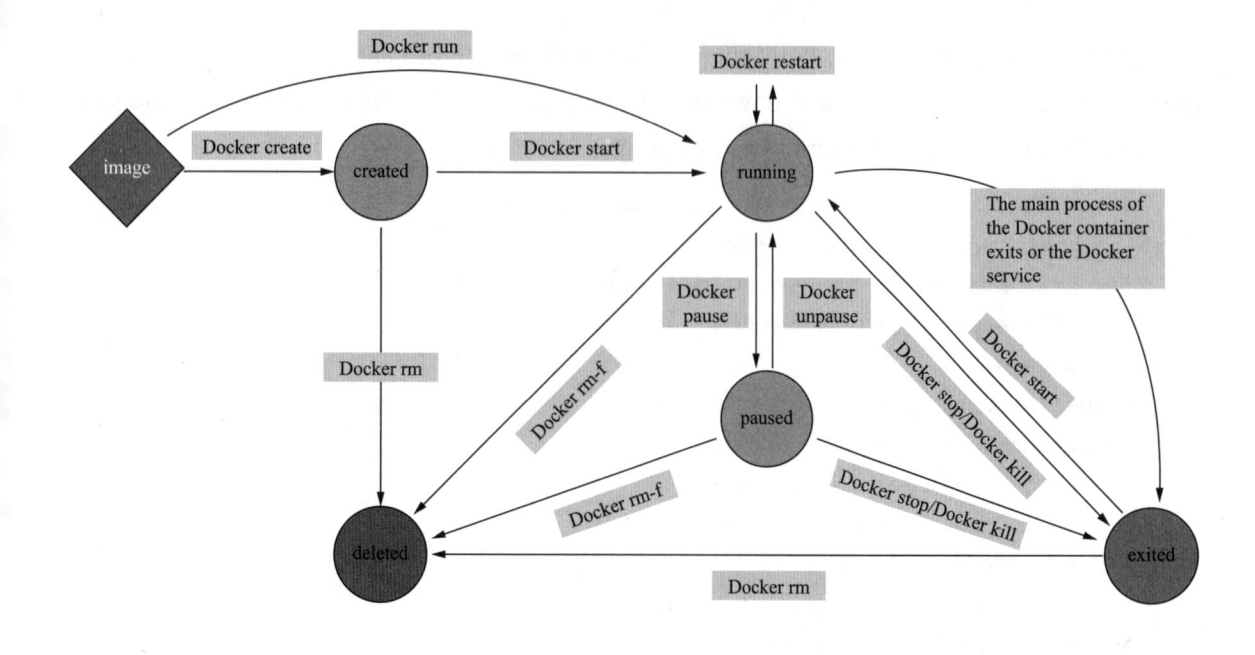

Fig. 4-5   The life cycle process of the entire container

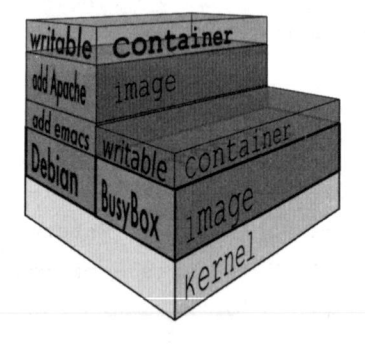

Fig. 4-6   The hierarchical structure of the container

4.3   Project practice—use Docker engine technology to manage containers

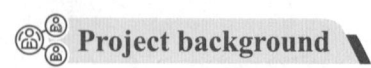 **Project background**

A company needs to migrate its business to the container cloud due to its business needs. It needs the operation and maintenance personnel to first build the container operation environment

and master the simple operation and maintenance methods involved in the image and container.

 **Project implementation**

Task 1　Install container engine

Build a yum source. First, build the base network source of centos, install the yum-utils tool using the network source, and download the configuration file docker-ce.repo for the yum warehouse source of the docker-ce software package.

```
[root@localhost ~]# curl -o /etc/yum.repos.d/CentOS-Base.repo https://mir-
rors.aliyun.com/repo/Centos-7.repo
[root@localhost ~]# yum -y install yum-utils
[root@localhost ~]# yum-config-manager \
--add-repo \
http://mirrors.aliyun.com/docker-ce/linux/centos/docker-ce.repo
[root@localhost ~]# yum repolist
Loaded plugins: fastestmirror
Loading mirror speeds from cached hostfile
* base: mirrors.aliyun.com
* epel: my.mirrors.thegigabit.com
* extras: mirrors.aliyun.com
* updates: mirrors.aliyun.com
docker-ce-stable                              | 3.5 kB  00:00:00
(1/2): docker-ce-stable/7/x86_64/primary_db                | 70 kB 00:00:00
(2/2): docker-ce-stable/7/x86_64/updateinfo                | 55 B 00:00:00
repo id                      repo name                        status
base/7/x86_64                CentOS-7 - Base - mirrors.aliyun.com  10,072
docker-ce-stable/7/x86_64    Docker CE Stable - x86_64             139
*epel/x86_64          Extra Packages for Enterprise Linux 7 - x86_64 13,712
extras/7/x86_64              CentOS-7 - Extras - mirrors.aliyun.com  500
local                        local                                3,971
updates/7/x86_64             CentOS-7 - Updates - mirrors.aliyun.com 3,252
repolist: 31,646
```

Install docker-ce and start the service.

```
[root@localhost ~]# yum -y install docker-ce
[root@localhost ~]# systemctl enable docker
[root@localhost ~]# systemctl start docker
```

After the installation is successful, use the docker info to view the docker information.

```
[root@localhost ~]# docker info
```

```
Client:
Context:    default
Debug Mode: false
Plugins:
app: Docker App (Docker Inc., v0.9.1-beta3)
buildx: Docker Buildx (Docker Inc., v0.7.1-docker)
scan: Docker Scan (Docker Inc., v0.12.0)

Server:
Containers: 0
Running: 0
Paused: 0
Stopped: 0
Images: 0
Server Version: 20.10.12
Storage Driver: overlay2
Backing Filesystem: xfs
Supports d_type: true
Native Overlay Diff: true
userxattr: false
Logging Driver: json-file
Cgroup Driver: cgroupfs
Cgroup Version: 1
Plugins:
Volume: local
Network: bridge host ipvlan macvlan null overlay
Log: awslogs fluentd gcplogs gelf journald json-file local logentries splunk
syslog
Swarm: inactive
Runtimes: io.containerd.runc.v2 io.containerd.runtime.v1.linux runc
Default Runtime: runc
Init Binary: docker-init
containerd version: 7b11cfaabd73bb80907dd23182b9347b4245eb5d
runc version: v1.0.2-0-g52b36a2
init version: de40ad0
Security Options:
seccomp
Profile: default
Kernel Version: 3.10.0-862.el7.x86_64
Operating System: CentOS Linux 7 (Core)
```

```
OSType: linux
Architecture: x86_64
CPUs: 2
Total Memory: 1.936GiB
Name: localhost.localdomain
ID: HWX4:KNYS:IDY6:T3LF:N6B4:GTXW:UAE5:UET4:QVFH:34GE:FFER:AZCK
Docker Root Dir: /var/lib/docker
Debug Mode: false
Registry: https://index.docker.io/v1/
Labels:
Experimental: false
Insecure Registries:
127.0.0.0/8
Live Restore Enabled: false
```

So far, docker-ce has been successfully installed. Next, configure the docker process configuration file daemon.json. In the configuration file, you can set data storage path graph, store driver type storage-driver, and add trust zone insecure-registries, docker network bridge IP address bip and other parameters.

```
[root@localhost ~]# vi /etc/docker/daemon.json
{
"graph": "/data/docker",
"storage-driver": "overlay",
"insecure-registries": ["registry.access.redhat.com","quay.io","harbor.
od.com"],
"bip": "172.7.21.1/24",
"exec-opts": ["native.cgroupdriver=systemd"]
}
[root@master ~]# systemctl daemon-reload
[root@localhost ~]# systemctl restart docker
```

Task 2　Start the simplest container "hello world"

Use the "docker run" command to run a container of hello-world images. This is the simplest container image, generally for testing. After the container starts, print a message "Hello from Docker!".

```
[root@localhost ~]# docker run hello-world
Unable to find image 'hello-world:latest' locally
latest: Pulling from library/hello-world
2db29710123e: Pull complete
```

```
Digest: sha256:2498fce14358aa50ead0cc6c19990fc6ff866ce72aeb5546e1d59caac3d0d60f
Status: Downloaded newer image for hello-world:latest
Hello from Docker!
This message shows that your installation appears to be working correctly.

To generate this message, Docker took the following steps:
1. The Docker client contacted the Docker daemon.
2. The Docker daemon pulled the "hello-world" image from the Docker Hub.
(amd64)
3. The Docker daemon created a new container from that image which runs the
executable that produces the output you are currently reading.
4. The Docker daemon streamed that output to the Docker client, which sent
it to your terminal.

To try something more ambitious, you can run an Ubuntu container with:
$ docker run -it ubuntu bash

Share images, automate workflows, and more with a free Docker ID:
https://hub.docker.com/

For more examples and ideas, visit:
https://docs.docker.com/get-started/
```

Task 3　Manage images

1. Connect the image warehouse

The container startup needs to be instantiated by an image. Docker uses the image warehouse "Docker Hub" by default. To use "Docker Hub", you must register first. Visit the official website of "Docker Hub" and complete the registration according to the prompts, as shown in Fig. 4-7.

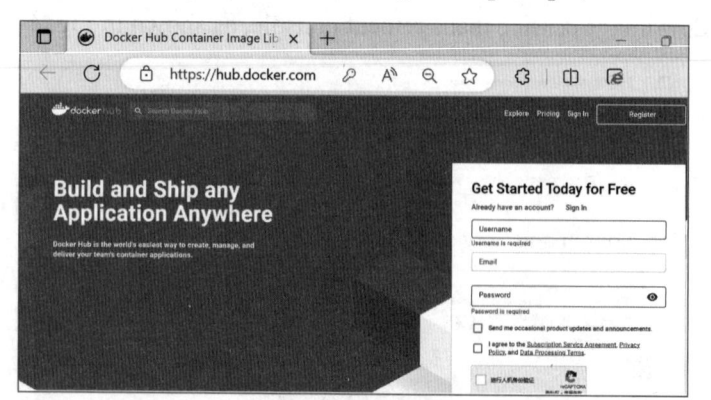

Fig. 4-7　Docker Hub registration interface

Log in to Docker Hub.

```
[root@localhost ~]# docker login docker.io
Login with your Docker ID to push and pull images from Docker Hub. If you
don't have a Docker ID, head over to https://hub.docker.com to create one.
Username: test123
Password:
WARNING! Your password will be stored unencrypted in /root/.docker/config.
json.
Configure a credential helper to remove this warning. See
https://docs.docker.com/engine/reference/commandline/login/#creden-
tials-store
Login Succeeded
```

It is slow to pull images from Docker Hub, so you can configure an image accelerator; moreover, you can choose to use Aliyun's accelerator product, that is, select "products" → "Containers and middleware" → "Container Image Service ACR" command, as shown in Fig. 4-8.

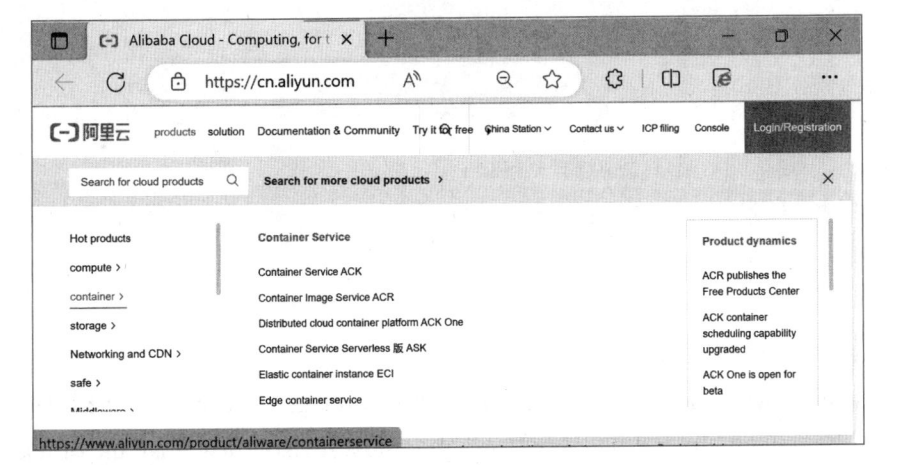

Fig. 4-8　Use Aliyun Accelerator

Enter the window image service management console, as shown in Fig. 4-9.

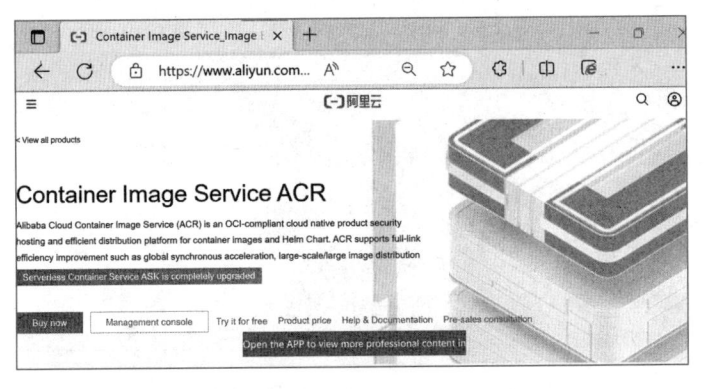

Fig. 4-9　Enter the container image service management console

Copy the accelerator address and add the warehouse image address in daemon.json to configure registry-mirrors, as shown in Fig. 4-10, and get the configuration information of the image accelerator.

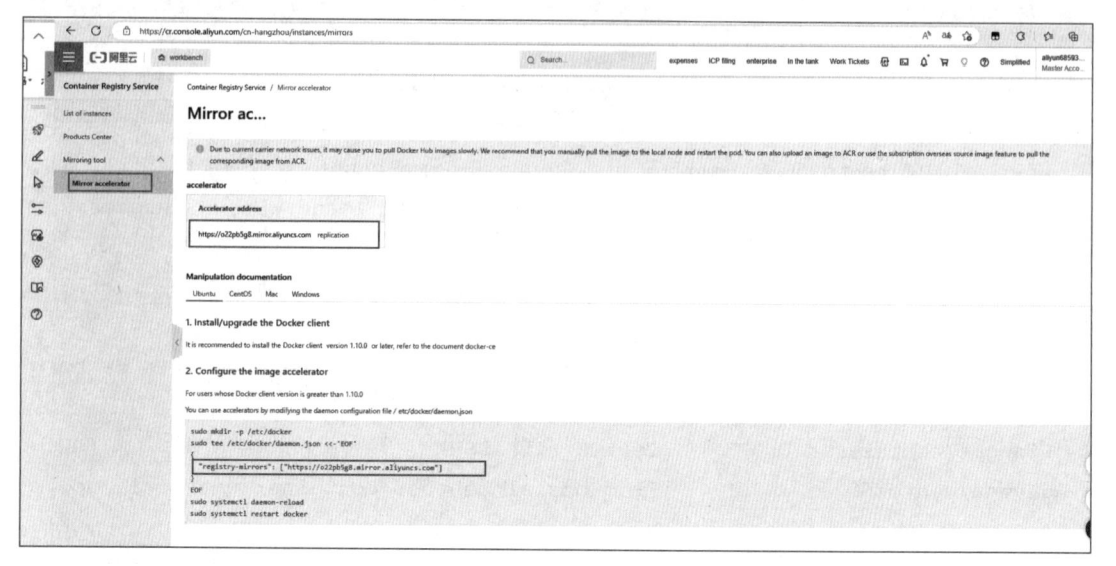

Fig. 4-10   Get image accelerator configurations

Follow the configuration information displayed on the page to write the configuration file daemon.json.

```
[root@localhost ~]# cat /etc/docker/daemon.json
{
"graph": "/data/docker",
"storage-driver": "overlay",
"insecure-registries": ["registry.access.redhat.com","quay.io","harbor.
od.com"],
"registry-mirrors": ["https://jctfhowk.mirror.aliyuncs.com"],
"bip": "172.7.21.1/24",
"exec-opts": ["native.cgroupdriver=systemd"],
"live-restore": true
}
[root@localhost ~]# systemctl restart docker
```

2. Search images

You can use "docker search" to search images and the search results will return the image name, descriptions and other information available in the warehouse.

```
[root@localhost ~]# docker search centos
NAME                          DESCRIPTION       STARS    OFFICIAL    AUTOMATED
centos                        The official build of CentOS.    6962    [OK]
```

```
ansible/centos7-ansible        Ansible on Centos7              135    [OK]
consol/centos-xfce-vnc         Centos container with "headless" VNC session…
                                                              132    [OK]
jdeathe/centos-ssh             OpenSSH / Supervisor / EPEL/IUS/SCL Repos - …
                                                              121    [OK]
centos/systemd                 systemd enabled base container. 105   [OK]
centos/mysql-57-centos7        MySQL 5.7 SQL database server   92
imagine10255/centos6-lnmp-php56    centos6-lnmp-php56          58    [OK]
tutum/centos                   Simple CentOS docker image with SSH access
                                                               48
centos/postgresql-96-centos7 PostgreSQL is an advanced Object-Relational …
                                                               45
centos/httpd-24-centos7        Platform for running Apache httpd 2.4 or bui…
                                                               41
kinogmt/centos-ssh             CentOS with SSH                 29    [OK]
guyton/centos6                 From official centos6 container with full up…
                                                               10    [OK]
centos/tools                   Docker image that has systems administration…
                                                                7    [OK]
drecom/centos-ruby             centos ruby                      6    [OK]
centos/redis                   Redis built for CentOS           6    [OK]
mamohr/centos-java             Oracle Java 8 Docker image based on Centos 7
                                                                4    [OK]
roboxes/centos8                A generic CentOS 8 base image.   4
darksheer/centos               Base Centos Image -- Updated hourly 3    [OK]
dokken/centos-7                CentOS 7 image for kitchen-dokken 2
miko2u/centos6                 CentOS6 日本语环境                  2    [OK]
amd64/centos                   The official build of CentOS.    2
mcnaughton/centos-base         centos base image                1    [OK]
blacklabelops/centos           CentOS Base Image! Built and Updates Daily!
                                                                1    [OK]
starlabio/centos-native-build  Our CentOS image for native builds  0    [OK]
smartentry/centos              centos with smartentry           0    [OK]
```

3. Download an image

Use "docker pull" to pull the image to the local memory. Use "docker images" and you can view the list of local images.

```
[root@localhost ~]# docker pull centos
Using default tag: latest
latest: Pulling from library/centos
```

```
a1d0c7532777: Pull complete
Digest: sha256:a27fd8080b517143cbbbab9dfb7c8571c40d67d534bbdee55bd6c473f432b177
Status: Downloaded newer image for centos:latest
[root@localhost ~]# docker.io/library/centos:latest
[root@localhost ~]#  docker pull centos:centos7
centos7: Pulling from library/centos
2d473b07cdd5: Pull complete
Digest: sha256:9d4bcbbb213dfd745b58be38b13b996ebb5ac315fe75711bd618426a630e0987
Status: Downloaded newer image for centos:centos7
docker.io/library/centos:centos7
[root@localhost ~]# docker images
REPOSITORY         TAG          IMAGE ID        CREATED          SIZE
hello-world        latest       feb5d9fea6a5    3 months ago     13.3KB
centos             centos7      eeb6ee3f44bd    3 months ago     204MB
centos             latest       5d0da3dc9764    3 months ago     231MB
[root@localhost ~]# docker tag eeb6ee3f44bd  docker.io/test123/centos:centos7
[root@localhost ~]# docker images
REPOSITORY         TAG          IMAGE ID        CREATED          SIZE
hello-world        latest       feb5d9fea6a5    3 months ago     13.3KB
test123/centos     centos7      eeb6ee3f44bd    3 months ago     204MB
centos             centos7      eeb6ee3f44bd    3 months ago     204MB
centos             latest       5d0da3dc9764    3 months ago     231MB
```

4. Push images

Use "docker push" to push an image to the warehouse. The name of the image follows the structure below: "${registry_name}/${repository_ name}/${image_name}: ${tag_name}". Where, ${registry_name}/${repository_name} is the warehouse name, ${image_name} is the image name, and ${tag_name} is the version number. For example, the name of an image is docker. io/library/ centos: latest, which refers to the centos image in docker. io/library, with the version name of "latest"; docker.io/test123/ ubuntu:v1.1.0 refers to the ubuntu image in docker.io/test123 with the version name of v1.1.0.

```
[root@localhost ~]# docker push docker.io/test123/centos:centos7
The push refers to repository [docker.io/test123/centos]
174f56854903: Layer already exists
centos7: digest: sha256:dead07b4d8ed7e29e98de0f4504d87e8880d4347859d839686a31da-
35a3b532f size: 529
```

After the image is pushed, log in "Docker Hub" and you can see that the local image has been pushed to the warehouse, as shown in Fig. 4-11 and Fig. 4-12.

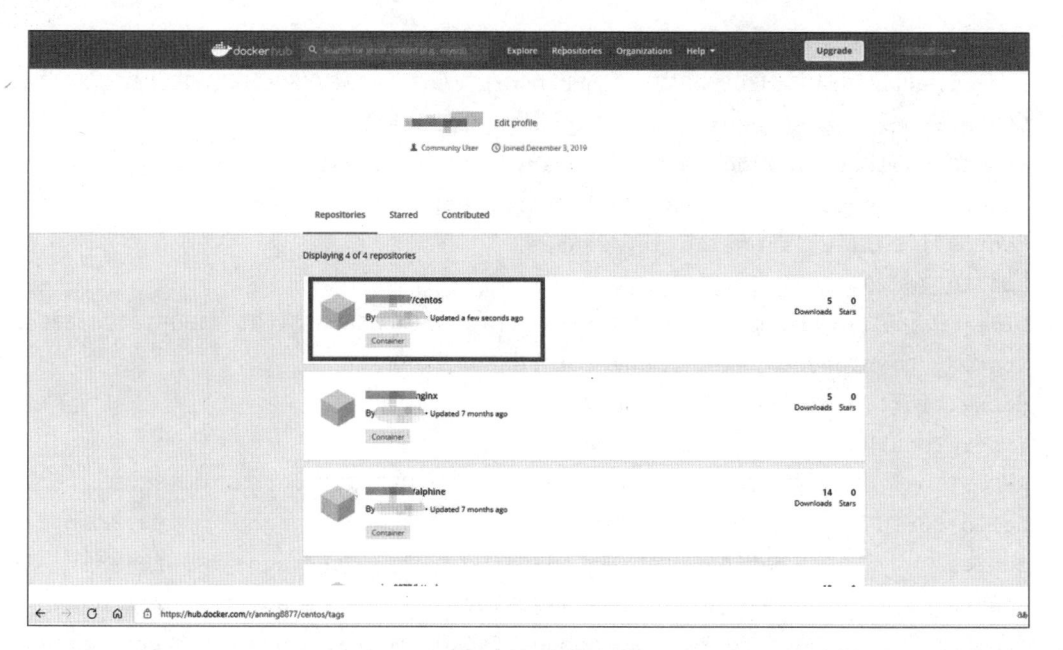

Fig. 4-11　List of images in the warehouse

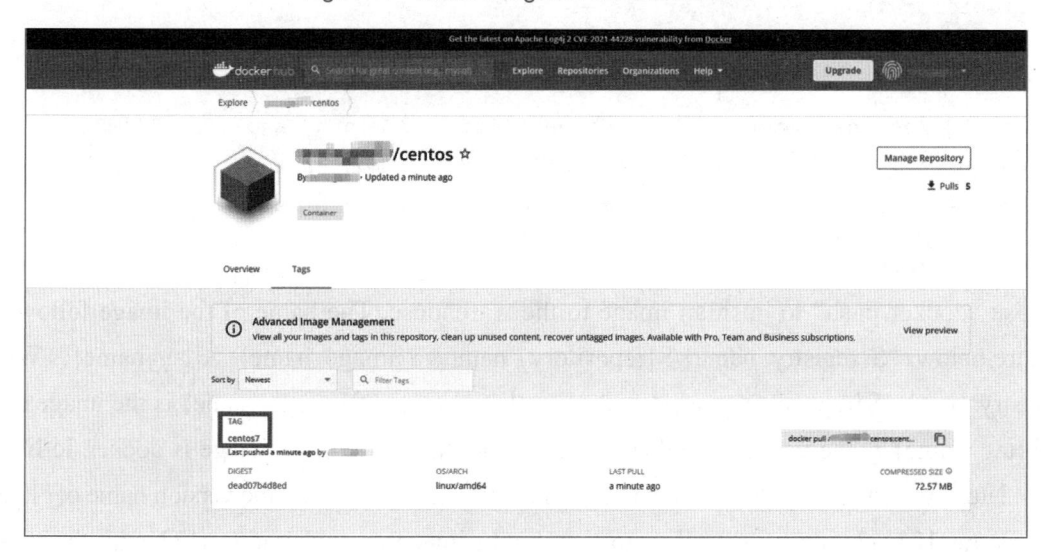

Fig. 4-12　Image details

5. Delete images

Use the "docker rmi" command to delete local images.

```
[root@localhost ~]# docker rmi docker.io/test123/centos:centos7
 Untagged: test123/centos:centos7
 Untagged: test123/centos@sha256:dead07b4d8ed7e29e98de0f4504d87e8880d-
4347859d839686a31da35a3b532f
 [root@localhost ~]# docker rmi -f eeb6ee3f44bd
 Untagged: test123/centos:centos7
```

```
Untagged: centos:centos7
Untagged: centos@sha256:9d4bcbbb213dfd745b58be38b13b996ebb5ac315fe75711b-
d618426a630e0987
Deleted: sha256:eeb6ee3f44bd0b5103bb561b4c16bcb82328cfe5809ab675bb17ab3a16c517c9
Deleted: sha256:174f5685490326fc0a1c0f5570b8663732189b327007e47ff13d2ca59673db024.
```

Task 4　Container management

1. View container process

```
[root@localhost ~]# docker ps
CONTAINER ID    IMAGE       COMMAND    CREATED    STATUS      PORTS      NAMES
[root@localhost ~]# docker ps -a
CONTAINER ID    IMAGE        COMMAND     CREATED     STATUS    PORTS     NAMES
5778103fc2bc    hello-world   "/hello"    20 hours ago    Exited (0) 20 hours
ago eloquent_hellman
```

2. Start container

The command to start the container is "docker run", which can be used as follows:

```
docker run [OPTIONS] IMAGE [COMMAND] [ARG...]
```

OPTIONS:

-i: Start the container interactively.

-t: Use the standard input and output associated with the terminal to the container.

-d: Run in the background.

--rm: Exit and delete the container.

--name: Define the container name.

COMMAND: The command to run when starting the container.

Here follows an example of starting the centos container interactively and then it will enter the bash interface of the system. Interactively, you can enter the container for operation.

```
[root@localhost ~]# docker run -it centos:latest /bin/bash
[root@7a1565402db7 /]# ls
bin  dev  etc  home  lib  lib64  lost+found  media  mnt  opt  proc  root
run  sbin  srv  sys  tmp  usr  var
[root@localhost ~]# docker ps -a
CONTAINER ID    IMAGE        COMMAND      CREATED      STATUS     PORTS      NAMES
ccbd2941c4f7    centos:latest  "/bin/bash"  6 seconds ago    Up 5 seconds
heuristic_gates
5778103fc2bc    hello-world  "/hello"  20 hours ago    Exited (0) 20 hours ago
eloquent_hellman
```

Use the "--rm" parameter to exit and delete the container. In many scenarios, only a simple process is temporarily running in the container. Then, it doesn't function and can be deleted automatically in this way.

```
[root@localhost ~]# docker run --rm centos /bin/echo hello
hello
[root@localhost ~]# docker ps -a
CONTAINER ID    IMAGE    COMMAND    CREATED    STATUS    PORTS    NAMES
ccbd2941c4f7    centos:latest "/bin/bash" About a minute ago  Exited (0) 34
seconds ago heuristic_gates
5778103fc2bc    hello-world  "/hello"    20 hours ago    Exited (0) 20 hours
ago eloquent_hellman
```

When the container needs to hang up in the background for continuous operation, you can use the "- d" parameter to start the container in the background.

```
[root@localhost ~]# docker run -d --name test1 centos /bin/sleep 300
7aee3bb74a5dc395a7a46262cfd7fd60c83044bf6f42aaf86f436a7842309973
[root@localhost ~]# docker ps
CONTAINER ID IMAGE    COMMAND    CREATED    STATUS    PORTS    NAMES
7aee3bb74a5d centos  "/bin/sleep 300"  4 seconds ago  Up 3 seconds    test1
[root@localhost ~]# ps aux|grep sleep
root   65969 0.1 0.0 23032  936 ?    Ss  23:45 0:00 /usr/bin/coreutils --coreuti-
ls-prog-shebang=sleep /bin/sleep 300
```

3. Enter the container

Use the "docker exec" command to enter the container.

```
[root@localhost ~]# docker ps
CONTAINER ID  IMAGE  COMMAND CREATED STATUS PORTS NAMES
7aee3bb74a5d centos "/bin/sleep 300" 3 minutes ago  Up 3 minutes    test1
[root@localhost ~]# docker exec -it 7aee3bb74a5d /bin/bash
[root@7aee3bb74a5d /]# ps aux
USER    PID %CPU %MEM    VSZ   RSS TTY    STAT START    TIME COMMAND
root    1  0.0 0.0  23032  936 ?      Ss 15:45 0:00/usr/bin/coreutils
--coreutils-prog-shebang=sleep /bin/sleep 300
root    8  0.5  0.1  12036  2096 pts/0      Ss    15:49 0:00 /bin/bash
root    22 0.0  0.0  44652  1776 pts/0      R+    15:49 0:00 ps aux
```

4. Stop the container

Use the "docker stop" command to stop the container.

```
[root@localhost ~]# docker ps
```

```
CONTAINER ID   IMAGE     COMMAND       CREATED      STATUS      PORTS      NAMES
7aee3bb74a5d   centos    "/bin/sleep 300"  4 minutes ago   Up 4 minutes        test1
[root@localhost ~]# docker stop 7aee3bb74a5d
7aee3bb74a5d
[root@localhost ~]# docker ps
CONTAINER ID   IMAGE     COMMAND       CREATED      STATUS      PORTS      NAMES
```

5. Restart the container

Use the "docker restart" command to restart the container.

```
[root@localhost ~]# docker ps
CONTAINER ID   IMAGE     COMMAND       CREATED      STATUS      PORTS  ·  NAMES
[root@localhost ~]# docker ps -a
CONTAINER ID   IMAGE       COMMAND       CREATED      STATUS        PORTS      NAMES
7aee3bb74a5d   centos     "/bin/sleep 300" 6 minutes ago  Exited (0) About a
minute ago              test1
5778103fc2bc   hello-world  "/hello"      21 hours ago   Exited (0) 21 hours
ago eloquent_hellman
[root@localhost ~]# docker restart 7aee3bb74a5d
7aee3bb74a5d
[root@localhost ~]# docker ps
CONTAINER ID   IMAGE     COMMAND       CREATED      STATUS      PORTS      NAMES
7aee3bb74a5d   centos    "/bin/sleep 300"  6 minutes ago  Up 2 seconds       test1
```

6. Delete the container

Use the "docker rm" command to delete a container. If you delete a running container, add - f for forced deletion.

```
[root@localhost ~]# docker rm test1
Error response from daemon: You cannot remove a running container 7aee3b-
b74a5dc395a7a46262cfd7fd60c83044bf6f42aaf86f436a7842309973. Stop the container before at-
tempting removal or force remove
[root@localhost ~]# docker rm -f test1
test1
```

To delete the exited containers in batches sometimes, you can use the text processing tool "awk" in the shell to obtain the ID of the exited container and further use circular statements to delete in batches.

```
[root@localhost ~]# docker ps -a
CONTAINER ID   IMAGE     COMMAND       CREATED      STATUS         PORTS      NAMES
dd05fc14a462   centos    "/bin/bash"   10 minutes ago  Exited (0) 10 minutes
```

```
ago test
    ccbd2941c4f7   centos:latest   "/bin/bash"   13 minutes ago     Exited (0) 12
minutes ago              heuristic_gates
    70cc518f6fc1   centos:latest   "pwd"      14 minutes ago    Exited (0) 14 minutes
ago            busy_feynman
    7a1565402db7   centos:latest "/bin/bash" 15 minutes ago    Exited (0) 14 min-
utes ago              beautiful_noether
    80bea167a41c   centos:latest "pwd"      15 minutes ago   Exited (0) 15 minutes
ago              reverent_hermann
    92db7da2f9cf   centos:latest "/bin/bash"   15 minutes ago     Exited (0) 15 min-
utes ago              objective_hugle
    306b4ab346cd   centos:latest "/bin/bash"   15 minutes ago     Exited (0) 15 min-
utes ago              blissful_lehmann
    5778103fc2bc   hello-world   "/hello"      21 hours ago     Exited (0) 21 hours
ago              eloquent_hellman
    [root@localhost ~]# for i in `docker ps -a|grep -i exit|awk '{print $1}'`;do
docker rm -f $i;done
    dd05fc14a462
    ccbd2941c4f7
    70cc518f6fc1
    7a1565402db7
    80bea167a41c
    92db7da2f9cf
    306b4ab346cd
    5778103fc2bc
    [root@localhost ~]# docker ps -a
    CONTAINER ID   IMAGE     COMMAND     CREATED     STATUS     PORTS     NAMES
```

7. Modify the container

Sometimes you want to solidify the changes to the basic image to the read-only layer and build a new image. At this time, you can rebuild the image through the "docker commit" command. In this instance, based on the centos image, the written txt document is solidified into the image. When a new image is used to start the container, you can see that the added txt document has been solidified to the read-only layer after the container is started.

```
[root@localhost ~]# docker run -it centos /bin/bash
[root@4c8eed52f1c5 /]# vi 1.txt
hello world!
[root@4c8eed52f1c5 /]# ls -l
total 4
```

```
-rw-r--r--      1 root root      13 Jan    8    17:38 1.txt
lrwxrwxrwx      1 root root       7 Nov    3    2020 bin -> usr/bin
drwxr-xr-x      5 root root     360 Jan    8    17:38 dev
drwxr-xr-x      1 root root      66 Jan    8    17:38 etc
drwxr-xr-x      2 root root       6 Nov    3    2020 home
lrwxrwxrwx      1 root root       7 Nov    3    2020 lib -> usr/lib
lrwxrwxrwx      1 root root       9 Nov    3    2020 lib64 -> usr/lib64
drwx------      2 root root       6 Sep   15    14:17 lost+found
drwxr-xr-x      2 root root       6 Nov    3    2020 media
drwxr-xr-x      2 root root       6 Nov    3    2020 mnt
drwxr-xr-x      2 root root       6 Nov    3    2020 opt
dr-xr-xr-x    127 root root       0 Jan    8    17:38 proc
dr-xr-x---      2 root root     162 Sep   15    14:17 root
drwxr-xr-x     11 root root     163 Sep   15    14:17 run
lrwxrwxrwx      1 root root       8 Nov    3    2020 sbin -> usr/sbin
drwxr-xr-x      2 root root       6 Nov    3    2020 srv
dr-xr-xr-x     13 root root       0 Jan    8    17:38 sys
drwxrwxrwt      7 root root     171 Sep   15    14:17 tmp
drwxr-xr-x     12 root root     144 Sep   15    14:17 usr
drwxr-xr-x     20 root root     262 Sep   15    14:17 var
[root@4c8eed52f1c5 /]# exit
exit
[root@localhost ~]# docker commit -p 4c8eed52f1c5 test123/centos_with_1.txt
sha256:5268290bcab95fabfbaaeb7a602395e885a3803c595db362418adfae485d3ad3
[root@localhost ~]# docker images
REPOSITORY                     TAG      IMAGE ID       CREATED          SIZE
test123/centos_with_1.txt      latest   5268290bcab9   18 seconds ago   231MB
hello-world                    latest   feb5d9fea6a5   3 months ago     13.3KB
centos                         latest   5d0da3dc9764   3 months ago     231MB
[root@localhost ~]# docker run -it test123/centos_with_1.txt /bin/bash
[root@3bf91a5992b2 /]# ls
1.txt  bin  dev  etc  home  lib  lib64  lost+found  media  mnt  opt  proc
root  run  sbin  srv  sys  tmp  usr  var
[root@3bf91a5992b2 /]# cat 1.txt
hello world!
```

8. Export images

You can export images by packaging and use "docker save" to save the images and package them as a tar file.

```
[root@localhost ~]# docker images
```

```
REPOSITORY                      TAG     IMAGE ID      CREATED          SIZE
test123/centos_with_1.txt       latest  5268290bcab9 6 minutes ago    231MB
hello-world                     latest  feb5d9fea6a5 3 months ago     13.3KB
centos                          latest  5d0da3dc9764 3 months ago     231MB
[root@localhost ~]# docker save 5268290bcab9 centos_with_1.txt.tar
cowardly refusing to save to a terminal. Use the -o flag or redirect
[root@localhost ~]# docker save 5268290bcab9 > centos_with_1.txt.tar
[root@localhost ~]# ll
total 233016
-rw-r--r--  1 root root        11 Sep 14 00:20 1
-rw-r--r--  1 root root         7 Sep 14 00:21 2
drwxr-xr-x  2 root root         6 Sep 13 18:15 abc
-rw-r--r--  1 root root         8 Sep 14 00:24 a.exe
-rw-------. 1 root root      1381 Apr  8  2021 anaconda-ks.cfg
-rw-r--r--  1 root root         0 Sep 14 00:18 a.txt
-rw-r--r--  1 root root         0 Sep 13 19:36 b.txt
-rw-r--r--  1 root root 238586880 Jan  9 01:47 centos_with_1.txt.tar
-rw-r--r--  1 root root         6 Sep 13 19:37 t5
```

If an image is deleted, you can restore it from the tar package.

```
[root@localhost ~]# docker rmi -f 5268290bcab9
Untagged: test123/centos_with_1.txt:latest
Deleted: sha256:5268290bcab95fabfbaaeb7a602395e885a3803c595db362418adfae485d3ad3
[root@localhost ~]# docker images
REPOSITORY       TAG       IMAGE ID        CREATED          SIZE
hello-world      latest    feb5d9fea6a5    3 months ago     13.3KB
centos           latest    5d0da3dc9764    3 months ago     231MB
[root@localhost ~]# docker load < centos_with_1.txt.tar
Loaded image ID: sha256:5268290bcab95fabfbaaeb7a602395e885a3803c595db362418adfae-
485d3ad3
[root@localhost ~]# docker images
REPOSITORY       TAG       IMAGE ID        CREATED          SIZE
<none>           <none>    5268290bcab9    8 minutes ago    231MB
hello-world      latest    feb5d9fea6a5    3 months ago     13.3KB
centos           latest    5d0da3dc9764    3 months ago     231MB
[root@localhost ~]# docker tag 5268290bcab9 test123/centos_with_1.txt
[root@localhost ~]# docker images
REPOSITORY                      TAG     IMAGE ID      CREATED          SIZE
test123/centos_with_1.txt       latest  5268290bcab9 8 minutes ago    231MB
hello-world                     latest  feb5d9fea6a5 3 months ago     13.3KB
```

```
centos                        latest  5d0da3dc9764 3 months ago    231MB
[root@localhost ~]# docker run -it 5268290bcab9 /bin/bash
[root@89190dbb29b9 /]# ls
1.txt  dev  home  lib64        media  opt   root  sbin  sys  usr
bin    etc  lib   lost+found  mnt    proc  run   srv   tmp  var
[root@89190dbb29b9 /]# cat 1.txt
hello world!
```

9. View container logs

You can view the container operation logs via the docker logs + container ID number.

```
[root@localhost ~]# docker ps -a
CONTAINER ID    IMAGE        COMMAND      CREATED      STATUS      PORTS      NAMES
89190dbb29b9    5268290bcab9     "/bin/bash" 40 hours ago  Exited (0) 40 hours
ago         funny_engelbart
3bf91a5992b2    test123/centos_with_1.txt   "/bin/bash"   41 hours ago   Exited
(130) 41 hours ago         goofy_gates
4c8eed52f1c5  centos      "/bin/bash"   41 hours ago  Exited (0) 41 hours ago
romantic_shockley
[root@localhost ~]# docker logs 89190dbb29b9
[root@89190dbb29b9 /]# ls
1.txt  dev  home  lib64            media  opt   root  sbin   sys    usr
bin    etc  lib   lost+found mnt   proc   run   srv   tmp    var
[root@89190dbb29b9 /]# cat 1.txt
hello world!
[root@89190dbb29b9 /]# exit
exit
```

10. Local system and container interaction

The container often undertakes to provide services, which means that it should expose its own port to provide outside services. For example, when a nginx container is running, its port 80 will be mapped to the port 81 of the host for access to the nginx service via "host IP: 81".

```
[root@localhost ~]# docker pull nginx
Using default tag: latest
latest: Pulling from library/nginx
a2abf6c4d29d: Pull complete
a9edb18cadd1: Pull complete
589b7251471a: Pull complete
186b1aaa4aa6: Pull complete
b4df32aa5a72: Pull complete
a0bcbecc962e: Pull complete
```

```
Digest: sha256:0d17b565c37bcbd895e9d92315a05c1c3c9a29f762b011a10c54a66cd53c9b31
Status: Downloaded newer image for nginx:latest
docker.io/library/nginx:latest
[root@localhost ~]# docker images
REPOSITORY                  TAG         IMAGE ID        CREATED        SIZE
test123/centos_with_1.txt   latest      5268290bcab9    41 hours ago   231MB
nginx                       latest      605c77e624dd    11 days ago    141MB
hello-world                 latest      feb5d9fea6a5    3 months ago   13.3KB
centos                      latest      5d0da3dc9764    3 months ago   231MB
[root@localhost ~]# docker tag 605c77e624dd test123/nginx
[root@localhost ~]# docker images
REPOSITORY                  TAG         IMAGE ID        CREATED        SIZE
test123/centos_with_1.txt   latest      5268290bcab9    41 hours ago   231MB
test123/nginx               latest      605c77e624dd    11 days ago    141MB
nginx                       latest      605c77e624dd    11 days ago    141MB
hello-world                 latest      feb5d9fea6a5    3 months ago   13.3KB
centos                      latest      5d0da3dc9764    3 months ago   231MB
[root@localhost ~]# docker run --rm --name mynginx -d -p81:80 test123/nginx
7bba3aa0b54aba442e639fd27f4218a9c0229a1767990fc64938c96008f832cb
[root@localhost ~]# docker ps -a
CONTAINER ID  IMAGE       COMMAND       CREATED      STATUS     PORTS      NAMES
7bba3aa0b54a  test123/nginx    "/docker-entrypoint.…"  4 seconds ago  Up 3
seconds            0.0.0.0:81->80/tcp, :::81->80/tcp  mynginx
89190dbb29b9  5268290bcab9      "/bin/bash"   41 hours ago    Exited (0) 41
hours ago     funny_engelbart
3bf91a5992b2  test123/centos_with_1.txt  "/bin/bash"   41 hours ago    Exited
(130) 41 hours ago    goofy_gates
4c8eed52f1c5  centos      "/bin/bash"   41 hours ago  Exited (0) 41 hours ago
romantic_shockley
```

When accessing the port 81 of the host, you can see the nginx access interface as shown in Fig. 4-13.

When operating a container, you can map local data volumes in the container, for example, if an existing container is providing nginx services and assuming that the website code is updated, you can create a directory locally as the site code storage directory and then map it into the container by means of data volume loading. In this way, if the site code is updated, only the files in the local directory need to be updated, and the container will automatically map synchronously and the site contents can be updated without stopping any service.

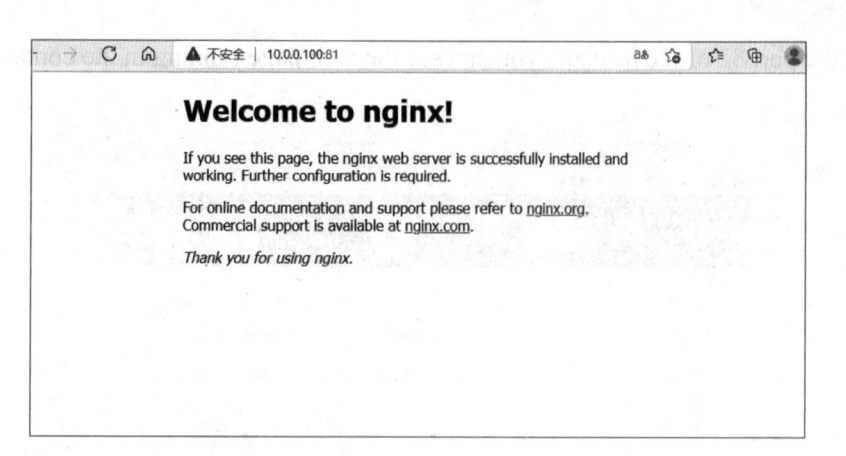

Fig. 4-13   Nginx access interface

```
[root@localhost ~]# mkdir html
[root@localhost ~]# cd html/
[root@localhost html]# wget www.taobao.com -O index.html
--2022-01-10 18:38:43--  http://www.taobao.com/
Resolving www.taobao.com (www.taobao.com)... 219.144.108.233, 219.144.108.232,
240e:bf:c800:2912:3::3fa, ...
Connecting to www.taobao.com (www.taobao.com)|219.144.108.233|:80... connected.
HTTP request sent, awaiting response... 301 Moved Permanently
Location: https://www.taobao.com/ [following]
--2022-01-10 18:38:43--  https://www.taobao.com/
Connecting to www.taobao.com (www.taobao.com)|219.144.108.233|:443... connected.
HTTP request sent, awaiting response... 200 OK
Length: unspecified [text/html]
Saving to: 'index.html'

    [ <=>                            ] 113,640      --.-K/s   in 0.03s

2022-01-10 18:38:43 (3.95 MB/s) - 'index.html' saved [113640]

[root@localhost html]# docker run -d --rm --name mytaobao -p82:80 -v/root/
html:/usr/share/nginx/html test123/nginx
660265c5e6adb442ef71bd4d3c34729cd98d2d11878da7f6aafc99c8feda5e16
[root@localhost html]# docker run -d --rm --name mytaobao -p82:80 -v/root/
html:/usr/share/nginx/html test123/nginx
660265c5e6adb442ef71bd4d3c34729cd98d2d11878da7f6aafc99c8feda5e16
[root@localhost html]# docker exec -it mytaobao /bin/bash
[root@660265c5e6ad:/# ls /usr/share/nginx/html/
index.html
```

Access the port 82 of the host and you can see the web page running in the container, as shown in Fig. 4-14.

Fig. 4-14   Web page running in container

4.4 Project practice—define and run multiple containers based on Docker Compose

 **Project background**

In a business scenario that uses Docker to run services, more than one container is usually required to complete the whole business. If you define and run containers individually, the workload will be relatively huge, but "Docker Compose" can provide a way to define and run multiple containers. To use "Docker Compose", firstly define the environment required by the application in Dockerfile; then write a yml file to configure the services required in the whole business scenario; finally, use the "docker-compose up" command to create and start all services at once.

Now the company needs to build a website to record the site traffic and use containers to realize the business scenarios.

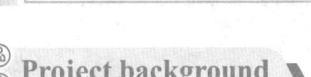 **Project implementation**

Task 1   Install docker-compose

Install the dependent packages required by docker-compose.

```
[root@localhost ~]# yum -y install epel-release
[root@localhost ~]# yum -y install python-pip python3-devel libffi-devel
openssl-dev gcc glibc-devel rust cargo make
```

Download the docker-compose application and grant executable permissions. Link the executable file to /usr/bin/. After the installation is successful, use docker-compose -- version to view docker-compose related information.

```
[root@localhost ~]# curl -L "https://github.com/docker/compose/releases/
download/v1.29.2/docker-compose-$(uname -s)-$(uname -m)" -o /usr/local/bin/dock-
er-compose
[root@localhost ~]# chmod +x /usr/local/bin/docker-compose
[root@localhost ~]# ln -s /usr/local/bin/docker-compose /usr/bin/docker-com-
pose
[root@localhost ~]# docker-compose --version
docker-compose version 1.29.2, build 5becea4c
```

Task 2　Build a website to record site traffic

1. Basic environment settings

Create a new cell directory composetest under the administrator directory.

```
[root@localhost ~]# mkdir composetest
[root@localhost ~]# cd composetest
```

Write the cell application app.py. Use the classic web framework "flask" and the database "redis".

```
[root@localhost composetest]# vi app.py
import time

import redis
from flask import Flask

app = Flask(__name__)
cache = redis.Redis(host='redis', port=6379)

def get_hit_count():
    retries = 5
    while True:
        try:
            return cache.incr('hits')
        except redis.exceptions.ConnectionError as exc:
```

```
             if retries == 0:
                  raise exc
             retries -= 1
             time.sleep(0.5)

@app.route('/')
def hello():
    count = get_hit_count()
    return 'Hello World! I have been seen {} times.\n'.format(count)
```

Set the dependent software packages "flask" and "redis" to be installed.

```
[root@localhost composetest]# vi requirements.txt
flask
redis
```

2. Create Dockerfile

Write Dockerfile, build it from python 3.7, set the working directory to /code, and set the environment variable to run "flask".

```
[root@localhost composetest]# vi Dockerfile
FROM python:3.7-alpine
WORKDIR /code
ENV FLASK_APP=app.py
ENV FLASK_RUN_HOST=0.0.0.0
RUN echo https://mirrors.ustc.edu.cn/alpine/latest-stable/main > /etc/apk/
repositories; \
     echo https://mirrors.ustc.edu.cn/alpine/latest-stable/community >> /etc/
apk/repositories
RUN apk add --update --no-cache gcc musl-dev linux-headers
COPY requirements.txt requirements.txt
RUN pip install -r requirements.txt
EXPOSE 5000
COPY . .
CMD ["flask", "run"]
```

3. Write a yml file to define the service

Define the initial image, port and other related configurations of the two services of "web" and "redis".

```
[root@localhost composetest]# vi docker-compose.yml

version: "3.9"
```

```
services:
  web:
    build: .
    ports:
      - "5000:5000"
  redis:
    image: "redis:alpine"
```

4. Use "docker-compose up" to run the application

```
[root@localhost composetest]# docker-compose up
Building web
Sending build context to Docker daemon  18.43kB
Step 1/11 : FROM python:3.7-alpine
 ---> a1034fd13493
Step 2/11 : WORKDIR /code
 ---> Using cache
 ---> 3d40488c8b7d
Step 3/11 : ENV FLASK_APP=app.py
 ---> Using cache
 ---> d34305afff76
Step 4/11 : ENV FLASK_RUN_HOST=0.0.0.0
 ---> Using cache
 ---> 2ad191618735
Step 5/11 : RUN echo https://mirrors.ustc.edu.cn/alpine/latest-stable/main
> /etc/apk/repositories;  echo https://mirrors.ustc.edu.cn/alpine/latest-stable/
community >> /etc/apk/repositories
 ---> Running in afac6d11bafd
Removing intermediate container afac6d11bafd
 ---> 8fe22683fb43
Step 6/11 : RUN apk add --update --no-cache gcc musl-dev linux-headers
 ---> Running in 100502fca72e
fetch https://mirrors.ustc.edu.cn/alpine/latest-stable/main/x86_64/APKINDEX.
tar.gz
fetch https://mirrors.ustc.edu.cn/alpine/latest-stable/community/x86_64/AP-
KINDEX.tar.gz
(1/13) Installing libgcc (10.3.1_git20211027-r0)
(2/13) Installing libstdc++ (10.3.1_git20211027-r0)
(3/13) Installing binutils (2.37-r3)
(4/13) Installing libgomp (10.3.1_git20211027-r0)
(5/13) Installing libatomic (10.3.1_git20211027-r0)
```

```
(6/13) Installing libgphobos (10.3.1_git20211027-r0)
(7/13) Installing gmp (6.2.1-r0)
(8/13) Installing isl22 (0.22-r0)
(9/13) Installing mpfr4 (4.1.0-r0)
(10/13) Installing mpc1 (1.2.1-r0)
(11/13) Installing gcc (10.3.1_git20211027-r0)
(12/13) Installing linux-headers (5.10.41-r0)
(13/13) Installing musl-dev (1.2.2-r7)
Executing busybox-1.34.1-r3.trigger
OK: 139 MiB in 48 packages
Removing intermediate container 100502fca72e
 ---> d0eb07ac0f8c
Step 7/11 : COPY requirements.txt requirements.txt
 ---> 13243085a68f
Step 8/11 : RUN pip install -r requirements.txt
 ---> Running in a7ab2f639ab5
Collecting flask
  Downloading Flask-2.0.2-py3-none-any.whl (95 KB)
Collecting redis
  Downloading redis-4.1.1-py3-none-any.whl (173 KB)
Collecting click>=7.1.2
  Downloading click-8.0.3-py3-none-any.whl (97 KB)
Collecting Jinja2>=3.0
  Downloading Jinja2-3.0.3-py3-none-any.whl (133 KB)
Collecting Werkzeug>=2.0
  Downloading Werkzeug-2.0.2-py3-none-any.whl (288 KB)
Collecting itsdangerous>=2.0
  Downloading itsdangerous-2.0.1-py3-none-any.whl (18 KB)
Collecting deprecated>=1.2.3
  Downloading Deprecated-1.2.13-py2.py3-none-any.whl (9.6 KB)
Collecting importlib-metadata>=1.0
  Downloading importlib_metadata-4.10.1-py3-none-any.whl (17 KB)
Collecting packaging>=20.4
  Downloading packaging-21.3-py3-none-any.whl (40 KB)
Collecting wrapt<2,>=1.10
  Downloading wrapt-1.13.3-cp37-cp37m-musllinux_1_1_x86_64.whl (78 KB)
Collecting typing-extensions>=3.6.4
  Downloading typing_extensions-4.0.1-py3-none-any.whl (22 KB)
Collecting zipp>=0.5
  Downloading zipp-3.7.0-py3-none-any.whl (5.3 KB)
```

```
Collecting MarkupSafe>=2.0
   Downloading MarkupSafe-2.0.1-cp37-cp37m-musllinux_1_1_x86_64.whl (30 KB)
Collecting pyparsing!=3.0.5,>=2.0.2
   Downloading pyparsing-3.0.6-py3-none-any.whl (97 KB)
Installing collected packages: zipp, typing-extensions, wrapt, pyparsing,
MarkupSafe, importlib-metadata, Werkzeug, packaging, Jinja2, itsdangerous, dep-
recated, click, redis, flask
Successfully installed Jinja2-3.0.3 MarkupSafe-2.0.1 Werkzeug-2.0.2 click-8.0.3
deprecated-1.2.13 flask-2.0.2 importlib-metadata-4.10.1 itsdangerous-2.0.1 packaging-
21.3 pyparsing-3.0.6 redis-4.1.1 typing-extensions-4.0.1 wrapt-1.13.3 zipp-3.7.0
WARNING: Running pip as the 'root' user can result in broken permissions and
conflicting behaviour with the system package manager. It is recommended to use a
virtual environment instead: https://pip.pypa.io/warnings/venv
WARNING: You are using pip version 21.2.4; however, version 21.3.1 is avail-
able.
You should consider upgrading via the '/usr/local/bin/python -m pip install
--upgrade pip' command.
Removing intermediate container a7ab2f639ab5
  ---> 9bc52916561f
Step 9/11 : EXPOSE 5000
  ---> Running in e74001bcf9e7
Removing intermediate container e74001bcf9e7
  ---> 0bd1e7c23eb2
Step 10/11 : COPY . .
  ---> b02f061ec235
Step 11/11 : CMD ["flask", "run"]
  ---> Running in 1bf7da5bd459
Removing intermediate container 1bf7da5bd459
  ---> 278691cb207f
Successfully built 278691cb207f
Successfully tagged composetest_web:latest
WARNING: Image for service web was built because it did not already exist.
To rebuild this image you must use `docker-compose build` or `docker-compose up
--build`.
Pulling redis (redis:alpine)...
alpine: Pulling from library/redis
59bf1c3509f3: Already exists
719adce26c52: Pull complete
b8f35e378c31: Pull complete
d034517f789c: Pull complete
```

```
3772d4d76753: Pull complete
211a7f52febb: Pull complete
Digest: sha256:4bed291aa5efb9f0d77b76ff7d4ab71eee410962965d052552db1fb80576431d
Status: Downloaded newer image for redis:alpine
Creating composetest_web_1   ... done
Creating composetest_redis_1 ... done
Attaching to composetest_redis_1, composetest_web_1
redis_1  | 1:C 18 Jan 2022 10:08:48.953 # oO0OoO0OoO0Oo Redis is starting
oO0OoO0OoO0Oo
redis_1  | 1:C 18 Jan 2022 10:08:48.954 # Redis version=6.2.6, bits=64, com-
mit=00000000, modified=0, pid=1, just started
redis_1  | 1:C 18 Jan 2022 10:08:48.954 # Warning: no config file specified,
using the default config. In order to specify a config file use redis-server /path/
to/redis.conf
redis_1  | 1:M 18 Jan 2022 10:08:48.955 * monotonic clock: POSIX clock_get-
time
redis_1  | 1:M 18 Jan 2022 10:08:48.956 * Running mode=standalone,
port=6379.
redis_1  | 1:M 18 Jan 2022 10:08:48.956 # WARNING: The TCP backlog setting
of 511 cannot be enforced because /proc/sys/net/core/somaxconn is set to the
lower value of 128.
redis_1  | 1:M 18 Jan 2022 10:08:48.956 # Server initialized
redis_1  | 1:M 18 Jan 2022 10:08:48.956 # WARNING overcommit_memory is set
to 0! Background save may fail under low memory condition. To fix this issue add
'vm.overcommit_memory = 1' to /etc/sysctl.conf and then reboot or run the com-
mand 'sysctl vm.overcommit_memory=1' for this to take effect.
redis_1  | 1:M 18 Jan 2022 10:08:48.956 * Ready to accept connections
web_1    | * Serving Flask app 'app.py' (lazy loading)
web_1    | * Environment: production
web_1    |   WARNING: This is a development server. Do not use it in a pro-
duction deployment.
web_1    |   Use a production WSGI server instead.
web_1    | * Debug mode: off
web_1    | * Running on all addresses.
web_1    |   WARNING: This is a development server. Do not use it in a pro-
duction deployment.
web_1    | * Running on http://172.17.0.3:5000/ (Press CTRL+C to quit)
web_1    | 192.168.100.1 - - [18/Jan/2022 10:09:20] "GET / HTTP/1.1" 200 -
web_1    | 192.168.100.1 - - [18/Jan/2022 10:09:20] "GET /favicon.ico
HTTP/1.1" 404 -
```

```
web_1     | 192.168.100.1 - - [18/Jan/2022 10:09:31] "GET / HTTP/1.1" 200 -
web_1     | 192.168.100.1 - - [18/Jan/2022 10:09:32] "GET / HTTP/1.1" 200 -
web_1     | 192.168.100.1 - - [18/Jan/2022 10:09:32] "GET / HTTP/1.1" 200 -
web_1     | 192.168.100.1 - - [18/Jan/2022 10:09:33] "GET / HTTP/1.1" 200 -
web_1     | 192.168.100.1 - - [18/Jan/2022 10:14:03] "GET / HTTP/1.1" 200 -
redis_1   | 1:M 18 Jan 2022 11:08:49.052 * 1 changes in 3600 seconds. Saving...
redis_1   | 1:M 18 Jan 2022 11:08:49.072 * Background saving started by pid 13
redis_1   | 13:C 18 Jan 2022 11:08:49.089 * DB saved on disk
redis_1   | 13:C 18 Jan 2022 11:08:49.089 * RDB: 4 MB of memory used by copy-
on-write
redis_1   | 1:M 18 Jan 2022 11:08:49.183 * Background saving terminated with
success
```

5. Test

Enter http://IP:5000/ in the browser and you can see the website page, as shown in Fig. 4-15.

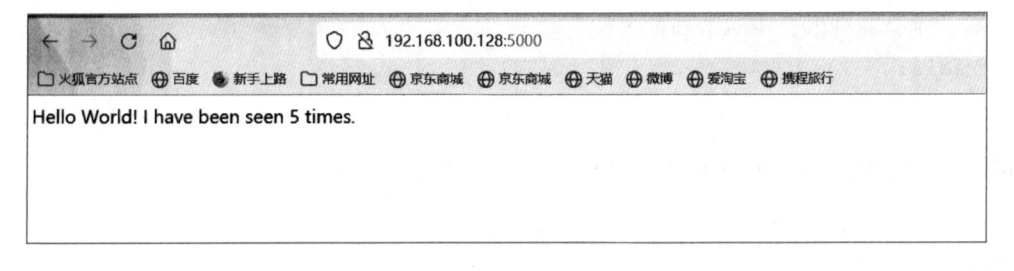

Fig. 4-15   Run web service page in container

Each time the page is refreshed, the counter will add 1, as shown in Fig. 4-16.

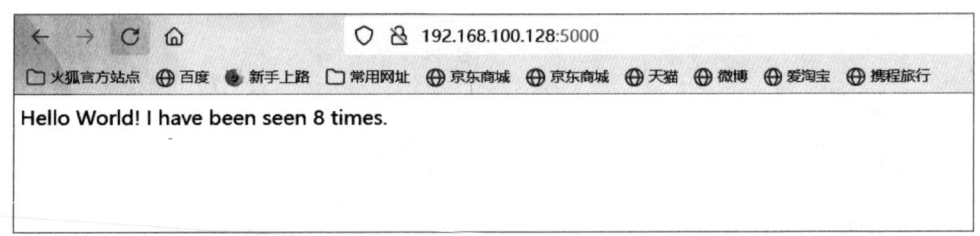

Fig. 4-16   Counter increment page is refreshed

6. Update service

When the service needs updating, you can update the application in real time by means of data volume loading.

Firstly, stop the service

```
[root@localhost composetest]# docker-compose down
Stopping composetest_redis_1 ... done
Stopping composetest_web_1   ... done
Removing composetest_redis_1 ... done
```

```
Removing composetest_web_1   ... done
Removing network composetest_default
```

Modify the yml file and load it to bind the service so that the modified application will take effect immediately without rebuilding the image.

```
[root@localhost composetest]# vi docker-compose.yml
version: "3.9"
services:
  web:
    build: .
    ports:
      - "5000:5000"
    volumes:
      - .:/code
    environment:
      FLASK_ENV: development
  redis:
    image: "redis:alpine"
```

Modify the application contents and view the immediate effect.

```
[root@localhost composetest]# vi app.py
import time

import redis
from flask import Flask

app = Flask(__name__)
cache = redis.Redis(host='redis', port=6379)


def get_hit_count():
    retries = 5
    while True:
        try:
            return cache.incr('hits')
        except redis.exceptions.ConnectionError as exc:
            if retries == 0:
                raise exc
            retries -= 1
            time.sleep(0.5)
```

```
@app.route('/')
def hello():
    count = get_hit_count()
    return 'Hello World@@@@@! I have been seen {} times.\n'.format(count)
```

Access the host port 5000 and you can see that the changes to the page contents take effect in real time, as shown in Fig. 4-17.

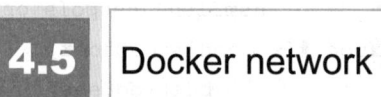

Fig. 4-17 Application page updated

4.5 Docker network

When using containers to build business, sometimes it is necessary to ensure the connectivity of the network between containers to achieve mutual access between containers, or to connect the containers with the workload in the non-container environment, and then it is necessary to deploy a container network environment for access to the containers.

Docker has such three network modes as follows:

1. None network mode

The none network mode, as the name implies, means that there is no network configured in the container, which is specified with "-- net=none". The container has an independent network namespace in the none network mode without network configuration. This network mode is generally applied to containers requiring high security but no communication. See Fig. 4-18 for the network relationship between the container and the host in the none network mode.

2. Host network mode

The host network mode means that the container and host share the same network stack and network namespace, which is specified with "-- net=host". The advantage of network stack sharing is that the docker network has high transmission efficiency, but the corresponding flexibility is poor, and the same network stack may cause mapping port conflicts. See Fig. 4-19 for the network relationship between the docker and the host in the host network mode.

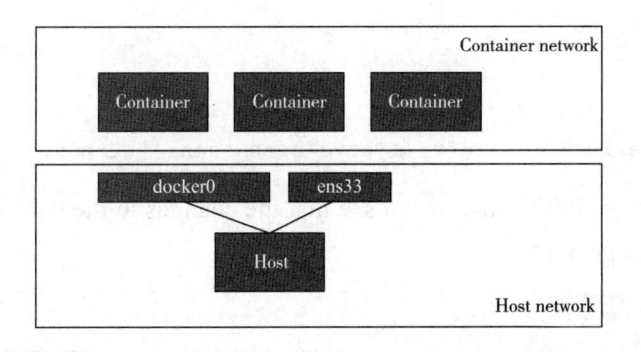

Fig. 4-18   Relationship between container and host in none network mode

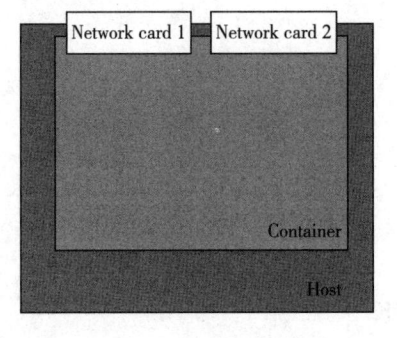

Fig. 4-19   Relationship between container and host in host network mode

3. Bridge network mode

The bridge network mode is the default service mode of the Docker network, which can be specified with "-- net=bridge". The bridge network mode means that the docker is bridged with the host network via the virtual bridge. After the docker is installed, a default virtual bridge docker0 will be created automatically in the host as the virtual bridge device connected to the container in the bridge network mode, with the address of 172.17.0.1/16 by default. Moreover, virtual bridges can also be created according to actual needs and the containers connected by different bridges are in different subnets. See Fig. 4-20 for the network relationship between the container and the host in the bridge network mode.

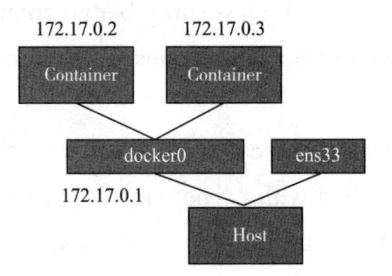

Fig. 4-20   Relationship between container and host in bridge network mode

4.6 Project practice—Docker network operation and maintenance management

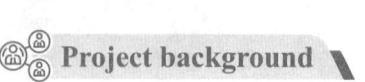

Project background

Setting an appropriate network mode for the container is an important basis for the container to provide external services and forward data flow inside the container. In order to ensure the functions of the business in the container, some company needs to configure different network modes for different types of containers and carry out operation and maintenance management accordingly.

Project implementation

Task 1　Build a docker of none network mode

Use the busybox image to build a container in the none network mode, check its network configuration, and find that there is no container or network configured.

```
root@localhost html]# docker run -it --network=none  busybox
Unable to find image 'busybox:latest' locally
latest: Pulling from library/busybox
5cc84ad355aa: Pull complete
Digest: sha256:5acba83a746c7608ed544dc1533b87c737a0b0fb730301639a0179f9344b1678
Status: Downloaded newer image for busybox:latest
/ # ip a
1: lo: <LOOPBACK,UP,LOWER_UP> mtu 65536 qdisc noqueue qlen 1000
    link/loopback 00:00:00:00:00:00 brd 00:00:00:00:00:00
    inet 127.0.0.1/8 scope host lo
      valid_lft forever preferred_lft forever
/ #
```

Task 2　Build a docker of host network mode

Use the busybox image to build a container in the host network mode. After the container is started successfully, check its network configuration and find that the network configuration in the container is completely consistent with that of the host.

```
[root@localhost html]# ip a
1: lo: <LOOPBACK,UP,LOWER_UP> mtu 65536 qdisc noqueue state UNKNOWN group de-
fault qlen 1000
      link/loopback 00:00:00:00:00:00 brd 00:00:00:00:00:00
```

```
        inet 127.0.0.1/8 scope host lo
           valid_lft forever preferred_lft forever
        inet6 ::1/128 scope host
           valid_lft forever preferred_lft forever
    2: ens33: <BROADCAST,MULTICAST,UP,LOWER_UP> mtu 1500 qdisc pfifo_fast state UP
group default qlen 1000
        link/ether 00:0c:29:6f:bf:6d brd ff:ff:ff:ff:ff:ff
        inet 10.0.0.137/24 brd 10.0.0.255 scope global noprefixroute dynamic ens33
           valid_lft 1356sec preferred_lft 1356sec
        inet6 fe80::20c:29ff:fe6f:bf6d/64 scope link
           valid_lft forever preferred_lft forever
    3: docker0: <BROADCAST,MULTICAST,UP,LOWER_UP> mtu 1500 qdisc noqueue state UP
group default
        link/ether 02:42:8b:e2:1b:29 brd ff:ff:ff:ff:ff:ff
        inet 172.17.0.1/16 brd 172.17.255.255 scope global docker0
           valid_lft forever preferred_lft forever
        inet6 fe80::42:8bff:fee2:1b29/64 scope link
           valid_lft forever preferred_lft forever
    27: vethd942024@if26: <BROADCAST,MULTICAST,UP,LOWER_UP> mtu 1500 qdisc noqueue
master docker0 state UP group default
        link/ether da:b0:db:95:26:bf brd ff:ff:ff:ff:ff:ff link-netnsid 0
        inet6 fe80::d8b0:dbff:fe95:26bf/64 scope link
           valid_lft forever preferred_lft forever
    29: veth2b63158@if28: <BROADCAST,MULTICAST,UP,LOWER_UP> mtu 1500 qdisc noqueue
master docker0 state UP group default
        link/ether be:27:26:fe:d9:f9 brd ff:ff:ff:ff:ff:ff link-netnsid 1
        inet6 fe80::bc27:26ff:fefe:d9f9/64 scope link
           valid_lft forever preferred_lft forever
[root@localhost html]# docker run -it --network=host busybox
/ # hostname
localhost.localdomain
/ # ip a
1: lo: <LOOPBACK,UP,LOWER_UP> mtu 65536 qdisc noqueue qlen 1000
        link/loopback 00:00:00:00:00:00 brd 00:00:00:00:00:00
        inet 127.0.0.1/8 scope host lo
           valid_lft forever preferred_lft forever
        inet6 ::1/128 scope host
           valid_lft forever preferred_lft forever
    2: ens33: <BROADCAST,MULTICAST,UP,LOWER_UP> mtu 1500 qdisc pfifo_fast qlen 1000
        link/ether 00:0c:29:6f:bf:6d brd ff:ff:ff:ff:ff:ff
```

```
        inet 10.0.0.137/24 brd 10.0.0.255 scope global dynamic noprefixroute ens33
            valid_lft 1336sec preferred_lft 1336sec
        inet6 fe80::20c:29ff:fe6f:bf6d/64 scope link
            valid_lft forever preferred_lft forever
    3: docker0: <BROADCAST,MULTICAST,UP,LOWER_UP> mtu 1500 qdisc noqueue
        link/ether 02:42:8b:e2:1b:29 brd ff:ff:ff:ff:ff:ff
        inet 172.17.0.1/16 brd 172.17.255.255 scope global docker0
            valid_lft forever preferred_lft forever
        inet6 fe80::42:8bff:fee2:1b29/64 scope link
            valid_lft forever preferred_lft forever
    27: vethd942024@if26: <BROADCAST,MULTICAST,UP,LOWER_UP,M-DOWN> mtu 1500 qdisc
noqueue master docker0
        link/ether da:b0:db:95:26:bf brd ff:ff:ff:ff:ff:ff
        inet6 fe80::d8b0:dbff:fe95:26bf/64 scope link
            valid_lft forever preferred_lft forever
    29: veth2b63158@if28: <BROADCAST,MULTICAST,UP,LOWER_UP,M-DOWN> mtu 1500 qdisc
noqueue master docker0
        link/ether be:27:26:fe:d9:f9 brd ff:ff:ff:ff:ff:ff
        inet6 fe80::bc27:26ff:fefe:d9f9/64 scope link
            valid_lft forever preferred_lft forever
    / #
```

Task 3 Build a container of bridge network mode

Use the busybox image to build a container in the bridge network mode. After the container is started successfully, check its network configuration and find that the IP address of the container is in the same subnet as docker0. Here, the docker is bridged to the virtual bridge docker0.

```
    3: docker0: <BROADCAST,MULTICAST,UP,LOWER_UP> mtu 1500 qdisc noqueue
        link/ether 02:42:8b:e2:1b:29 brd ff:ff:ff:ff:ff:ff
        inet 172.17.0.1/16 brd 172.17.255.255 scope global docker0
            valid_lft forever preferred_lft forever
        inet6 fe80::42:8bff:fee2:1b29/64 scope link
            valid_lft forever preferred_lft forever
[root@localhost html]# docker run -it busybox
/ # ip a
1: lo: <LOOPBACK,UP,LOWER_UP> mtu 65536 qdisc noqueue qlen 1000
    link/loopback 00:00:00:00:00:00 brd 00:00:00:00:00:00
    inet 127.0.0.1/8 scope host lo
        valid_lft forever preferred_lft forever
30: eth0@if31: <BROADCAST,MULTICAST,UP,LOWER_UP,M-DOWN> mtu 1500 qdisc noqueue
```

```
            link/ether 02:42:ac:11:00:04 brd ff:ff:ff:ff:ff:ff
            inet 172.17.0.4/16 brd 172.17.255.255 scope global eth0
                valid_lft forever preferred_lft forever
    / #
    [root@localhost html]#  yum install bridge-utils
    [root@localhost html]# brctl show
    bridge name      bridge id              STP enabled        interfaces
    docker0          8000.02428be21b29      no                 veth2b63158
                                                               vethd942024
```

Task 4   Build a user-defined network

In addition to the default docker0, users can also add virtual bridges according to their specific needs. After a virtual bridge is added successfully, use the "ip a" command to view it and you can see that an additional network device br-9291c369eef2 appears in the host. This is a virtual network bridge, whose default address is 172. 18.0. 1/16 if no IP is specified.

```
    [root@localhost html]# docker network create --driver bridge mynet
    9291c369eef252bb602a32f20cea7d443c7f092725322bec4afffbd24e48d345
    [root@localhost html]# ip a
    1: lo: <LOOPBACK,UP,LOWER_UP> mtu 65536 qdisc noqueue state UNKNOWN group de-
    fault qlen 1000
            link/loopback 00:00:00:00:00:00 brd 00:00:00:00:00:00
            inet 127.0.0.1/8 scope host lo
                valid_lft forever preferred_lft forever
            inet6 ::1/128 scope host
                valid_lft forever preferred_lft forever
    2: ens33: <BROADCAST,MULTICAST,UP,LOWER_UP> mtu 1500 qdisc pfifo_fast state UP
    group default qlen 1000
            link/ether 00:0c:29:6f:bf:6d brd ff:ff:ff:ff:ff:ff
            inet 10.0.0.137/24 brd 10.0.0.255 scope global noprefixroute dynamic ens33
                valid_lft 1592sec preferred_lft 1592sec
            inet6 fe80::20c:29ff:fe6f:bf6d/64 scope link
                valid_lft forever preferred_lft forever
    3: docker0: <BROADCAST,MULTICAST,UP,LOWER_UP> mtu 1500 qdisc noqueue state UP
    group default
            link/ether 02:42:8b:e2:1b:29 brd ff:ff:ff:ff:ff:ff
            inet 172.17.0.1/16 brd 172.17.255.255 scope global docker0
                valid_lft forever preferred_lft forever
            inet6 fe80::42:8bff:fee2:1b29/64 scope link
```

```
         valid_lft forever preferred_lft forever
   27: vethd942024@if26: <BROADCAST,MULTICAST,UP,LOWER_UP> mtu 1500 qdisc noqueue
master docker0 state UP group default
       link/ether da:b0:db:95:26:bf brd ff:ff:ff:ff:ff:ff link-netnsid 0
       inet6 fe80::d8b0:dbff:fe95:26bf/64 scope link
         valid_lft forever preferred_lft forever
   29: veth2b63158@if28: <BROADCAST,MULTICAST,UP,LOWER_UP> mtu 1500 qdisc noqueue
master docker0 state UP group default
       link/ether be:27:26:fe:d9:f9 brd ff:ff:ff:ff:ff:ff link-netnsid 1
       inet6 fe80::bc27:26ff:fefe:d9f9/64 scope link
         valid_lft forever preferred_lft forever
   32: br-9291c369eef2: <NO-CARRIER,BROADCAST,MULTICAST,UP> mtu 1500 qdisc noqueue
state DOWN group default
       link/ether 02:42:73:b2:9b:1e brd ff:ff:ff:ff:ff:ff
       inet 172.18.0.1/16 brd 172.18.255.255 scope global br-9291c369eef2
         valid_lft forever preferred_lft forever
```

If necessary, you can use the "--gateway" parameter to specify an IP address. For example, if the IP addresses of the new bridge and the subnet are specified as 172.100.100.1 and 172. 100. 100.0/24 respectively, you can see that a new virtual bridge of br-c1bf0d60c7d0 is added to the host, and its address is 172.100.100.1 specified.

```
   [root@localhost html]# docker network create --driver bridge --subnet
172.100.100.0/24 --gateway 172.100.100.1 mynet2
   c1bf0d60c7d02bb74ee4eea60f6b3500707e06068cd77be1bf5411477de92079
   [root@localhost html]# ip a
   1: lo: <LOOPBACK,UP,LOWER_UP> mtu 65536 qdisc noqueue state UNKNOWN group de-
fault qlen 1000
       link/loopback 00:00:00:00:00:00 brd 00:00:00:00:00:00
       inet 127.0.0.1/8 scope host lo
         valid_lft forever preferred_lft forever
       inet6 ::1/128 scope host
         valid_lft forever preferred_lft forever
   2: ens33: <BROADCAST,MULTICAST,UP,LOWER_UP> mtu 1500 qdisc pfifo_fast state UP
group default qlen 1000
       link/ether 00:0c:29:6f:bf:6d brd ff:ff:ff:ff:ff:ff
       inet 10.0.0.137/24 brd 10.0.0.255 scope global noprefixroute dynamic ens33
         valid_lft 1475sec preferred_lft 1475sec
       inet6 fe80::20c:29ff:fe6f:bf6d/64 scope link
         valid_lft forever preferred_lft forever
   3: docker0: <BROADCAST,MULTICAST,UP,LOWER_UP> mtu 1500 qdisc noqueue state UP
```

```
group default
        link/ether 02:42:8b:e2:1b:29 brd ff:ff:ff:ff:ff:ff
        inet 172.17.0.1/16 brd 172.17.255.255 scope global docker0
           valid_lft forever preferred_lft forever
        inet6 fe80::42:8bff:fee2:1b29/64 scope link
           valid_lft forever preferred_lft forever
    27: vethd942024@if26: <BROADCAST,MULTICAST,UP,LOWER_UP> mtu 1500 qdisc noqueue
master docker0 state UP group default
        link/ether da:b0:db:95:26:bf brd ff:ff:ff:ff:ff:ff link-netnsid 0
        inet6 fe80::d8b0:dbff:fe95:26bf/64 scope link
           valid_lft forever preferred_lft forever
    29: veth2b63158@if28: <BROADCAST,MULTICAST,UP,LOWER_UP> mtu 1500 qdisc noqueue
master docker0 state UP group default
        link/ether be:27:26:fe:d9:f9 brd ff:ff:ff:ff:ff:ff link-netnsid 1
        inet6 fe80::bc27:26ff:fefe:d9f9/64 scope link
           valid_lft forever preferred_lft forever
    32: br-9291c369eef2: <NO-CARRIER,BROADCAST,MULTICAST,UP> mtu 1500 qdisc noqueue
state DOWN group default
        link/ether 02:42:73:b2:9b:1e brd ff:ff:ff:ff:ff:ff
        inet 172.18.0.1/16 brd 172.18.255.255 scope global br-9291c369eef2
           valid_lft forever preferred_lft forever
    33: br-c1bf0d60c7d0: <NO-CARRIER,BROADCAST,MULTICAST,UP> mtu 1500 qdisc noqueue
state DOWN group default
        link/ether 02:42:50:8b:47:3b brd ff:ff:ff:ff:ff:ff
        inet 172.100.100.1/24 brd 172.100.100.255 scope global br-c1bf0d60c7d0
           valid_lft forever preferred_lft forever
```

After the custom network is created successfully, use the busybox image to start a container and specify its network as the custom network mynet2. After successful startup, check its IP address and you can see that the IP address obtained by the container is in the mynet2 network.

```
[root@localhost html]# docker run -it --network=mynet2 busybox
/ # ip a
1: lo: <LOOPBACK,UP,LOWER_UP> mtu 65536 qdisc noqueue qlen 1000
        link/loopback 00:00:00:00:00:00 brd 00:00:00:00:00:00
        inet 127.0.0.1/8 scope host lo
           valid_lft forever preferred_lft forever
    36: eth0@if37: <BROADCAST,MULTICAST,UP,LOWER_UP,M-DOWN> mtu 1500 qdisc noqueue
        link/ether 02:42:ac:64:64:02 brd ff:ff:ff:ff:ff:ff
        inet 172.100.100.2/24 brd 172.100.100.255 scope global eth0
           valid_lft forever preferred_lft forever
```

Task 5   Realize communication between different containers

In the container, different networks are mutually isolated, so if you want to achieve network connectivity between different containers, you must ensure that different dockers have the same network card to communicate with each other. Therefore, if the docker with docker0 as the gateway wants to communicate with the docker in the mynet2 network, a network card of the mynet2 network must be added to the docker, which can be realized by using "docker network connect mynet2".

```
# docker network connect mynet2 3086a330c2e9
[root@localhost html]# docker exec -it 3086a330c2e9 /bin/sh
/ # ip a
1: lo: <LOOPBACK,UP,LOWER_UP> mtu 65536 qdisc noqueue qlen 1000
    link/loopback 00:00:00:00:00:00 brd 00:00:00:00:00:00
    inet 127.0.0.1/8 scope host lo
       valid_lft forever preferred_lft forever
38: eth0@if39: <BROADCAST,MULTICAST,UP,LOWER_UP,M-DOWN> mtu 1500 qdisc noqueue
    link/ether 02:42:ac:11:00:04 brd ff:ff:ff:ff:ff:ff
    inet 172.17.0.4/16 brd 172.17.255.255 scope global eth0
       valid_lft forever preferred_lft forever
40: eth1@if41: <BROADCAST,MULTICAST,UP,LOWER_UP,M-DOWN> mtu 1500 qdisc noqueue
    link/ether 02:42:ac:64:64:02 brd ff:ff:ff:ff:ff:ff
    inet 172.100.100.2/24 brd 172.100.100.255 scope global eth1
       valid_lft forever preferred_lft forever
```

4.7   Project practice—build container private warehouse

 **Project background**

In the production environment, in order to improve security and facilitate the unit group to share images, it is now decided to build a private warehouse. There are two kinds of commonly used private warehouses: one is the Registry provided by Docker officially and the other is the more popular Docker Harbor.

Project implementation

Task 1   Build Docker Registry

1. Use the official Registry image to run the container

Expose the container port 5000.

```
[root@localhost ~]# docker run -d -p 5000:5000 --restart=always --name regis-
try registry
```

By default, the warehouse will be created in the /var/lib/registry directory of the container. You can use the "-v" parameter to store the image file in the local specified path. For example, in the following example, store the uploaded image file in the local directory "/opt/data/registry".

```
[root@localhost ~]# docker run -d \
    -p 5000:5000 \
    -v /opt/data/registry:/var/lib/registry \
    registry
```

2. Upload, search and download images in the private warehouse

After creating a private warehouse, you can use the docker tag to mark an image and then push it to the warehouse. For example, push to the registry warehouse with the address of 127.0.0 1:5000.

First view the existing images on the local host.

```
[root@localhost ~]# docker images
REPOSITORY      TAG        IMAGE ID        CREATED         SIZE
<none>          <none>     4565a1830fd7    12 days ago     231MB
busybox         latest     beae173ccac6    3 months ago    1.24MB
nginx           latest     605c77e624dd    3 months ago    141MB
httpd           latest     dabbfbe0c57b    3 months ago    144MB
registry        latest     b8604a3fe854    4 months ago    26.2MB
mariadb         latest     e2278f24ac88    5 months ago    410MB
hello-world     latest     feb5d9fea6a5    6 months ago    13.3KB
centos          latest     5d0da3dc9764    6 months ago    231MB
[root@localhost ~]# docker tag busybox:latest 127.0.0.1:5000/busybox:latest
[root@localhost ~]# docker images
REPOSITORY               TAG        IMAGE ID        CREATED        SIZE
<none>                   <none>     4565a1830fd7    12 days ago    231MB
127.0.0.1:5000/busybox   latest     beae173ccac6    3 months ago   1.24MB
busybox                  latest     beae173ccac6    3 months ago   1.24MB
nginx                    latest     605c77e624dd    3 months ago   141MB
httpd                    latest     dabbfbe0c57b    3 months ago   144MB
registry                 latest     b8604a3fe854    4 months ago   26.2MB
mariadb                  latest     e2278f24ac88    5 months ago   410MB
hello-world              latest     feb5d9fea6a5    6 months ago   13.3KB
centos                   latest     5d0da3dc9764    6 months ago   231MB
```

Use "docker push" to upload the image marked.

```
[root@localhost ~]# docker push 127.0.0.1:5000/busybox:latest
The push refers to repository [127.0.0.1:5000/busybox]
01fd6df81c8e: Pushed
latest: digest: sha256:62ffc2ed7554e4c6d360bce40bbcf196573dd27c4ce080641a2c59867e732dee
size: 527
```

Use "curl" to view the image in the warehouse.

```
[root@localhost ~]# curl 127.0.0.1:5000/v2/_catalog
{"repositories":["busybox"]}
```

Here you can see {"repositories": ["busybox"]}, indicating that the image has been uploaded successfully.

Next, verify whether we can remove it from the private warehouse and use the image we uploaded. Delete the local image first and then try to download the image from the private warehouse.

```
[root@localhost ~]# docker image rm 127.0.0.1:5000/busybox
Untagged: 127.0.0.1:5000/busybox:latest
Untagged: 127.0.0.1:5000/busybox@sha256:62ffc2ed7554e4c6d360bce40bbcf196573dd27c-
4ce080641a2c59867e732dee
[root@localhost ~]# docker pull 127.0.0.1:5000/busybox:latest
latest: Pulling from busybox
Digest: sha256:62ffc2ed7554e4c6d360bce40bbcf196573dd27c4ce080641a2c59867e732dee
Status: Downloaded newer image for 127.0.0.1:5000/busybox:latest
127.0.0.1:5000/busybox:latest
[root@localhost ~]# docker images
REPOSITORY               TAG       IMAGE ID        CREATED          SIZE
<none>                   <none>    4565a1830fd7    12 days ago      231MB
busybox                  latest    beae173ccac6    3 months ago     1.24MB
127.0.0.1:5000/busybox   latest    beae173ccac6    3 months ago     1.24MB
nginx                    latest    605c77e624dd    3 months ago     141MB
httpd                    latest    dabbfbe0c57b    3 months ago     144MB
registry                 latest    b8604a3fe854    4 months ago     26.2MB
mariadb                  latest    e2278f24ac88    5 months ago     410MB
hello-world              latest    feb5d9fea6a5    6 months ago     13.3KB
centos                   latest    5d0da3dc9764    6 months ago     231MB
```

3. Configure non-https warehouse address

To provide the private warehouse to the cluster in the LAN, the intranet address (for example, 192.168.199.100:5000) should be treated as the private warehouse address and you will find that the image cannot be pushed successfully.

```
[root@localhost ~]# docker tag busybox:latest 10.0.0.137:5000/busybox:latest
[root@localhost ~]# docker push 10.0.0.137:5000/busybox:latest
The push refers to repository [10.0.0.137:5000/busybox]
Get "https://10.0.0.137:5000/v2/": http: server gave HTTP response to HTTPS
client
```

The reason is as follows: Docker does not allow non-https image push by default. You can cancel the limit through Docker's configuration options.

```
[root@localhost ~]# cat /etc/docker/daemon.json
{
    "registry-mirrors": ["https://jctfhowk.mirror.aliyuncs.com"],
    "insecure-registries": [
        "10.0.0.137:5000"
    ]
}
[root@localhost ~]# systemctl daemon-reload
[root@localhost ~]# systemctl restart docker
[root@localhost ~]# docker push 10.0.0.137:5000/busybox:latest
The push refers to repository [10.0.0.137:5000/busybox]
01fd6df81c8e: Layer already exists
latest: digest: sha256:62ffc2ed7554e4c6d360bce40bbcf196573dd27c4ce080641a2c59867e732dee
size: 527
```

Task 2　Build Harbor

1. Issue a certificate for the server

Harbor access needs https, so it is necessary to issue a certificate for the server first.

```
[root@localhost data]# mkdir -pv /data/ssl
mkdir: created directory '/data/ssl'
[root@localhost data]# cd ssl/
[root@localhost ssl]# openssl genrsa -out server.key 1024
Generating RSA private key, 1024 bit long modulus
.................................++++++
.............++++++
e is 65537 (0x10001)
[root@localhost ssl]# chmod 600 server.key
[root@localhost ssl]# openssl req -new -key server.key -out server.csr
You are about to be asked to enter information that will be incorporated
into your certificate request.ser
What you are about to enter is what is called a Distinguished Name or a DN.
```

```
There are quite a few fields but you can leave some blank
For some fields there will be a default value,
If you enter '.', the field will be left blank.
-----
Country Name (2 letter code) [XX]:CN
State or Province Name (full name) []:Sichuan
Locality Name (eg, city) [Default City]:Chengdu
Organization Name (eg, company) [Default Company Ltd]:cdp
Organizational Unit Name (eg, section) []:cloud
Common Name (eg, your name or your server's hostname) []:harbor.cdp.com
Email Address []:

Please enter the following 'extra' attributes
to be sent with your certificate request
A challenge password []:
An optional company name []:
[root@localhost ssl]# cd /etc/pki/CA/
[root@localhost CA]# openssl genrsa -out private/cakey.pem 2048
Generating RSA private key, 2048 bit long modulus
.............................................................+++
...............+++
e is 65537 (0x10001)
[root@localhost CA]# chmod 600 private/cakey.pem
[root@localhost CA]# openssl req -new -x509 -key private/cakey.pem -out ca-
cert.pem -days 365
You are about to be asked to enter information that will be incorporated
into your certificate request.
What you are about to enter is what is called a Distinguished Name or a DN.
There are quite a few fields but you can leave some blank
For some fields there will be a default value,
If you enter '.', the field will be left blank.
-----
Country Name (2 letter code) [XX]:CN
State or Province Name (full name) []:Sichuan
Locality Name (eg, city) [Default City]:Chengdu
Organization Name (eg, company) [Default Company Ltd]:cdp
Organizational Unit Name (eg, section) []:cloud
Common Name (eg, your name or your server's hostname) []:
Email Address []:
[root@localhost ~]# cd /etc/pki/CA/
```

```
[root@localhost CA]# touch index.txt
[root@localhost CA]# echo "01" > serial
[root@localhost CA]# cd /data/ssl/
[root@localhost ssl]# openssl ca -in server.csr -out server.crt -days 365
Using configuration from /etc/pki/tls/openssl.cnf
Check that the request matches the signature
Signature ok
Certificate Details:
        Serial Number: 1 (0x1)
        Validity
            Not Before: Jan 10 16:47:20 2022 GMT
            Not After : Jan 10 16:47:20 2023 GMT
        Subject:
            countryName = CN
            stateOrProvinceName = Sichuan
            organizationName = cdp
            organizationalUnitName = cloud
            commonName = harbor.cdp.com
    X509v3 extensions:
        X509v3 Basic Constraints:
            CA:FALSE
        Netscape Comment:
            OpenSSL Generated Certificate
        X509v3 Subject Key Identifier:
            6D:15:28:38:4B:E1:6A:24:FB:C3:ED:C1:BF:F2:69:72:89:89:5E:15
        X509v3 Authority Key Identifier:
            keyid:FA:F2:CE:1A:C5:CA:D7:47:1A:AE:07:89:8B:52:5F:15:8F:CF:43:6
Certificate is to be certified until Jan 10 16:47:20 2023 GMT (365 days)
Sign the certificate? [y/n]:y


1 out of 1 certificate requests certified, commit? [y/n]y
Write out database with 1 new entries
Data Base Updated
```

2. Install docker-compose

```
[root@localhost ssl]# yum -y install docker-compose
```

3. Download the Harbor offline installation package

Edit harbor.cfg to configure the server address and use the protocol, certificate path, etc. to

deploy Harbor via docker-compose.

```
[root@localhost ssl]# cd /opt/
[root@localhost opt]# wget https://storage.googleapis.com/harbor-releases/
release-1.7.0/harbor-offline-installer-v1.7.1.tgz
   --2022-01-11 00:50:22--  https://storage.googleapis.com/harbor-releases/re-
lease-1.7.0/harbor-offline-installer-v1.7.1.tgz
   Resolving storage.googleapis.com (storage.googleapis.com)... 172.217.160.112,
142.251.43.16, 172.217.160.80, ...
   Connecting to storage.googleapis.com (storage.googleapis.com)|172.217.160.112|:
443... connected.
   HTTP request sent, awaiting response... 200 OK
   Length: 597857483 (570M) [application/x-tar]
   Saving to: 'harbor-offline-installer-v1.7.1.tgz'


   100%[=======================================================================
===================================================>] 597,857,483 10.5MB/s    in 53s


   2022-01-11 00:51:15 (10.8 MB/s) - 'harbor-offline-installer-v1.7.1.tgz' saved
[597857483/597857483]
[root@localhost opt]# tar -xvf harbor-offline-installer-v1.7.1.tgz -C /opt/
[root@localhost opt]# ll
total 583848
drwx--x--x 4 root root         28 Jan  8 00:20 containerd
drwxr-xr-x 3 root root        270 Jan 11 00:54 harbor
-rw-r--r-- 1 root root 597857483 Jan  7  2019 harbor-offline-installer-v1.7.1.tgz
[root@localhost harbor]# cd harbor
[root@localhost harbor]# vi harbor.cfg
hostname = 10.0.0.100
ui_url_protocol = https
ssl_cert = /data/ssl/server.crt
ssl_cert_key = /data/ssl/server.key
[root@localhost harbor]# ./prepare
Generated and saved secret to file: /data/secretkey
Generated configuration file: ./common/config/nginx/nginx.conf
Generated configuration file: ./common/config/adminserver/env
Generated configuration file: ./common/config/core/env
Generated configuration file: ./common/config/registry/config.yml
Generated configuration file: ./common/config/db/env
Generated configuration file: ./common/config/jobservice/env
```

```
Generated configuration file: ./common/config/jobservice/config.yml

Generated configuration file: ./common/config/log/logrotate.conf

Generated configuration file: ./common/config/registryctl/env

Generated configuration file: ./common/config/core/app.conf

Generated certificate, key file: ./common/config/core/private_key.pem, cert
file: ./common/config/registry/root.crt

The configuration files are ready, please use docker-compose to start the ser-
vice.

[root@localhost harbor]# ./install.sh


[Step 0]: checking installation environment ...


Note: docker version: 20.10.12


Note: docker-compose version: 1.18.0


[Step 1]: loading Harbor images ...

Loaded image: goharbor/registry-photon:v2.6.2-v1.7.1

Loaded image: goharbor/harbor-migrator:v1.7.1

Loaded image: goharbor/harbor-adminserver:v1.7.1

Loaded image: goharbor/harbor-core:v1.7.1

Loaded image: goharbor/harbor-log:v1.7.1

Loaded image: goharbor/harbor-jobservice:v1.7.1

Loaded image: goharbor/notary-server-photon:v0.6.1-v1.7.1

Loaded image: goharbor/clair-photon:v2.0.7-v1.7.1

Loaded image: goharbor/harbor-portal:v1.7.1

Loaded image: goharbor/harbor-db:v1.7.1

Loaded image: goharbor/redis-photon:v1.7.1

Loaded image: goharbor/nginx-photon:v1.7.1

Loaded image: goharbor/harbor-registryctl:v1.7.1

Loaded image: goharbor/notary-signer-photon:v0.6.1-v1.7.1

Loaded image: goharbor/chartmuseum-photon:v0.7.1-v1.7.1


[Step 2]: preparing environment ...

Clearing the configuration file: ./common/config/adminserver/env

Clearing the configuration file: ./common/config/core/env

Clearing the configuration file: ./common/config/core/app.conf

Clearing the configuration file: ./common/config/core/private_key.pem
```

```
Clearing the configuration file: ./common/config/db/env
Clearing the configuration file: ./common/config/jobservice/env
Clearing the configuration file: ./common/config/jobservice/config.yml
Clearing the configuration file: ./common/config/registry/config.yml
Clearing the configuration file: ./common/config/registry/root.crt
Clearing the configuration file: ./common/config/registryctl/env
Clearing the configuration file: ./common/config/registryctl/config.yml
Clearing the configuration file: ./common/config/nginx/cert/server.crt
Clearing the configuration file: ./common/config/nginx/cert/server.key
Clearing the configuration file: ./common/config/nginx/nginx.conf
Clearing the configuration file: ./common/config/log/logrotate.conf
loaded secret from file: /data/secretkey
Generated configuration file: ./common/config/nginx/nginx.conf
Generated configuration file: ./common/config/adminserver/env
Generated configuration file: ./common/config/core/env
Generated configuration file: ./common/config/registry/config.yml
Generated configuration file: ./common/config/db/env
Generated configuration file: ./common/config/jobservice/env
Generated configuration file: ./common/config/jobservice/config.yml
Generated configuration file: ./common/config/log/logrotate.conf
Generated configuration file: ./common/config/registryctl/env
Generated configuration file: ./common/config/core/app.conf
Generated certificate, key file: ./common/config/core/private_key.pem, cert
file: ./common/config/registry/root.crt
```

The configuration files are ready, please use docker-compose to start the service.

```
[Step 3]: checking existing instance of Harbor ...

Note: stopping existing Harbor instance ...
Stopping harbor-jobservice   ... done
Stopping harbor-portal       ... done
Stopping harbor-core         ... done
Stopping redis               ... done
Stopping harbor-db           ... done
Stopping harbor-adminserver ... done
Stopping registryctl         ... done
Stopping registry            ... done
```

```
Stopping harbor-log         ... done
Removing nginx              ... done
Removing harbor-jobservice  ... done
Removing harbor-portal      ... done
Removing harbor-core        ... done
Removing redis              ... done
Removing harbor-db          ... done
Creating harbor-log ... done
Removing registryctl        ... done
Removing registry           ... done
Removing harbor-log         ... done
Removing network harbor_harbor
Creating harbor-adminserver ... done
Creating harbor-core ... done
[Step 4]: starting Harbor ...
Creating harbor-portal ... done
Creating nginx ... done
Creating registry ...
Creating registryctl ...
Creating redis ...
Creating harbor-adminserver ...
Creating harbor-db ...
Creating harbor-core ...
Creating harbor-portal ...
Creating harbor-jobservice ...
Creating nginx ...


√ ----Harbor has been installed and started successfully.----


Now you should be able to visit the admin portal at https://10.0.0.100.
For more details, please visit https://github.com/goharbor/harbor .
```

After deployment, use https://IP/harbor to access the web interface of Harbor in the browser with the default user of admin and the password Harbor12345, as shown in Fig. 4-21.

In the Harbor management platform, you can manage cells, images, etc., and monitor the warehouse status information, as shown in Fig. 4-22.

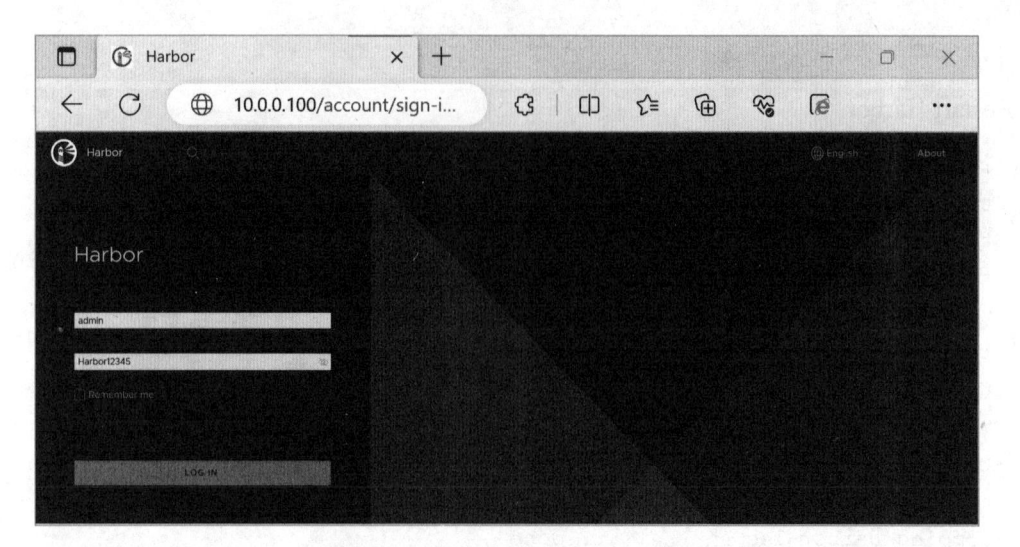

Fig. 4-21   Harbor login interface

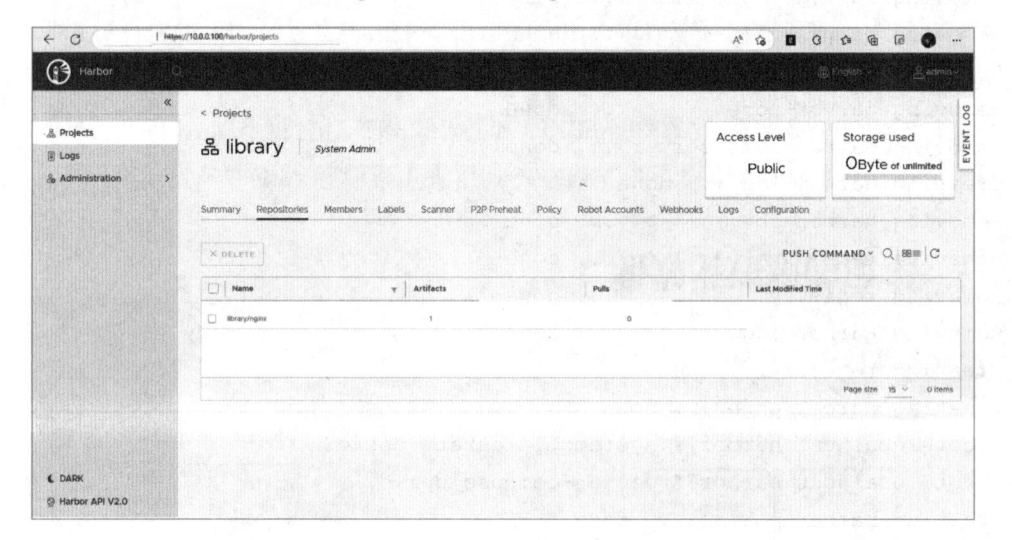

Fig. 4-22   Harbor management interface

4. Use Harbor to set Harbor as a local private warehouse

Edit daemon.json and add a private image warehouse.

```
[root@localhost harbor]# cat /etc/docker/daemon.json
{
    "graph": "/data/docker",
    "storage-driver": "overlay",
    "insecure-registries": ["registry.access.redhat.com","quay.
io","10.0.0.100","harbor.od.com"],
    "registry-mirrors": ["https://jctfhowk.mirror.aliyuncs.com"],
    "bip": "172.7.21.1/24",
    "exec-opts": ["native.cgroupdriver=systemd"],
    "live-restore": true
```

```
}
```

Restart Harbor.

```
[root@localhost harbor]# docker-compose down
Stopping harbor-jobservice  ... done
Stopping nginx              ... done
Stopping harbor-portal      ... done
Stopping harbor-core        ... done
Stopping harbor-adminserver ... done
Stopping registryctl        ... done
Stopping harbor-db          ... done
Stopping redis              ... done
Stopping registry           ... done
Stopping harbor-log         ... done
Removing harbor-jobservice  ... done
Removing nginx              ... done
Removing harbor-portal      ... done
Removing harbor-core        ... done
Removing harbor-adminserver ... done
Creating harbor-log ... done
Removing harbor-db          ... done
Removing redis              ... done
Removing registry           ... done
Removing harbor-log         ... done
Creating registry ... done
Creating harbor-core ... done
[root@localhost harbor]# systemctl restart docker
[root@localhost harbor]# docker-compose up -d
Creating harbor-portal ... done
Creating nginx ... done
Creating registry ...
Creating harbor-db ...
Creating harbor-adminserver ...
Creating redis ...
Creating registryctl ...
Creating harbor-core ...
Creating harbor-portal ...
Creating harbor-jobservice ...
Creating nginx ...
```

Log in Harbor.

```
[root@localhost harbor]# docker login https://10.0.0.100
Username: admin
```

```
Password:
WARNING! Your password will be stored unencrypted in /root/.docker/config.
json.
Configure a credential helper to remove this warning. See
https://docs.docker.com/engine/reference/commandline/login/#creden-
tials-store


Login Succeeded
```

Push the image to the Harbor warehouse.

```
[root@localhost harbor]# docker tag 605c77e624dd 10.0.0.100/library/nginx
[root@localhost harbor]# docker push 10.0.0.100/library/nginx
Using default tag: latest
The push refers to repository [10.0.0.100/library/nginx]
d874fd2bc83b: Pushed
32ce5f6a5106: Pushed
f1db227348d0: Pushed
b8d6e692a25e: Pushed
e379e8aedd4d: Pushed
2edcec3590a4: Pushed
latest: digest: sha256:ee89b00528ff4f02f2405e4ee221743ebc3f8e8dd0bfd5c4c20a-
2fa2aaa7ede3 size: 1570
```

The image pushed can be seen in the web management interface of Harbor, as shown in Fig. 4-23.

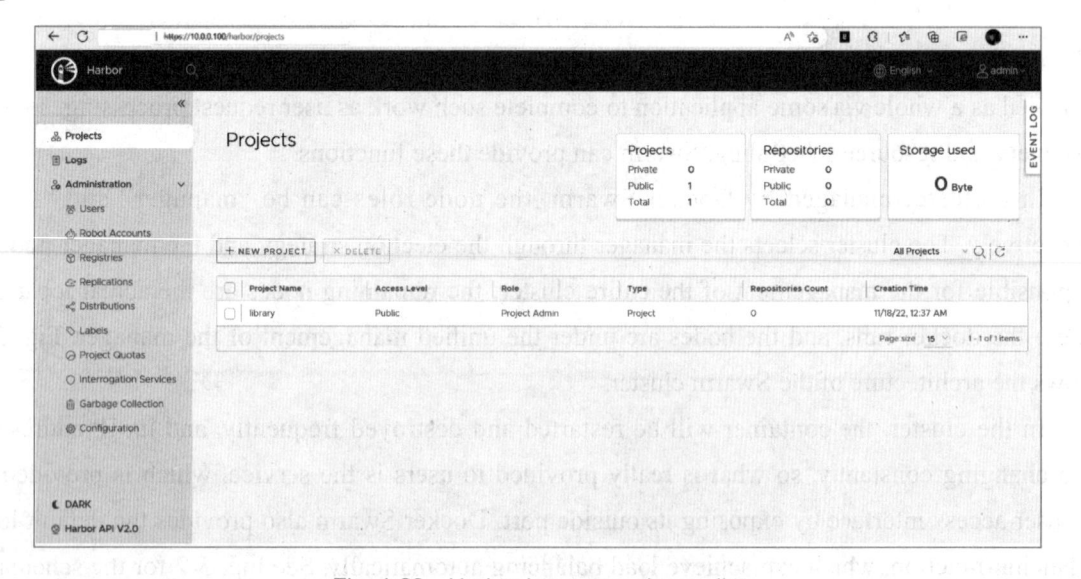

Fig. 4-23   Harbor image warehouse list

Unit 5

Container cluster management

5.1   Docker Swarm

As an open-source unit of Docker, Swarm is currently a service scheduling tool supported by Docker engine by default. Docker Swarm is a technology developed to manage Docker clusters. Like OpenStack, single-node computing resources will have single-point failures, insufficient resources and other problems when providing services. Therefore, in the container cloud, clusters will also be built to solve such problems. The cluster provides services as a whole, so it needs to be managed as a whole via some application to complete such work as user request processing, service discovery and resource scheduling. Swarm can provide these functions.

In a cluster managed by Docker Swarm, the node roles can be "manager" and "node" respectively. The cluster selects the manager through the election strategy and the manager node is responsible for the management of the entire cluster; the remaining nodes are the actual locations where the docker runs, and the nodes are under the unified management of the manager. Fig. 5-1 shows the architecture of the Swarm cluster.

In the cluster, the container will be restarted and destroyed frequently, and its IP address is also changing constantly, so what is really provided to users is the service, which is provided to the user access interface by exposing its outside port. Docker Swarm also provides the default load balancing function, which can achieve load balancing automatically. See Fig. 5-2 for the schematic diagram.

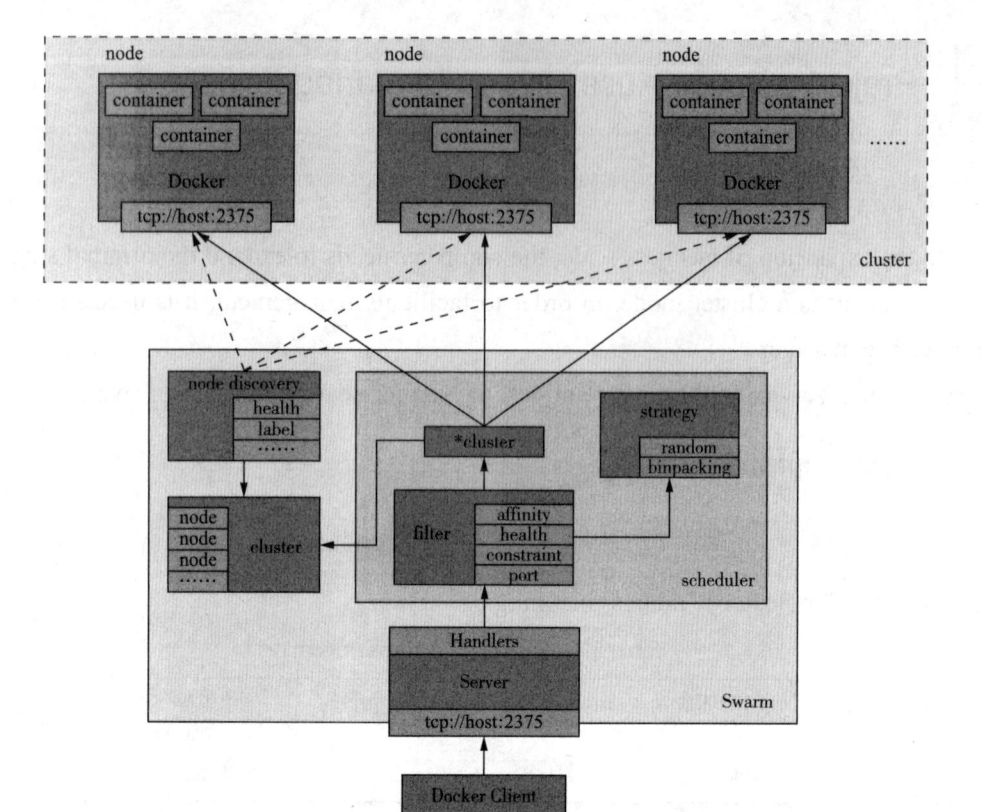

Fig. 5-1　Swarm cluster architecture

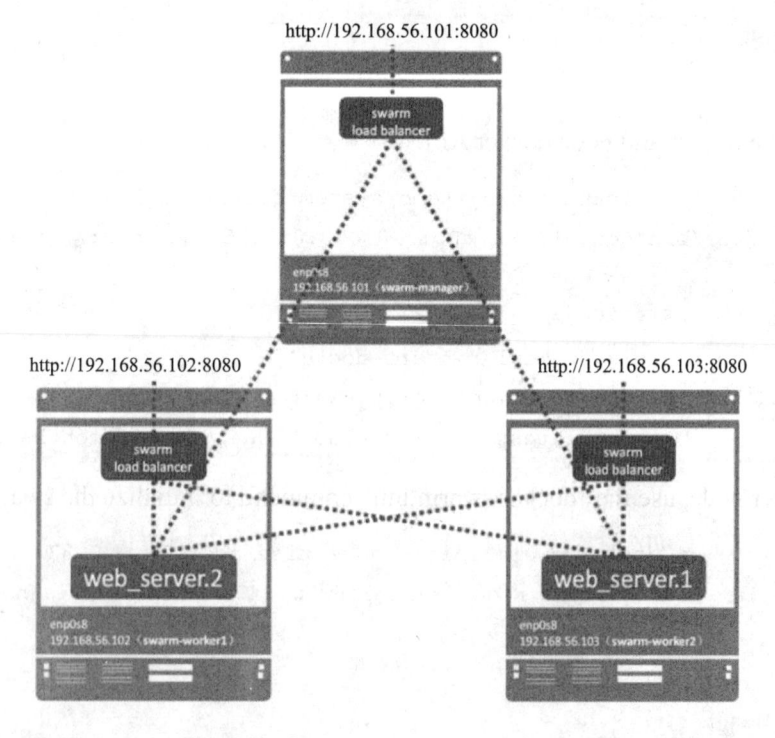

Fig. 5-2　Swarm load balancing functions

5.2 Project practice—use Swarm to manage clusters

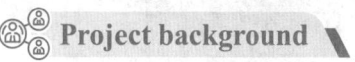 **Project background**

Due to the expansion of business scale, the company needs to expand the original single-node docker environment to a cluster mode. In order to facilitate management, it is necessary to install the web interface manager.

Start the nginx service in the cluster and test the service scaling function of swarm.

 **Project implementation**

Task 1 Basic environment construction

See Table 5-1 for cluster IP address planning.

Table 5-1 Cluster IP address planning

Node name	IP address
master	10.0.0.37
node	10.0.0.8

Modify the hosts resolution file of each host in the cluster.

```
[root@localhost ~]# cat /etc/hosts
10.0.0.37 master
10.0.0.8 node
```

Configure the master and node docker API.

```
[root@localhost ~]# cat /lib/systemd/system/docker.service|grep ExecStart
    ExecStart=/usr/bin/dockerd -H tcp://0.0.0.0:2375 -H unix:///var/run/docker.
sock
[root@localhost ~]# systemctl daemon-reload
[root@localhost ~]# systemctl restart docker
[root@localhost ~]# docker images|grep swarm
swarm               latest              1a5eb59a410f    16 months ago    12.7MB
```

At the master node, use the "docker swarm inti" command to initialize the swarm cluster.

```
[root@master ~]# docker swarm init --advertise-addr 10.0.0.37
Swarm initialized: current node (x4rzm3l6wkwyod717t3xmhc4u) is now a manager.
```

```
To add a worker to this swarm, run the following command:
```

```
docker swarm join --token SWMTKN-1-3abvt79csrclmad83q384otm7p3udsupvmurj4r9vp-
fqbssoj9-0bfn4wv5zeqsc1zjq7m5m8o8i 10.0.0.37:2377
```

After initialization, use the cluster token to add the node to the cluster as prompted above.

```
[root@node ~]# docker swarm join --token SWMTKN-1-3abvt79csrclmad83q384otm-
7p3udsupvmurj4r9vpfqbssoj9-0bfn4wv5zeqsc1zjq7m5m8o8i 10.0.0.37:2377
This node joined a swarm as a worker.
```

After the cluster is created, use the "docker node ls" command to verify the swarm cluster. You can see that there are two nodes in the current cluster, namely master and node, both of which are ready.

```
[root@master ~]# docker node ls
ID            HOSTNAME    STATUS    AVAILABILITY    MANAGER STATUS    ENGINE VERSION
x4rzm3l6wkwyod717t3xmhc4u *  master   Ready    Active      Leader      20.10.12
vlh80zt9f3vijilv5d06sm8hv    node     Ready    Active                  20.10.12
```

Task 2　Install the web interface management tool portainer

Start the portainer service in the docker. Firstly, create a data volume to be used by the portainer, then create a service named portainer, map the working port of the service to the host port 9000, define the number of replicas as 1, and let the container work on the manager node.

```
[root@master ~]# docker volume create portainer_data
portainer_data
[root@master ~]# docker service create --name portainer --publish 9000:9000
--replicas=1 --constraint 'node.role==manager' --mount type=bind,src=/var/
run/docker.sock,dst=/var/run/docker.sock --mount type=volume,src=portainer_
data,dst=/data portainer/portainer -H unix:///var/run/docker.sock
b2xwx0yrmh0hazf9o2048eiw3
overall progress: 1 out of 1 tasks
1/1: running   [==================================================>]
verify: Service converged
```

After the container is started successfully, access http://host IP: 9000 with a browser and you can see the home page interface of the portainer. On the home page interface of the portainer, you can manage the entire cluster through interface operation, as shown in Fig. 5-3 and Fig. 5-4.

Fig. 5-3   Portainer home page interface

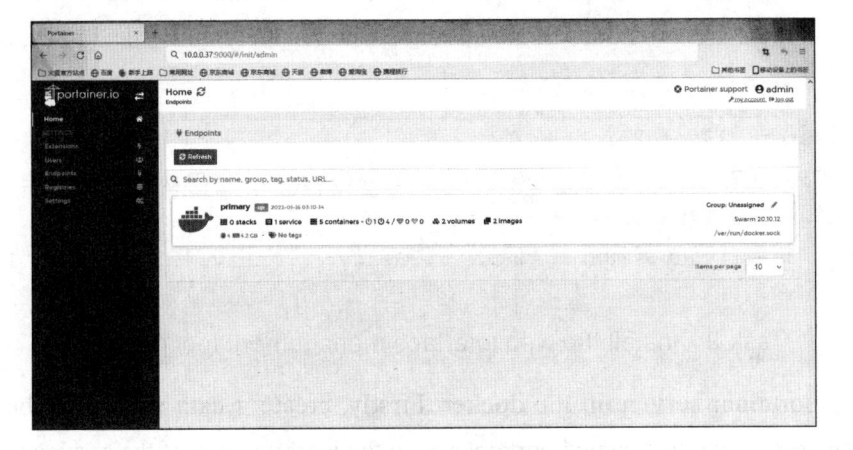

Fig. 5-4   Portainer management page

Task 3   Manage Docker Service

1. Service scalability

Start the nginx service. The default number of replicas is 2.

```
[root@master ~]# docker service create --name nginx --replicas 2 nginx
z7myxrt4qfxckuc5fmfd66b8f
overall progress: 2 out of 2 tasks
1/2: running
2/2: running
verify: Service converged
[root@master ~]# docker service ls
ID              NAME           MODE           REPLICAS      IMAGE          PORTS
z7myxrt4qfxc    nginx          replicated     2/2           nginx:latest
xmbiumsjtyxl    protainer      replicated     0/1           portainer/portainer:
latest   *:9000->9000/tcp
[root@master ~]# docker service ps nginx
ID            NAME       IMAGE            NODE DESIRED STATE CURRENT STATE ERROR PORTS
```

```
1frin78pprg6 nginx.1 nginx:latest node Running Running 52 seconds ago
3081v2i8bxro nginx.2 nginx:latest node Running Running 52 seconds ago
```

Expand the service scale to five replicas and check the service status. You can see that the number of dockers has expanded from 2 to 5.

```
[root@master ~]# docker service scale nginx=5
nginx scaled to 5
overall progress: 5 out of 5 tasks
1/5: running
2/5: running
3/5: running
4/5: running
5/5: running
verify: Service converged
[root@master ~]# docker service ls
ID                    NAME         MODE          REPLICAS        IMAGE
PORTS
    z7myxrt4qfxc    nginx       replicated    5/5         nginx:latest
    xmbiumsjtyxl    protainer    replicated    0/1           portainer/portainer:lat-
est    *:9000->9000/tcp
[root@master ~]# docker service ps nginx
ID            NAME        IMAGE        NODE       DESIRED STATE    CURRENT
STATE            ERROR      PORTS
    1frin78pprg6    nginx.1    nginx:latest    node       Running         Running 2
minutes ago
    3081v2i8bxro    nginx.2    nginx:latest    node       Running         Running 2
minutes ago
    me8aig85vx0v    nginx.3    nginx:latest    master       Running         Running
33 seconds ago
    fw9tf8jdxtg5    nginx.4    nginx:latest    node       Running         Running 33
seconds ago
    63w2yl6aooxs    nginx.5    nginx:latest    node       Running         Running 33
seconds ago
```

If it is hoped that the docker is dispatched only to non-manager nodes, you can set to prohibit the docker from running on the manager node.

```
[root@master ~]# docker node update --availability active master
master
[root@master ~]# docker node ls
ID                              HOSTNAME    STATUS    AVAILABILITY    MANAGER
```

```
STATUS    ENGINE VERSION
    gordckzydp01ig1lzhxbabh92 *    master      Ready       Drain          Leader
20.10.12
    tffdxg32b63990ep360wdwc6t               node        Ready        Active
20.10.12
    [root@master ~]# docker service ps nginx
    ID            NAME        IMAGE         NODE      DESIRED STATE    CURRENT
STATE           ERROR       PORTS
    1frin78pprg6   nginx.1     nginx:latest   node      Running         Running 3
minutes ago
    3081v2i8bxro   nginx.2     nginx:latest   node      Running         Running 3
minutes ago
    me8aig85vx0v   nginx.3     nginx:latest   node       Running          Running
about a minute ago
    fw9tf8jdxtg5   nginx.4     nginx:latest   node       Running          Running
about a minute ago
    63w2yl6aooxs   nginx.5     nginx:latest   node       Running          Running
about a minute ago
```

Expose the service port outside to access the web site page through the browser and use http://manager IP address: 8080 to access the site. At this time, Dockers Swarm will automatically provide load balancing function, as shown in Fig. 5-5.

```
    [root@master ~]# docker service update --publish-add 8080:80 nginx
    nginx
    overall progress: 5 out of 5 tasks
    1/5: running
    2/5: running
    3/5: running
    4/5: running
    5/5: running
    verify: Service converged
```

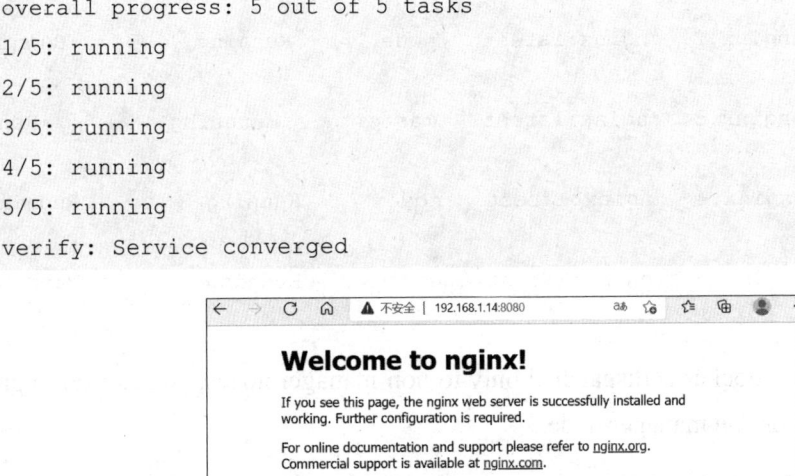

Fig. 5-5　Docker cluster web service page